River Life

The Natural and Cultural History of a Northern River

by John Bates

Illustrated by Terry Daulton

Manitowish River Press
Mercer, Wisconsin

River Life:
The Natural and Cultural History of a Northern River

Printed in the United States on 50% recycled paper.

Editor: Greg Linder
Illustrations: Terry Daulton
Cover photography: Jeff Richter
Book design: Katie Miller
Cover design: Katie Miller

Publisher's Cataloging in Publication Data
Bates, John, 1951-
River Life: The Natural and Cultural History of a Northern River /
written by John Bates ; illustrated by Terry Daulton
Includes bibliographical references and index.
ISBN 0-9656763-3-1 (softcover)
1. Stream Ecology
2. Rivers
3. Freshwater Ecology
4. Nature Study
5. Natural History
Library of Congress Catalog Card Number: 00-193568
0 9 8 7 6 5 4 3 2 1

Published by: Manitowish River Press
4245N Hwy. 47
Mercer, WI 54547
Phone: (715) 476-2828
Fax: (715) 476-2818
E-mail: manitowish@centurytel.net
Web site: www.manitowish.com

I am indebted to the following natural resource professionals for supplying essential information to me along the way: Ron Eckstein, Jeff Roth, Tim Kratz, Susan Knight, Sue Brisk, Carl Watras, Karen Wilson, Dennis Scholl, Kate Barrett, Katherine Webster, Paul Schultz, Mark Bruhy, and Katie Egan-Bruhy.

My deepest thanks go to the following people for making the production of this book possible: Terry Daulton, Katie Miller, Greg Linder, Mary Burns, and Stephany Freedman.

"All nature is my bride," said Thoreau.
My desire is to experience all nature with my bride.
This book is to you Mary.

Grateful acknowledgment is made to the following for permission to reprint previously published material:

Brooks, Paul. From SPEAKING FOR NATURE by Paul Brooks. Copyright © 1980. Reprinted by permission of the Sierra Club.

Curtis, John. From VEGETATION OF WISCONSIN by John Curtis. Copyright © 1959. Reprinted by permission of the publisher, The University of Wisconsin Press.

Dennis, Jerry. From THE BIRD IN THE WATERFALL by Jerry Dennis. Copyright © 1996. Used by permission of Jerry Dennis.

Eastman, John. From "The Ghost Forest," with permission from Natural History Magazine, January, 1986. Copyright © 1986 The American Museum of Natural History.

Eggers, Steve. From WETLAND PLANTS AND PLANT COMMUNITIES OF MINNESOTA AND WISCONSIN, Second Edition. Copyright © 1997. Used by permission of Steve Eggers.

Heat-Moon, William Least. From RIVER HORSE. Copyright © 1999. Reprinted by permission of Houghton Mifflin Company.

Hildebrand, John. From READING THE RIVER by John Hildebrand. Copyright © 1997. Reprinted by permission of the publisher, The University of Wisconsin Press.

Isherwood, Justin. From "Of River, Time and Steerage," Wisconsin Natural Resources Magazine, Copyright © 1982 by Justin Isherwood. Used by permission of Justin Isherwood.

Kappel-Smith, Diana. From WINTERING. Copyright © 1979, 1980, 1982, 1983, 1984 by Diana Kappel-Smith. Reprinted by permission of Diana Kappel-Smith.

Leopold, Aldo. From A SAND COUNTY ALMANAC: AND SKETCHES HERE AND THERE SPECIAL COMMEMORATIVE EDITION by Aldo Leopold. Copyright © 1949, 1977 by Oxford University Press, Inc. Used by permission of Oxford University Press, Inc.

Leopold, Luna. From WATERS, RIVERS AND CREEKS by Luna Leopold. Copyright © 1997. Used by permission of University Science Books.

Middleton, Harry. From RIVERS OF MEMORY by Harry Middleton. Copyright © 1993. Used by permission of Pruett Books.

Murray, John. From THE RIVER READER, edited by John Murray. Copyright © 1998. Used by permission of the publisher, The Lyons Press, New York, New York.
Oliver, Mary. From HOUSE OF LIGHT by Mary Oliver. Copyright © 1990 by Mary Oliver. Reprinted by permission of Beacon Press, Boston.

Olson, Sigurd. From LISTENING POINT, reprinted by permission of Alfred A. Knopf. Copyright © 1958 by Sigurd Olson.

Stafford, William. "Ask Me" Copyright © 1977, 1998 by the Estate of William Stafford. Reprinted from THE WAY IT IS: NEW AND SELECTED POEMS with the permission of Graywolf Press.

Stains, Bill. From THE WHISTLE OF THE JAY by Bill Stains. Copyright © 1979. Used by permission of Bill Stains.

Steinhart, Peter. From "A Vision of Lakes," Audubon Magazine, July, 1987, by Peter Steinhart. Used by permission of Peter Steinhart.

Steinhart, Peter. From "Trusting Water," Audubon Magazine, November, 1986, by Peter Steinhart. Used by permission of Peter Steinhart.

Steinhart, Peter. From "The Meaning of Creeks," Audubon Magazine, May, 1989, by Peter Steinhart. Used by permission of Peter Steinhart.

Zwinger, Ann. From RUN, RIVER, RUN by Ann Zwinger. Copyright © 1975. Used by permission of Ann Zwinger.

The following poems of John Bates' have been previously published in Wisconsin Academy Review: "Pine Voices" (Winter 1992), "Manitowish" (Fall 1994).

Table of Contents

Introduction

People spend a lot time trying to figure out who they are. The real question is not who am I, but where is here?" — Northrop Frye

A Dictionary of the Ojibway Language, written by Father Frederic Baraga and published by the Minnesota Historical Society in 1878, gives three words that could be used to define Manitowish:

manitowish — "small animal, a marten or weasel"

manitow — "I am a spirit"

manitowis — "I am looked upon (or considered) as a spirit"

Given the French and American penchant for mispronouncing and distorting American Indian words, who really knows how close "Manitowish" may be to the original Ojibwa word? The earliest European reference to the Manitowish River that I am aware of is from Thomas Jefferson Cram in 1840, an engineer given the job of delineating the border between the state of Michigan and Wisconsin Territory. He referred to the river as the "Manitouish River." A few years later in 1847, geologist J.G. Norwood wrote in his journal of paddling on the "Manidowish River."

I asked an Ojibwa elder from the Bad River Reservation in northern Wisconsin what the "wish" part of Manitowish might mean, and he said "tricky, kind of playful." "Like an otter?" I asked. "Yes, just like that!" he said. So, Manitowish most likely means "playful spirit." That fits the river as I know it.

I also admit a desire to say the real meaning of Manitowish is something like "Leech" or "Black Fly" or "Evil Smell." After all, I'd prefer not to share the river with too many of you. Critical mass comes quickly on small rivers. If you must come, you should be required to sign a vow guaranteeing your passionate protection of the river's inhabitants, and its feeling of wildness—upon penalty of permanent exile to urban malls.

≈ ≈ ≈ ≈ ≈

Every river speaks with its own voice, but like humans, each river generates its expression from a general template. The diversity of river environments runs the spectrum from torrential mountain brooks to quiet streams, and from large lowland rivers to great rivers whose watersheds consume subcontinents. It's fair to say the Manitowish River no more represents all northern rivers than any one forest stand represents all forests. However, it's also fair to say that exploring the ecological, cultural, hydrological, and geomorphological processes of the Manitowish River can provide significant insight into how many northern rivers work. The knowledge needed to try to understand a small river like the Manitowish is similar to the knowledge necessary to fathom the Brule or the Vermilion or the Two-Hearted rivers.

Because every river occurs in widely varying conditions of climate, vegetation, topography, and geology, each has its own constellation of remarkable attributes: a certain wildness to its shoreline, a particular color, a varying but unique speed and tumble of current, an unmistakable layout of islands or rocks in the streambed, a sound and a scent and a feel that is all its own. Like friends, we get to know individual rivers and can recognize each instantly from all others. Still, these rivers have more in common than what sets them apart, and those commonalties form the basis and intent of this book—to begin to understand in laypersons' terms the life of northern rivers.

≈ ≈ ≈ ≈ ≈

When you look upon a river, what first strikes you? Likely you notice the dimensions of the river, the strength of the current, how high or low the water level appears, the shape of the channel, the presence of boulders or islands in the river, the shoreline topography and vegetation, the aquatic vegetation, and the presence of wildlife.

In the case of the Manitowish, you would likely observe that its watershed is predominately wooded and wild, that flatwater dominates, that its water is stained a light brown and is shallow. If you asked a few locals about the fishing, you'd hear that the fishery includes smallmouth bass, largemouth bass, northern pike, muskellunge, walleye, panfish, and in some areas, lake sturgeon. A birder or hunter might tell you that waterfowl utilize the river as a spring and fall migration stopover site. Campers would say that the 38 campsites along the river make canoe-camping easy. An environmentalist might note that three dams alter the flow over the river's 44 mile course.

That's an accurate but limited snapshot. It says little about the river's personality. The Manitowish isn't lineal like a brochure overview would show it. There's nothing direct about it or easily defined. It takes its sweet time, wandering and meandering. The word "meander" derives from the old Roman name, the *Maeander*. The name was originally given to Turkey's Menderes River, a river so convoluted that in ancient times it was thought it turned and flowed backward.

Meander fits the Manitowish. I'm surprised that the Ojibwa didn't incorporate their word for "meander" into the river's name.

Readers should know that like the *Maeander* and the Manitowish, this book loops and turns back upon itself at times. Often the writing slows down to look at something and contemplate its life. Numerous sidebars offer a closer look at a topic. Think of the sidebars as sandbars—places to get hung up, sit a spell, and think. Note that some sandbars are placed at the end of chapters so as not to interrupt the flow of your trip.

As with any river trip, there will be things you wish to spend time looking at and thinking about, and there will be things you wish to ignore. Do so freely. Skip around. Focus on what you want to know about, and leave the rest for another time.

Rivers trips also have a randomness to them. They're a potpourri, a wandering through an environment, as well as through our own mind. This book follows the Manitowish in its wandering. I've read many books that attempt to organize a river into precise systematic chapters. That works well for text books. But that's not how any of us actually experiences a river.

Finally, this book combines science and poetry, a purposeful mix, which recognizes that science gives us more and more information and often less and less understanding, while poetry talks about why we are on the river in the first place. Mary Oliver, Pulitzer Prize-winning poet, says in one of her poems, "Science is only the golden boat on the dark river." All the gold that science uncovers means nothing without poetry to also explore the dark waters.

I hope you come to see rivers as sacred places where sacred experiences are possible. Like most humans, rivers are ever-changing. They are worthy of our life-long involvement and respect, while equally deserving of our playfulness and simple joy. Loving a river is like any love affair. Our love deepens the more we understand and appreciate the complexity and individuality of our partner.

With our love for rivers will come the commitment to protect them. That's what we humans do for those we love—we honor, cherish, and safeguard them.

The Manitowish River

Chapter 1
Our Course: From High Lake to Fishtrap Dam

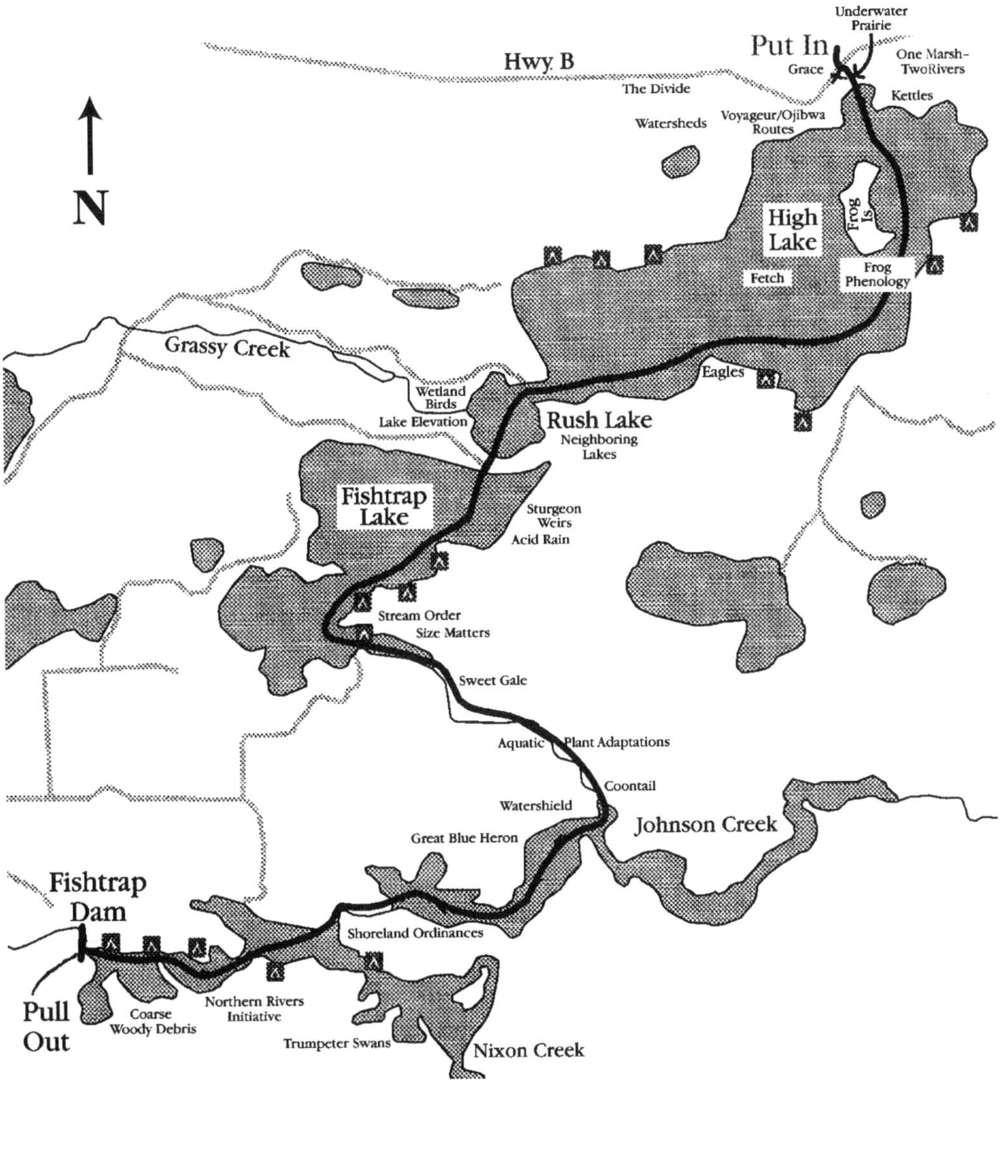

The river is old. Old as few things are old; old like orbits and dust and gravity.

—Justin Isherwood

Chapter 1

Of Grace, Watersheds, Time Travel, Eagles, and Swans

Quick or smooth, broad as the Hudson or narrow enough to scrape your gunwales, every river is a world of its own, unique in pattern and personality. Each mile on a river will take you further from home than a hundred miles on a road. You will see more in an hour than a motorist will in a week. — Paul Brooks

Why Paddle Rivers?

In many spiritual texts, we are told to put ourselves in the way of grace. I think we intuitively know this when we paddle—putting-in on a river is a way of putting ourselves in the way of grace.

But there are many forms of grace. Fortunately, the river seems to embody them all. Consider the gracefulness of the dancer: the swirl, the glide, the floating-on-air otherworldliness, the effortless strength and subtlety, the symmetry, the elegance of simple beauty. The river is all of these.

Consider the saying of grace before a meal: the request of a blessing for what we have been given, the thankfulness for being given anything at all, the gratitude to the animals and plants whose lives we take into our own. The river offers blessings too—the gift of seeing the world in action as it should be; the gift of free travel through an ever-changing landscape; the gift of quiet.

Consider receiving the grace of God, the hand of the divine cupped in offering, the rapture of beauty revealed, the understanding of what it means to be ALIVE, NOW, HERE. The river offers that.

Consider grace as charity and forgiveness. "There but for the grace of God, go I," we say. The river washes away sins in many faiths.

Consider being graced with beauty—adorned, crowned. The river is crowned with otters, yellow warblers, dazzling dragonflies, regal water lilies.

Consider gracious people—kind, good-hearted, good-natured. Rivers are my definition of good nature.

Consider what it means to be graced by someone in your presence—to be honored, to be exalted. The river graces you in the same way.

So, to float on a river is to put yourself in the way of grace.

Too often though, we work at cross-purposes to finding the way of grace. We impose ourselves on the world, on others, driven by a pleasure-seeking that is a taking rather than a receiving, a blindness among so many possible visions. I admit pleasure-seeking drives much of my river travel, too. And, I admit I've done my share of imposing, though some of my impositions I think the river laughed right along with—the playful soaking water-fights, the creeping up behind another canoe to catch a free ride, the burst-out, surprise attack from the tall sedges in the crease of the shoreline. Playful, joyful acts, something otters might do. I think that's grace too.

We'd be fools not to put ourselves in the way of the river's chorus of voices, of experiences, of philosophies, of ways of living life, fools not to quietly reflect on the quiet reflection.

Mary Oliver says about writing poetry, "One learns the craft, and then casts off. One hopes for gifts. One hopes for direction . . . It is intimate and inapprehensible."

Well, on the river, one learns about one's craft too—the canoe, the kayak, the raft—and then casts off, hoping for direction, for intimacy, for revelation of the inapprehensible.

Why Begin Here?

Because it's all downhill, or downriver, from here.

Rivers are all smooth glide. Riverwater must be the most intelligent substance on earth. It alone finds a way to go downhill all its life.

And what could be a better downhill than starting from the top of the Continental Divide (or Sub-continental Divide as some geographers refer to it)? In fact, that's where you are right now. The Manitowish River originates in High Lake in the Northern Highlands State Forest just south of the Continental Divide.

The Northern Highlands and the Continental Divide . . . the words might lead you to think of 12,000-foot-high mountains, Lewis and Clark, precipices, avalanches. But finding mountains in this area requires pre-glacial time travel. The Northern Highlands present themselves instead as a high ridge of land, which to the naked eye appears as little more than

Sub-continental divide

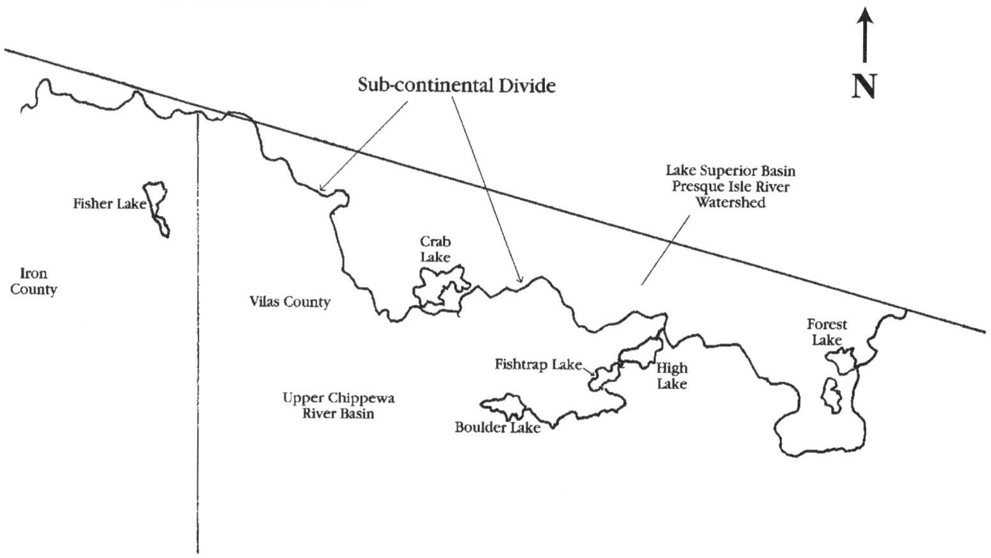

a low, undulating plain with occasional small hills. The elevation of High Lake peaks out at only 1,640 feet, while the Continental Divide tumbles along at around 1,700 feet.

Some 1,200 miles downriver from here, as the crow flies, and likely twice that far as the river flows, some of the water running under your boat right now will finally find level in the Gulf of Mexico at zero feet of elevation.

By my advanced math, that's less than a foot of descent per mile, a decline so imperceptible as to nearly give definition to the word "flat." But downhill it is. For comparison sake, plumbers minimally pitch pipes in a house at a drop of 1/4 inch per 10 feet (or 11 feet per mile) in order to keep the water moving. Plumbers wouldn't bet a plugged nickel on this river flowing.

Like the river, the surrounding land imperceptibly slopes down from the Continental Divide—from this point, no one could look around and say with accuracy that the landscape leans any one particular direction. But it does, and that makes all the difference.

Take a look at the watershed map. Wisconsin divides into two major watersheds. North and east lies the St. Lawrence Drainage, which leads to the Atlantic Ocean. Waters within this watershed flow by way of Lake Superior or Lake Michigan, and via the Fox River-Wolf River-Lake Winnebago system, the Menomonee River, and others.

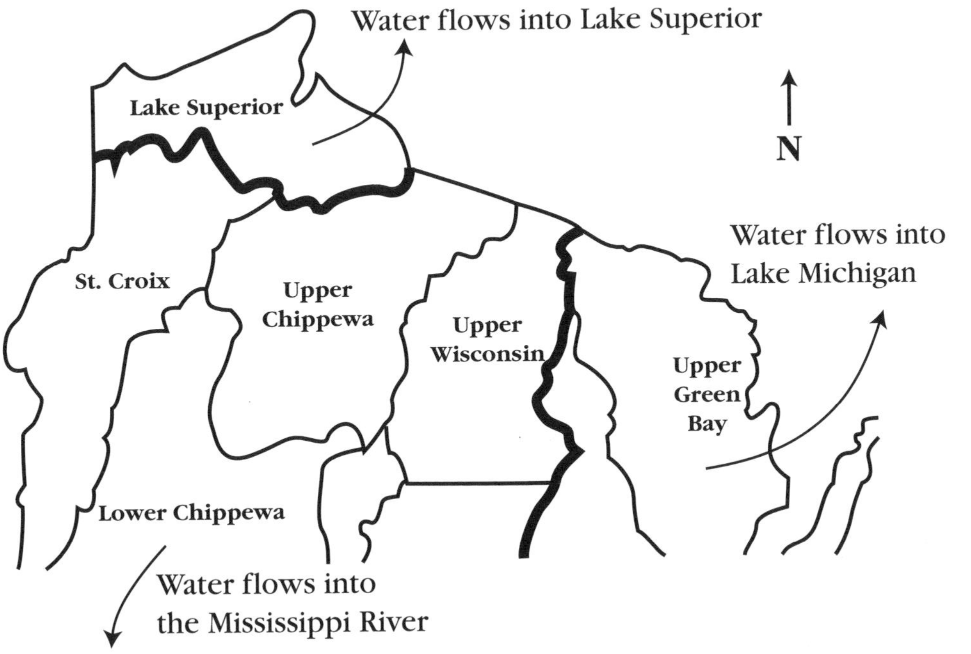

Watersheds of Northern Wisconsin

NORTHERN WISCONSIN WATERSHED CHART

DIVISION OF WATERSHEDS	RIVERS INCLUDED IN WATERSHED
Gulf of Mexico	
Mississippi Drainage:	
St. Croix Watershed	St. Croix, Yellow, Namekagon, Clam
Chippewa Watershed	Chippewa, Flambeau, Eau Claire
Wisconsin Watershed	Wisconsin, Tomahawk, and many smaller rivers & streams: Pelican, Somo, Prairie
St. Lawrence/Atlantic Ocean	
Lake Michigan Drainage:	
Fox River Watershed	Fox, Wolf, Embarrass
Menominee Watershed	Menominee, Brule, Pine, Pike
Other Streams	Peshtigo, Oconto
Lake Superior Drainage:	
Lake Superior Streams	(Bois) Brule, Montreal, Bad; and Ontonagon and Iron in Michigan.

South and west lies the Gulf of Mexico Drainage, which courses toward the Gulf by way of America's arterial river, the Mississippi. The Mississippi is fed by the St. Croix, Chippewa, Black, and Wisconsin Rivers, among many others. (Note that the words "drainage" and "watershed" can be used interchangeably.)

What this means to you is that because you're just a mile or so south of the Continental Divide, all rivers now flow south into the Gulf of Mexico. If you had a few months, you could pull out in New Orleans, with the only required walking being portages around dams and around rapids that look a bit too dicey.

Wander a mile east of here and you could pick up the nearby Ontonagon River and paddle to the Atlantic Ocean, coming out north of Maine by way of the St. Lawrence Seaway.

The fact that the Ontonagon rises so near to the Manitowish is hardly surprising. Along the Continental Divide, headwaters of separate rivers often begin in close proximity to one another. Sometimes two rivers rise from opposite ends of the same marsh. The Brule and St. Croix rivers may be the best-known example of the "one marsh-two rivers" phenomenon in northern Wisconsin. Both rivers originate in a large marsh off of Highway P near Solon Springs. The original portage trail connecting the rivers is signed and may still be walked.

In Iron County, the north-flowing West Fork of the Montreal and the south-flowing East Fork of the Chippewa head within two miles of one another. In Ashland County, the south-flowing West Fork of the Chippewa and the north-flowing Bad River (via the Brunsweiler River) head within a few miles of one another.

The Indians and voyageurs knew about these watershed divides, and their knowledge turned this area into a sort of river-trader's switchyard. A relatively short portage permitted them to pick up a river that would eventually go just about anywhere in the eastern U.S. Historically, this was a hub, like the big terminal at O'Hare Airport, the Greyhound station in St. Louis, the Grand Central train station in New York . . . well, I get carried away. Nonetheless, river routes radiated out from here, and with the right geographical wisdom, you could live a long life and seldom leave the water.

Northern Wisconsin's "Keynote": Water, Water, Everywhere

Headwater rivers make up only a portion of the geographic story here. Rivers often flow into lakes, forming chains of lakes linked by river sections. Over 15 percent of Vilas County's surface area is covered by lakes and ponds—140 square miles worth. Add in another 21 percent of the county that's covered by streams and wetlands, and for every three steps you take here, one step will likely squish.

To visualize how and why this landscape became so pockmarked by water, imagine the receding edge of the glaciers breaking into pieces, calving like current-day icebergs, and leaving behind huge chunks of ice buried in the sandy glacial drift. Over time, the ice blocks melted, and the sandy soil subsided into the hollows left by the melted ice. As it rained, or as groundwater flowed in through springs, some of these holes eventually filled with water.

Today, in the sandbox we call northern Wisconsin, we describe these hollows as ice block depressions or kettle holes, and if they're filled with water, as kettle lakes or pothole lakes. Whatever the name, ice block depressions formed the lakes and bogs of this area after the glaciers hightailed it back to Canada.

Kettle lakes are characteristically small and shallow. In Vilas and Oneida counties, over half the lakes are smaller than 10 acres, while another quarter are between 10 and 49 acres. In Oneida County, 88 percent of the lakes are less than 25 feet deep.

The density of kettle lakes here is rivaled only by northern Ontario, Alaska, and northern Europe, defining this ecoregion as one of the most concentrated lake districts in the world, a globally important land/lakescape. In other words, we have enough kettles here to set up one helluva tea party.

Rivers as Fluid Archaeological Trails: How Life Once Was

All of this connected water was the key to life centuries ago. River trade decreased dramatically with longer distances and greater difficulties in travel. Those 12,000-foot peaks along the Continental Divide in the

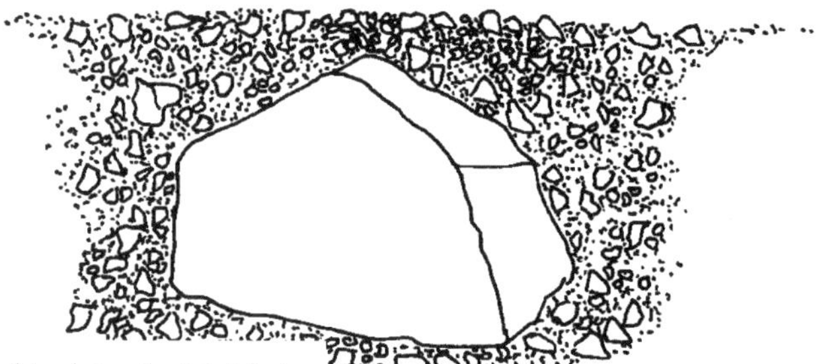

Ice block in glacial debris

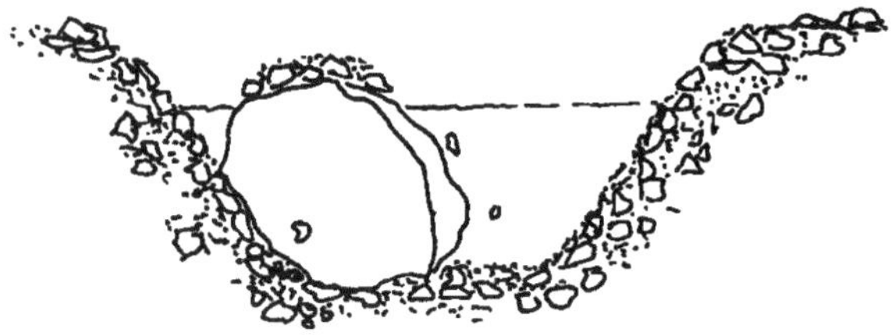

Ice block half melted

Ice block depression or kettle lake

Rockies were a daunting barrier to trade. But in northern Wisconsin, trade moved with comparative ease across the Divide. Copper from the Keweenaw Peninsula in the Upper Peninsula of Michigan commonly was traded with tribes from far-flung places like Louisiana, and archaeological remains from sites in Wisconsin include seashells from the Gulf Coast, obsidian from Wyoming, and pipestone from western Minnesota.

While we have no written records that go back thousands of years, it's fair to assume since the retreat of the last glacier that American Indians traveled the Manitowish River route through High Lake. In the mid-1600s, the Indians were joined by the voyageurs, and for nearly 200 years the Manitowish likely served the same purpose as its present-day concrete counterpart, Highway 51, joining topography, hydrography, and people over time and place.

A general accounting of the many water and portage trails utilized by the American Indians and voyageurs can be made—see the "Voyageur and American Indian Routes and Watersheds" table.

Only 150 years ago, this region was still true wilderness. The reality of wilderness life for an explorer of the time was far more arduous than the romantic ideal we tend to impose on our history. James Doty, who was to become the second governor of the Wisconsin Territory, traveled through the Upper Midwest with the Cass/Schoolcraft expedition in 1820. At age 21, he was selected as the secretary of the expedition, and his *Official Journal, 1820* provides an in-depth look at the hardships of such a life.

Let's take a quick side trip.

The expedition began in Detroit on May 24, 1820. Its assigned written purposes were to obtain information about:

"Northern Indians, their divisions, names, important men, of what numbers the tribes are composed, the country and its extent which they inhabit, and their history, manners and customs, as also what their feelings and dispositions are as respects the United States and Great Britain.

"The topography of the country is to be accurately observed and noted, and collect all the information possible necessary to form a complete map of this section of the Union.

"Eligible sites for forts are to be selected and purchased, and particularly one at Sault de Ste. Marie.

"The geological and mineralogical aspects of the country are to be examined, especially the copper mines, lead mines and gypsum quarries, their quality and quantity, and the facilities of obtaining them.

"... To ascertain the sources of the Mississippi may be considered another object."

No problem.

The expedition was manned by Lewis Cass, governor of Michigan Territory; Captain David Douglass of the Corps of Engineers and a mathematics professor; Henry Schoolcraft, a mineralogist and geologist; Dr. Alexander Woolcott, an Indian agent at Chicago; Doty; and then a variety of others, including an interpreter, a lieutenant of the Artillery, two assistants to Captain Douglass, as well as 10 Canadian voyageurs, 10 U.S. soldiers to serve as escorts, and 10 Indians (Ottawa, Chippewa, and Shawnee) to act as hunters. One of the Indians was Kewaygooshkum, an Ottawa chief who played an important part at the Treaty of Chicago.

To give you a flavor of the time, here are some topical excerpts from Doty's journal.

Waves and weather: "In crossing [Huron Bay on the Keweenaw], a violent gale of wind arose, and separated our little fleet. I never experienced so heavy a wind—the swells ran high dashing over our canoe—two or three from stem to stern. We were in imminent danger of breaking across them."

Storms: "At 10 o'clock last night (6/20) the most tremendous storm arose I ever witnessed. It came on from the N.W. and directly on the shore where we were camped [Pictured Rocks National Lakeshore in Michigan]. From the continued flashes of lightning the Lakes

ROUTE AND WATERSHED CHART

NAME OF ROUTE	ORIGINATES; WATERSHED (WS.)	CONNECTION WATERSHED (WS.)
1. Ashland Overland to the White River	Chequemegon Bay; Lake Superior ws.	White River, branch of the Bad River; Lake Superior ws.
2. Bad River to the Namekagon River	Mouth of the Bad River; Lake Superior ws.	Namekagon River; St. Croix drainage; Mississippi ws.
3. Bayfield-St. Croix Trail	Chequamegon Bay; Lake Superior ws.	St. Croix River; Mississippi ws.
4. Brule-St. Croix	Mouth of the Brule River; Lake Superior ws.	Headwaters of the St. Croix River, upper St. Croix Lake; Miss. ws.
5. Flambeau Trail	Mouth of the Montreal River; Lake Superior ws.	Lac du Flambeau; Chippewa drainage; Mississippi ws.
6. Iron River to Lac Vieux Desert	Mouth of the Iron River; Lake Superior ws.	Lac Vieux Desert, Wisconsin River headwaters; Miss ws.
7. Lac Courte Oreilles to the West Fork of the Chippewa River	Lac Courte Oreilles; Chippewa drainage; Mississippi ws.	West Fork of the Chippewa; Mississippi ws.
8. Lac du Flambeau to the Chippewa River	Lac du Flambeau; Chippewa drainage; Mississippi ws.	Chippewa River; Mississippi ws.
9. Lac du Flambeau to the Wisconsin River and Lac Vieux Desert	Lac du Flambeau; Chippewa Drainage, Mississippi ws.	Wisconsin River; Mississippi ws.
10. Lac Vieux Desert to the Menominee	Lac Vieux Desert; Mississippi ws.	Menominee River; Lake Michigan ws.

Continued

ROUTE AND WATERSHED CHART (Continued)

NAME OF ROUTE	ORIGINATES; WATERSHED (WS.)	CONNECTION WATERSHED (WS.)
11. Lac Vieux Desert, trails/military roads in the vicinity	Two trails/roads: 1) Green Bay; Lake Michigan ws.; 2) Wausau; Wisconsin River; Mississippi ws.	Lac Vieux Desert; Mississippi ws.
12. L'Anse-Lac Vieux Desert	Base of the Keweenaw Peninsula; Lake Superior ws.	Lac Vieux Desert; Mississippi ws.
13. Lake Superior to Lake Nebagamon	South shore of Lake Superior on the Poplar River; Lake Superior ws.	Lake Nebagamon; Lake Superior ws.
14. Long Lake Trail	Chequamegon Bay; Lake Superior ws.	Long Lake; Lake Superior ws.
15. Marengo River to English Lake	Marengo River (branch of the Bad River); Lake Superior ws.	English Lake; Lake Superior ws.
16. Montreal River to Lac Vieux Desert	Mouth of the Montreal River; Lake Superior ws.	Lac Vieux Desert; Mississippi ws.
17. Montreal River to Trout Lake, and Lac Vieux Desert	Mouth of the Montreal River; Lake Superior ws.	Lac Vieux Desert; Mississippi ws.
18. Odanah to Lac Courte Oreilles	Odanah, Bad River; Lake Superior ws.	Lac Courte Oreilles; Chippewa R. Drainage; Mississippi ws.
19. Ontonagon Route	Mouth of the Ontonagon River; Lake Superior ws.	Lac Vieux Desert; Mississippi ws.
20. St. Croix Trail	Base of Chequamegon Bay; Lake Superior ws.	St. Croix River; Mississippi ws.

appeared on fire. The wind was so strong it was with the utmost difficulty we kept up our tent—the Governor's was blown down and also one or two others. Lightning struck several times near us. The waves rolled up to the mouth of our tent, and completely over the Governor's wetting all his baggage and the gentlemen's with him."

Fighting rapids upstream: "For 10 minutes our canoe, with all of the men at the oars and paddles, did not stir 8 feet either way [on St. Mary's River at Sault Ste. Marie]."

Paddling downstream: "For miles it [the Mississippi River above the Twin Cities] is but a continued succession of rapids, and if the waters had not been at more than their usual height we could not have run them as easily and with the security we did, nor have went from 140 to 150 miles in a day with birch canoes."

Things we'll never see again: "Almost unlimited numbers of [passenger] pigeons came flying over the banks into the gulf, from whence they ascended to the opposite. They flew so low that 30 or 40 were killed by our men with clubs and sticks thrown into the flocks as they passed."

Wilderness life: "I very much doubt whether the desire to accumulate wealth could ever so strongly predominate in me as to induce me to forsake the pleasures, the comfort, and elegancies of civilized life for a residence in this dreary wilderness . . . Starvation has few horrors which are not at some season or other felt in a greater or less degree by the resident here. Even at this season these people are living on dried buffalo meat and Labrador tea! without bread or vegetables."

≈ ≈ ≈ ≈ ≈

Time to Put In

Put in at the boat landing on Highway B, about 16 miles west of Land O'Lakes, or 12 miles east of Presque Isle.

The channel into High Lake is a short paddle, and the water is cold and clear, even in July. Springwaters at the height of the landscape tend to be that way.

The underwater plant life is clearly visible. River ecologists and writers have often referred to the aquatic plant zone as an underwater prairie, a concept clearly visible here in the headwaters of the Manitowish.

Anchored in the sediments lives a wonderfully diverse array of plants, which has evolved to grow in three zones: under water, on the water, and well above the water. Submersed plants like coontail and milfoil, grow to varying heights completely under the water (though their flowers may reach the surface). Surface plants, like water lilies and watershield, emerge from the depths to float their leaves and flowers on top of the water. Tall emergents like pickerelweed stand a foot above the water, while cattail sometimes grows seven feet above the water, high enough to create lengthy shadows.

Some plants like duckweed don't fit neatly into any of the three categories. They simply drift rootless in the water column, their free-floating existence lending new definition to the notion of "going with the flow."

The ecology of aquatic plant communities is often analogous to upland plant communities like forests and prairies. Forests tend to be richer with the more layers they have, rising from low herbaceous species along the forest floor to tall herbs to shrubs to short trees to tall trees. Insects, birds, and mammals feed, mate, nest, and find cover at varying levels of the forest. So, as a general rule, the more layers in a forest, the higher the diversity of animals, and the greater the overall health of the forest.

The same holds true for underwater plant communities. A rich array of vertical layers in an underwater prairie presents the best overall habitat option for a host of creatures from snails to minnows to dragonflies to ducks.

As important as the ecological value of these underwater meadows is their sheer beauty. Many people see weeds when they look in the water, likely because they're used to nutrient-loaded lakes and rivers overloaded with plants. Too much of a good thing can be a bad thing. Seen in the right balance, though, aquatic plant communities equal the beauty and intrigue of any upland community.

≈ ≈ ≈ ≈ ≈

Notice along the channel the undisturbed forested uplands and wetlands leading down into the river. That bodes well, because if you want to know what the health of a river is likely to be, you look to the landscape first.

Here's why. Water gets into a river by three routes: drainage downhill over the surface of the land, groundwater seepage, or from precipitation.

Relative to water quality, the most important of the three routes is typically water pouring downhill because of its potential for carrying soil and chemicals to the river. Gravity does all the work—water drains down from

Underwater prairie

ridgelines. A watershed acts like a funnel to catch all the rain and snow that falls within its boundaries. It cradles the water within its ridgelines, and then funnels the water down to the river. In a hilly landscape, the watershed would look like a tall narrow funnel. In our landscape, the Northern Highlands, the watershed looks like a very broad, shallow funnel.

Or think of a watershed like a human circulatory system. Small capillaries run into larger veins, which run into arteries. The same is true here in a watershed, except a stream network is typically full of curves and rapids and maybe a waterfall or two. Hopefully, your circulatory system doesn't look quite like this, or you're in trouble.

Each watershed looks like the most creative of corrugated washboards, wrinkling into a highly complex landscape that may contain mountains, ridges, canyons, streams, old-growth forests, clearcuts, towns, farms, mills, you name it. Water heads down to the river carrying what it can over the landscape. There is no away; there's only below, into the river. The bottom line, a purposeful pun, is that whatever happens in the watershed is inevitably mirrored in the river.

Watershed quality makes all the difference in the water quality of the rivers that are at the bottom of watersheds. If the landscape remains natural, the drainage of rainwater and snowmelt down and across the land is usually slow. Most of the water percolates down into the soil, like coffee through a filter, before it reaches the

Manitowish River Watershed

The Manitowish River Watershed lies in the northeastern-most tip of the Upper Chippewa River Basin (or watershed—take your pick of terms), a 4,860-square-mile area divided into 23 watersheds. The basin includes portions of 11 counties. The northern area consist mostly of forests and wetlands, while the southern part of the basin tends toward agricultural use.

The two major rivers that drain the basin, the Chippewa and the Flambeau, are subdivided into the West Fork and East Fork of the Chippewa, and the North Fork and South Fork of the Flambeau.

Where does the Manitowish fit into this model? The Manitowish River meets the Bear River to form the beginning of the North Fork of the Flambeau.

Most of the Upper Chippewa River Basin is in county, state, or federal ownership, and represents some of the most remote and beautiful land and water in the state. Nearly the entire Manitowish River Watershed lies within the public ownership of the Northern Highlands/American Legion State Forest. The Manitowish does run through several very small communities—Boulder Junction, Manitowish Waters, and Manitowish—none of which currently poses known threats to water quality.

A Water Quality Management Plan was written for the upper Chippewa River Basin in 1996 as required by the Federal Clean Water Act. While not light bedtime reading, the plan is loaded with details on what needs to be done to maintain and enhance the overall health of the streams and rivers in this basin. If you live in Wisconsin, a similar plan should exist for your area. Contact the WDNR's Department of Water Quality Management or your local library to see a copy of the plan for where you live.

river. The result is usually clean water entering the river. Plug up the coffee filter—or compact the soil so that little rain percolates through—and you have bad coffee. Or a poor quality river system. Healthy watersheds help ensure healthy rivers.

> *Like us, rivers spring from obscure sources and flow toward unavoidable destinations. If the sea represents eternity, the rivers that flow into it are the twisting, bold, and unstoppable currents of time.* — Jerry Dennis

≈ ≈ ≈ ≈ ≈

Just before entering High Lake, a beaver lodge projects skyward on the left side of the channel. Or should I say, as of the year 2000, an active beaver lodge could be seen here. Beavers come and go over time, their demise most often created by their own appetites, or by those of trappers. Beavers lend veracity to the cliché of eating yourself out of house and home. So, this lodge my still be active at this site upon your passage-by, it may be slowly settling and decomposing, or it may have been washed away in the high water of spring snowmelt. Little is permanent along the banks of a river.

High Lake opens up to the southwest, perfect for the prevailing winds of summer to nail you head-on. The last time my wife Mary and I paddled High Lake, we fought our way through whitecaps beating hard against us, a dubious beginning to any trip. The lake covers over 734 acres, stretching nearly a mile and a half in length, thus providing a good, long fetch for waves to build up. Hopefully, your paddling luck is better than ours was that day.

Read more about fetch on the next page.

Note the shorelines in either direction—to your right (west), there are lots of homes, as well as some homes initially to your left (east). In the distance, natural shoreline beckons to the south and east. Less than half of High Lake is in state ownership. With a maximum depth of only 31 feet, waters heat up rather quickly in High Lake, resulting in a typical warmwater fishery of muskie, walleye, northern pike, and largemouth and smallmouth bass. Note that a mercury warning exists for fish caught in High Lake.

If you're an angler, check out the Fishing Hot Spots map of High Lake for a detailed look at where to wet your line. In 1933, the state record walleye, weighing in at 18 pounds, was caught on High Lake.

A large island, promisingly called Frog Island, forms a crescent-shaped wedge about a half mile out. I say promisingly, because frog populations are a good indicator of undisturbed shorelines, and the promise of a natural, undisturbed shoreline attracts me. Paddle out to the southern end of the island and enjoy the windbreak if the waves are buffeting you. A substantial stand of bulrush projects out from the southernmost edge of the island.

Frogs serve as indicators of river health much like the proverbial canary in the coal mine. They're certainly not the only indicator—the presence of eagles at the top of the food chain, and certain aquatic insects near the bottom of the food chain tells us a great deal in a quick glance too.

Fetch

Fetch, in lake terms, means the distance the wind travels over open water. The longer the fetch, the bigger that waves can build.

Wave height depends on three things: the strength of the wind, the length of time it has been blowing, and the fetch. Given how limited inland lakes are in size, lake waves are never as big as ocean waves, which have been known to reach more than 100 feet high!

For example, if we had a gale wind of 35 miles per hour blowing across a deep lake that had a fetch of 6 miles across it, the average wave height (crest to trough) would be only about 2 feet. Increase the fetch to 30 miles, and the average wave height would grow to over 7 1/2 feet. Let the gale blow for an entire day across a fetch of 300 miles (Lake Superior would work for this), and waves would build to nearly 9 feet.

These are averages. About 10 percent of the waves in the gale on Lake Superior would be higher than 18 feet, an exceptional height, and I think fair to say, a good day to keep the canoe on top of the car.

These numbers apply to deep-water waves, waves whose wave length (the distance from crest to crest) is more than twice that of the water's depth.

Shallow lakes are notorious for quickly building breaking whitecap waves. Because of the physics of waves, steep waves become unstable at a water depth equal to half of a wave's length. So, if a wave's length is only 10 feet, and if water depths are less than 5 feet, the wave crest travels faster than the bottom of the wave can. It then falls forward, "breaking." Waves only 10 feet apart are darn close, so when a wind squall hits a shallow, broad lake, it seems to the paddler like the lake suddenly churns itself into a dynamo.

In 1847, geologist J.G. Norwood described just such a moment when he attempted to cross Ike Walton Lake on the Lac du Flambeau Reservation about 30 miles southwest of here. Ike Walton covers 1,424 acres, but only averages 10 feet deep:

"We waited some time on the shore of this lake for the wind to subside, and at noon started across. By the time we had made two-thirds of the passage, the wind increased to a perfect gale, and wave after wave, which ran almost as high as I have ever seen them in Lake Superior, broke over our canoe, until it was more than half full of water, and in momentary danger of sinking. By great exertions, the men succeeded in reaching the borders of a small island, and we dragged the canoe into a marsh . . . The lake is about two and a half miles long and one mile and a half wide, a very small sheet of water to afford so heavy a swell."

During the day, frogs are relatively quiet, often keeping their own counsel until dusk, though I have heard frogs in full mating chorus at noon, too. The power of the mating urge frequently overcomes the desired intimacy of darkness.

What frogs might you hear? To read a site for potential frog populations, look at the date, the water temperature, the temporary or permanent nature of the waterbody, and the human development along the shoreline.

Frogs adhere to a regular timing and cycle of natural events, much like wildflowers or songbirds or other animals or plants. Every species of frog regularly breeds within a certain window of time, singing to attract its mates, then mating and laying its eggs, and finally withdrawing and going quietly about its way the rest of the year.

The first frogs of the year pipe up in late April to mid-May in northern Wisconsin. Wood frogs, spring peepers, and chorus frogs use temporary, or vernal, waterbodies in which to breed. Their choruses explode from forest and bog sites.

The crescendo of chorusing frogs and toads can amaze you. At these times, love knows no bounds. The male who sings loudest may have the best luck attracting females, so volume control isn't part of the mix.

All Wisconsin frogs and toads require water for breeding. Note how the early breeders call from ephemeral waters while later breeders use permanent waters. Deeper, permanent waters like High Lake warm up much more slowly than temporary, shallow ponds and marshes. Given that water temperature greatly affects breeding, it makes sense that early breeders would utilize the shallowest waters.

But vernal ponds represent a real roulette wheel. When wood frogs, spring peepers, chorus frogs, and leopard frogs move to temporary ponds to breed, they engage in a race to breed, hatch their eggs, and have their tadpoles transform into terrestrial frogs before the water dries up. It's a gamble that eliminates the danger of predators like fish. However, in a drought year, tadpoles may be left high and dry before they have developed lungs.

Development of shorelines greatly impacts frog populations. Given Frog Island's intact shorelines and undis-

Northwoods Frog Phenology

Species	Breeding Period	Habitat
Wood Frog	Late April-early May	vernal pond
Spring Peeper	late April-late May	vernal pond, marsh
Chorus Frog	late April-mid-May	vernal pond, marsh
Leopard Frog	early May-early June	lakes, streams, marsh
American Toad	early May-early June	shallow waters
Eastern Gray Treefrog	late May-June	trees/shrubs near water
Mink Frog	June	cool, permanent water
Green Frog	June-mid-July	all permanent water
Bullfrog	mid-June-July	all permanent water

turbed aquatic plant communities, frogs should find a good home here, albeit one that is limited in size.

≈ ≈ ≈ ≈ ≈

A large boulder rests on the southwest shoreline of Frog Island. Called glacial erratics by geologists, boulders were infrequently strewn upon the landscape by the retreating glaciers.

American Indians sometimes saw these boulders as spirit rocks, stopping whenever paddling past to confer prayers and offerings upon the site. Spirit Rock, situated in Flambeau Lake on the Lac du Flambeau Reservation, may be the most notable example of a spirit rock in our area. Many people still stop near it to offer thanks and prayers.

These glacial erratics are the natural equivalents of the sculptures we so commonly build near fountains in our cities. I always try to take note of the artistry of the boulders, and their sentinel permanence upon the Manitowish. While the plants and animals come and go, the boulders will forever watch and be watched.

≈ ≈ ≈ ≈ ≈

A lake is the landscape's most beautiful and expressive feature. It is the Earth's eye, looking into which the beholder measures the depth of his own nature. — Henry David Thoreau

Paddle southwest from Frog Island toward a rounded headland on the southern shore of High Lake. Easily visible, an eagle's nest protrudes a fourth of the way down a large pine. Two immature eaglets kept an eye on our proceeding the last time we paddled by this nest, and hopefully adults and/or immatures will follow your progress as you pass by.

An eagle pair has nested on High Lake since at least 1973, though within the overall territorial area, the nesting site has moved between Frog Island and two different trees on the south shoreline.

The current nest is located on private land within a few hundred yards of buildings, amply demonstrating how eagles have acclimated to limited human development.

Unlike osprey, eagles like some shade on their nests, typically building these enormous structures a fourth or so of the way down from the top of a tree. Still, they seem to prefer a panoramic view of the landscape, so in Wisconsin, eagles choose large white pines nearly three-fourths of the time for their nests. These nests average 75 feet above the ground, and about 4 feet wide and 3 feet deep. The largest active nest in Vilas County is on Dunn Lake, and measures 11 feet deep by nearly 4 feet wide. The nest towers 105 feet above the landscape in a huge old white pine with a 42-inch diameter.

While eagles are opportunists, foraging on everything from dead deer to muskrat to ducklings, their main dietary focus remains fish. The eagles' presence on High Lake indicates that the lake has a good fishery, at least by eagle standards. In Wisconsin, typically 95 percent of an eagle's diet consists of fish, with suckers, bullheads, and northern pike representing the top three species. An occasional duck, baby muskrat or baby beaver, and carrion supplement the cuisine.

One north-central Minnesota study of prey remains collected in nests found that fish species comprised 90 percent of eagles' diets. Bullheads made up 39 percent of the fish consumed, suckers 32 percent, small northern pike 15 percent, and other fish 4 percent. Mallards, ring-necked ducks, lesser scaup, and wood ducks represented nearly 5 percent of their total diet, with ring-billed and herring gulls providing a tasty sidedish of 2 percent. Muskrat were the only mammalian species found in the nests, and at 1 percent of the total diet, represented little more than a culinary diversion.

The likely bias in this study, and in most others, is that prey remains were only gathered during the summer months, and only from the nests. Adult eagles don't always eat at the nest; some prey, like carrion deer, are too large to be hauled back to the nest. Diet also changes with the season—road-killed deer, for instance, comprise a large share of an eagle's diet in early spring before ice-out occurs in northern Wisconsin.

Other historic studies illustrate the opportunistic nature of eagles. Murie's 1940 study of eagles on the Aleutian Islands found 80 percent of their diet to be birds, the dominant fauna of the Islands. Retfailvi's 1970 study on San Juan Island in Washington found that 21 percent of the eagles' diet consisted of mammals, particularly road-killed rabbits. A 1984-85 study in the Apostle Islands of food habits, foraging areas, and chemical contaminants in eagle foods found fish represented 52 percent of prey remains and birds 47 percent. Three species of suckers comprised 28 percent of the total diet, while herring gulls made up another 20 percent.

So, what are these eagles on High Lake likely to be eating? The fishery surely provides the bulk of the eagles' diet. High Lake is well-known for its walleye, and while eagles will eat walleye, the majority of their diet here is likely suckers, bullheads, and northern pike, all of which populate the lake in good numbers. Unfortunately, time and money constraints make it impossible for researchers to gather study data that separates out nest-to-nest feeding differences, so it's impossible to say what the exact diet may be of this eagle pair.

≈ ≈ ≈ ≈ ≈

High Lake flows out into a short channel that opens into Rush Lake, which in high water appears as a very shallow 44-acre lake, and in low water presents itself as a wetland with a stream running through it. At a maximum depth of 7 feet, Rush Lake typically grows aquatic plants from end to end, although boat traffic and the gentle Manitowish current tend to keep the narrow river channel open.

A rough birchbark sign hand-lettered with the greeting "Wildlife Sanctuary" appears at the head of the lake, a nonofficial but accurate posting. If water levels are high enough, take time to explore away from the channel and into the sloughs for wetland birds.

Lakes are frequently classified by their water circulation (hydrology) into four main groups:

1- seepage
2- drainage
3- drained
4- spring

Wetland Birds

Not all wetlands are created equal. Birds apparently understand this. A black ash swamp offers very different nesting, feeding, and cover opportunities than does an open bog. Some of our most distinctive bird communities are associated with specific wetland types. Still, a disclaimer should be noted: Exact correlations between bird species and habitat preferences seldom exist. Most species don't have strong habitat preferences (see Janet Green's book *Birds and Forests* for a very thorough look at the complex relationships between birds and habitats).

Nevertheless, some birds are clearly associated with wetland communities, and as a paddler, you should have a search vision that includes these species as "most likely" to be seen when entering their characteristic habitat. Be looking for the following (adapted from Green, 1995):

Sedge Meadow/Fen
Sedge Wren
Bobolink
Savannah Sparrow
Le Conte's Sparrow

Open Bog/Muskeg
Tennessee Warbler
Nashville Warbler
Palm Warbler
Connecticut Warbler
Lincoln's Sparrow

Closed Bog
Spruce Grouse
Yellow-bellied Flycatcher
Gray Jay
Boreal Chickadee
Golden-crowned Kinglet
Ruby-crowned Kinglet
Swainson's Thrush
Hermit Thrush
Cape May Warbler
Yellow-rumped Warbler

Alder/Willow Thickets
Alder Flycatcher
Gray Catbird
Golden-winged Warbler
Yellow Warbler
Northern Waterthrush
Common Yellowthroat
Clay-colored Sparrow
Song Sparrow
Swamp Sparrow
American Goldfinch

Many of the species listed above, like gray catbird and song sparrow, are good examples of why we can't absolutely habitat-type most birds. On our property along the Manitowish River, gray catbirds and song sparrows sing every spring morning from the willow shrubs. So, they must be wetland birds, right? Well, no. Both species are apparently more attracted to the structure of the vegetation than the actual species of plants. Catbirds are equally likely to nest in ornamental shrubs in the suburbs, as well as early regenerating clearcuts. Song sparrows seem to most prefer edges, whether they be hedgerows of farm fields, forest edges, residential shrubs, or river shorelines.

The bottom line? Birds often have preferred habitats, but few species are restricted to only one habitat.

Rush and High lakes are drainage lakes. They have a surface inlet and outlet stream—the Manitowish River. Nearly all the lakes along the Manitowish Chain are drainage lakes, given that the Manitowish has to flow in and has to flow out. Drainage lakes make up two-thirds of the lake area of the Northern Highlands/American Legion State Forest (NH/AL), but comprise less than a one-third of the total number of lakes.

The most common lake type in the NH/AL is seepage lakes. Seepage lakes seep—they have no surface outlet and no surface inlet. Landlocked, the lakes can only receive water via rain and groundwater inflow, and lose it via evaporation and groundwater outflow. Some 60 percent of the lakes over 10 acres in the NH/AL are seepage lakes. Typically, small seepage lakes sit at the top of the topography, while larger drainage lakes gather water at the low end.

Drained lakes have an intermittent outlet, but no inlet, while spring lakes have a permanent outlet, but no inlet. Both are much less common than seepage and drainage lakes.

The source and circulation of water in a lake significantly influences water quality, but the elevation of a lake in the landscape may be of even greater importance. The water is so clear in High and Rush lakes, in large part because they are so high in the landscape, so near the Continental Divide. Landscape position makes a big difference in the chemistry and biology of a lake.

Picture this area as a big sandbox. Given that most of our soils here resemble your favorite beach, the sandbox picture should work pretty well. Now, recalling your childhood, mentally make like a bulldozer (or glacier) and grade your sandbox. Create a topography by making one end higher than the other, scoop out holes at different elevations in your box, make narrow depressions and ridges for water to run—some from hole to hole, while leaving other holes with no connecting ditches. Consider the wooden bottom of your sandbox to be bedrock. Have a good time and feel free to make funny bulldozer noises.

Now, imagine a cloud coming over and rain falling on this landscape, watching all the time how the water moves. Water will flow visibly over the sand surface, as well as invisibly under the sand. Holes at the top of the sandbox will collect water, and represent lakes that are at the top of the water table—these typically are clear seepage lakes. Nearly all the water that comes into them is through rain, while very little seeps in from the ground.

Water seeps from High Lake into the ground, and flows downhill underground just as surface water flows downhill over the ground. Like surface water, the groundwater picks up minerals from the soil that it flows through, depositing these minerals, like calcium and silica, in the next lake or river that the waters encounter. The lower the next lake in the landscape, the more groundwater it typically receives compared to precipitation. Since minerals feed animals and plants, lakes and rivers lower in the landscape tend to have more aquatic life. The richer water produces more growth.

For over a decade, the University of Wisconsin's Trout Lake Station has studied three lakes that illustrate the sandbox/landscape position concept—Crystal Lake, Big Muskellunge Lake, and Trout Lake, all just southwest of Rush Lake near Boulder Junction. A narrow ridge of land several hundred feet wide separates Crystal and Big Muskie. But Big Muskie lies 5 feet lower in the landscape, and receives nearly 15 percent of its water from groundwater while Crystal receives only 5 percent from groundwater.

The differences in aquatic communities are startling, even though these lakes are right next to each other. Crystal Lake traditionally supports far fewer fish, and until recently, was dominated by yellow perch. Now exotic smelt have invaded and dominate the fish community. Big Muskie has a far more diverse fish community, supporting muskie, walleye, and northern pike, and, not surprisingly, a healthy number of anglers.

Lower yet in the landscape lies Trout Lake, which receives much more groundwater, and is also connected to other lakes through rivers and creeks. Trout has an even more diverse collection of fish, invertebrates, and plants than Big Muskie because of the productivity of its waters.

So whenever paddling on a stretch of water, take a look at the elevation of the water relative to the rest of the watershed. While many other factors influence water quality, position in the sandbox plays a highly influential role.

Read more about neighboring lakes at the end of the chapter.

≈ ≈ ≈ ≈ ≈

Rush Lake's channel edges are dominated by expanses of pickerelweed and water lily that flower in great beds in July. In high water, you might try to paddle up Grassy Creek into Grassy Lake. However, beware that paddling creeks in the Northwoods is often a lesson in entangling oneself in the arms of alders, or meeting up-close-and-personal the logging handiwork of beavers. Still, Grassy Creek flows mostly through bog, so in spring you stand some chance of success given the lack of woody vegetation along the banks.

A culvert under a narrow two-lane road soon connects Rush Lake with Fishtrap Lake. A take-out is on the immediate left if you wish a quick exit off the water. A small parking area here provides room for a couple of cars.

Two islands loom immediately ahead upon entering Fishtrap Lake. The one farther west (on the right) has a home on it, but the other is uninhabited. The western shoreline is dominated by homes, so stay along the eastern (left) shore—much of it is state-owned and still relatively wild.

The name conferred upon this lake, Fishtrap Lake, leads some to wonder if this was a historical site for trapping fish. The native tribes were well aware when fish would spawn and were known to use weirs to trap and then spear fish as they moved en masse to their spawning grounds. When sturgeon spawn in late May/early June, they can easily be picked up along the shoreline (you still need a strong back though). A weir would help direct the sturgeon to a site easily accessible for spearing or capture.

Sturgeon were particularly prized for their size and their oil content. The

Grassy Lake

Grassy Lake, an infertile, softwater, spring-fed drainage lake, covers 220 acres with a maximum of only 4 feet of water. It's more of a frying-pan lake than one of the pothole lakes we talk so much about having in northern Wisconsin. Aquatic plants including watershield, pickerelweed, and pondweed dominate the surface of the lake, offering truth-in-advertising to the name, though none of these are true grasses.

Muck comprises most of the bottom material, so swimming is off anyone's to-do list here. It's plant life that's the draw. The shorelines offer a diverse array of wetland communities to explore, including sedge meadows, shrub-carr, muskeg, alder thicket and spruce-tamarack swamp. Look for plants like sweet gale, marsh cinquefoil, blue flag iris, swamp milkweed, and Canada bluejoint grass along the edges, as well as larger shrubs like speckled alder, leatherleaf, bog birch, and meadowsweet.

Wild rice may be present here as well, given that the lake provides the right physical characteristics—a minimum water depth, a muck bottom, and some moving water.

Look for an eagle's nest at the northern end of lake.

A small parking area off High Lake Road on the east side of the lake provides quick canoe access.

George L. Brown Cultural Museum in Lac du Flambeau houses the mount of a sturgeon over 7 feet long and weighing 200 pounds, speared on Lake Pokegama in Lac du Flambeau. Sturgeon commonly live over a century and reach astounding sizes; the *Fond du Lac Journal* in Wisconsin reported the capture of a 9-foot-long, 297 pounder in the spring of 1881.

While sturgeon are noted for tasting good, a spawning female also often produces 50,000-700,000 eggs. The eggs are a highly valued caviar today, and a food very likely coveted historically.

Sturgeon no longer live in this section of the river, victims of the Rest Lake and Fishtrap dams, which do not provide a fish ladder for migrating fish. But a remnant population of sturgeon does populate lower stretches of the river.

Read more about sturgeon weirs on the next page.

≈ ≈ ≈ ≈ ≈

Fish populations rise and fall in relation to many factors, often from pollutants like industrial chemicals, farm fertilizers, and human sewage. However, one pollutant that does not appear to be causing current direct damage to northern Wisconsin lakes is acid rain (though it may be contributing to mercury availability). Forty percent of the state's northern lakes are considered vulnerable to acid rain. When sulfur dioxide is pumped into the air from burning fossil fuels, it

Sturgeon Weirs

The Cass Expedition, which included Michigan Territorial Governor Lewis Cass, Henry Schoolcraft, and James Doty, camped on June 27, 1820, at the mouth of the Ontonagon River on Lake Superior. Here, in view of the Porcupine Mountains, Cass, Schoolcraft, some local Chippewa guides, and a few others ascended the Ontonagon in search of copper. On their route upstream, they came to a sturgeon weir. Doty described it as follows:

"At the distance of four miles we reached a sturgeon fishery, which the Indians have established in the river by means of a weir extending from bank to bank. This weir is constructed of saplings and small trees, sharpened and drove into the clayey bottom of the river with an inclination down stream, and supported by crotched stakes bracing against the current. Against the sides of these inclined stakes, long poles are placed horizontally and secured by hickory withes in such a manner as to afford the Indians a passage from one end to the other, and at the same time allow them to sit and fish upon any part of it. The sturgeon are caught with an iron hook, fixed at the end of a long, slender pole, which the Indian, setting on the weir holds to the bottom of the river, and when he feels the fish pressing against the slender pole, jerks it up with a sudden and very dexterous motion, and seldom fails to bring up the sturgeon. On one side of the weir, an opening is left for the fish to pass up, which they do at this season in vast numbers, but in their descent they are hurried by the current against the hooks of the savages, who are thickly planted on every part of the weir. The number of sturgeon caught at this place is astonishing, and the Indians rely almost entirely upon this fishery for a subsistence. What is not wanted for immediate consumption is cut into thin slices and dried or smoked . . . The sturgeon are generally from two feet to four feet in length . . . This fishery is of great importance to the Indians of the region, and appears to have been known to them for the earliest times, and has been constantly resorted to without any apparent diminution in the quantity taken. Henry [Alexander Henry, a fur trader during the 1760s and 70s, and the first Englishman to leave a record of the upper Great Lakes] says in 1765 that a month's subsistence for a regiment could have been taken in a few hours time."

The Ojibwa probably would have caught sturgeon wherever and whenever possible. I can find no references that list the Manitowish River as a site of a sturgeon weir—it's pure conjecture on my part. But the Manitowish River still supports a remnant sturgeon population, so it's highly probable that the Ojibwa would have caught sturgeon on the river. The sturgeon weirs would have been placed in the rocky riffles where sturgeon typically spawn. Riffle areas supposedly existed between the lakes, but the water today is artificially raised by the Fishtrap Dam.

forms sulfuric acid, a major component of acid rain.

Legislative efforts to reduce acid rain began in 1985 when Wisconsin Act 296 mandated a 50 percent reduction in sulfur dioxide emissions by the year 1993. As sulfur dioxide emissions declined in the ensuing decade, the acidity of rain and snow collected at monitoring stations across northern Wisconsin showed a steady and substantial decline, too. By the late 1990s, the pH of precipitation typically exceeded 4.7, a level that limnologists feel protects northern lakes from further acidification.

Lakes that were monitored in three regions of northern Wisconsin since 1987 increased from an average pH of 5.5 to 6.0 or higher. This 0.5 change in pH amounts to a threefold decrease in acidity (pH is measured on a log scale). Most importantly, lakes below a pH of 6.0 often lose or have reduced fish and plant communities. So, the very positive news is that as the 21st century begins, acid rain deposition continues to decline, and impacts appear minimal in northern Wisconsin lakes.

Lakes highest in the landscape are typically most sensitive to acidification because they receive little groundwater. Groundwater contains substances from the soil and rocks that neutralize, or buffer, acids. Thus, along the Manitowish River, those lakes farther downriver are least susceptible to acidification.

≈ ≈ ≈ ≈ ≈

The first peninsula jutting into Fishtrap Lake has a DNR campsite and an eagle's nest on it. A second campsite is located on a narrow finger of land marking the start of the river channel, though the topo maps refer to the next several miles of water as "Wide Creek." When Wide Creek meets Johnson Creek, according to the United States Geological Survey (USGS) map, the Manitowish begins. But the naming of a beginning and ending to many rivers is often arbitrary and may be hydrologically inaccurate. While I have conferred headwater status upon High Lake, one could just as easily place the mantle upon Johnson Creek.

For that matter, why is Wide Creek a creek and not a river? When does a creek become a river, a river become a lake? At nearly a quarter-mile wide, Wide Creek could easily be called a lake, as could Johnson Creek where it widens before meeting Wide Creek. For waterbodies, size and assigned status are often a matter of our biases, not unlike most human endeavor.

Read more about river and lake size on the next page.

Ecologists classify streams in a hierarchical system called stream order. The smallest, permanently flowing stream without any tributaries is termed a first order stream. The union of two first order streams creates a second order stream. The union of two second order streams creates a third order stream and so on (for algebra lovers, when two streams of order n merge, a stream of order $n + 1$ results.) The venation, or branching pattern, of the various orders of a river system looks like a genealogical family tree or a diagram of the blood stream.

Obviously, smaller order streams far outnumber larger order streams—the rule of thumb is there are typically

Size Matters—Sort Of

The Random House Unabridged Dictionary says a brook is the smallest natural stream of freshwater, while a creek is larger than a brook, but smaller than a river. A river is defined as larger than a creek and flowing in a definite channel, and a stream is any flowing body of water whether a brook, creek, or river. Got that?

Ecologists often classify water flow by saying rivulets flow into brooks, which flow into creeks, which flow into rivers. They see a stream as any mass of water moving through a defined course. So, when in doubt, call any moving water a "stream," and you'll have the highest odds of accuracy.

I like one of the definitions of "river" in my College Edition Random House Dictionary: "any stream or copious outpouring."

In his book *The Bird in the Waterfall*, Jerry Dennis describes the youthfulness of creeks and brooks as "the kicking colts of the aquatic world," an apt description of their playfulness and spontaneity.

Lakes and ponds have a definition problem, too. Limnologists, the scientists who study freshwater, define a lake as any standing body of nonmarine water. So, if it's not an ocean, it's a lake. But like most scientific definitions and terms, different regions of the country have their own take on the terms. Walden Pond in Massachusetts, likely the most famous pond in all of America, is 103-feet deep, clearly a lake by most everyone else's standards.

In Wisconsin, we're much more liberal than New Englanders with our definition of a lake. We're willing to call nearly anything that holds water a lake. Thus, Wisconsin totals 14,973 documented lakes, so sayeth the Wisconsin DNR publication "Wisconsin Lakes." About four-fifths of those, or 11,821, are less than 24 acres, and nearly 9,000 of these are so small that they're unnamed.

Almost one million of Wisconsin's nearly 36 million acres are classified as lakes, making our surface area about 3 percent water. Vilas County, home to most of the length of the Manitowish River, has the greatest surfacewater area of any county, its 1,237 lakes covering nearly 15 percent of the county.

three to four times as many streams of order $n - 1$ as of order n, each of which is roughly half as long, and drains about one-fifth as large an area. The relatively constant ratio of three to four times as many streams of a given order to the number of the next higher order shows the highly organized character of river systems.

An easy way to draw out stream order is to place tracing paper over a topographic map, and trace only the blue lines. The resulting network of streams can then be analyzed. If you want to get fancy, use different colors to highlight each order of stream.

There are a few problems with the stream order system. Maps may not accurately identify all streams and their courses, while wet and dry years may either change the determination of a stream's order or create an intermittent stream that might not run in a dry year.

Stream order also ignores the entry of streams of order n into order $n + 1$. Only when streams of equal order merge do they create a higher order stream. It's significant to note that nearly half the streams of a given order enter directly into streams two or more orders higher.

The Manitowish begins as a first order stream if you use High Lake as its source. Or it begins as a third order stream if you use Johnson Creek as its source—Garland Creek and Syphon Creek join to form a second order stream, while Salsich Creek and an unnamed creek also form a second

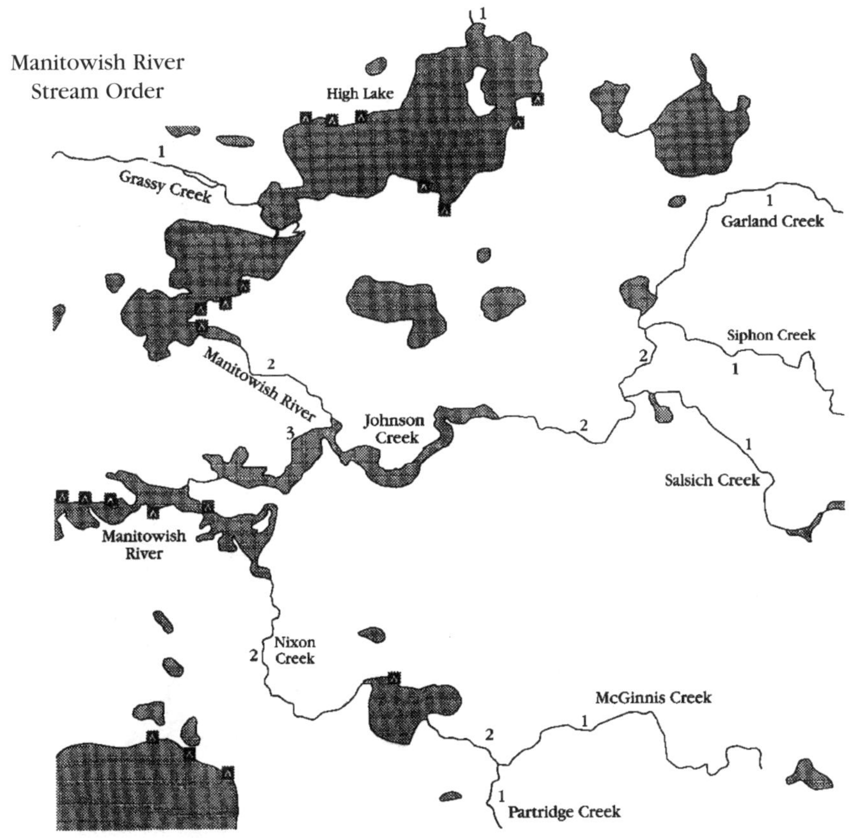

order stream; the two second order streams come together in section 8 to form the third order Johnson Creek. Where the third order Trout River and third order Manitowish River come together in Spider Lake, the Manitowish becomes a fourth order stream.

At the extreme end of stream order, tenth order streams are mighty rare; the Mississippi appears to be the only one in North America, draining an area of about 1.8 million square miles over its 3,600-mile-long course.

≈ ≈ ≈ ≈ ≈

"Wide Creek" represents the first lengthy river section of the Manitowish. Excellent examples of bog and sedge meadow box in the river on either side for nearly a half mile. Leatherleaf, bog rosemary, and sweet gale dominate the shoreline, all aquatic shrubs that provide thick root systems to stabilize the erodable interface of land and water.

Sweet gale *(Myrica gale)* may be the most common shrub along the Manitowish, yet in all the trips I've led over the years along the river, I've only met a few people who knew the species. A shame too. Crush one of its leaves, and a fragrant sage-like aroma emanates. Use the fresh leaves to season the next fish you catch along the river, just as the American Indians and early Europeans did. If you have a hand magnifying lens along, take a look at the resin dots on the leaves—they furnish the aroma.

To get a feel for how well sweet gale anchors the shoreline, try to pull out one of the plants. Unless you wrestle for a living, the plant won't budge. But you might, so be careful not to flip your canoe in the effort.

Sweet gale's cousin, sweet fern *(Comptonia peregrina)*, also belongs to the bayberry family, and scents the open upland sands so common in the Northwoods. Both species are nitrogen-fixers, improving the soil quality while keeping it in one place. Especially noticeable on hot, sunny days, sweet fern's heavy aroma also "fixes" your location—many people associate the smell of the Northwoods with sweet fern.

The topo map shows islands in Wide Creek as you progress southeast, but later in the summer when the aquatic vegetation is high, these islands can't be differentiated from the overall wetland complex.

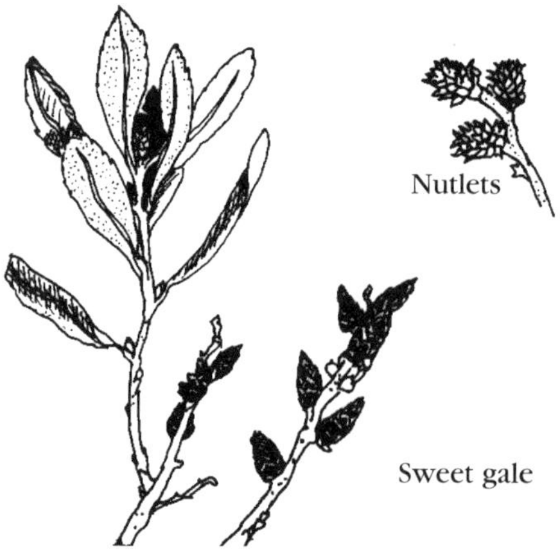

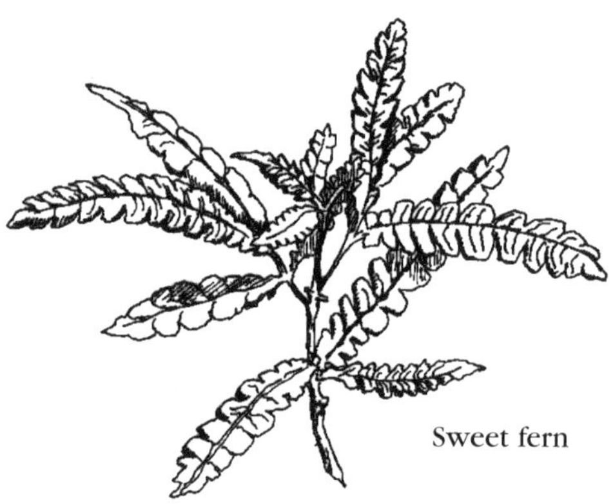

Huge clonal beds of pickerelweed and water lilies emerge from the shallow waters. Water lilies grow large floating leaves on stems that rise through water usually about waist deep. Often a few large roots of water lilies also float on top of the water, likely the result of the foraging efforts of a muskrat or beaver. These huge rootstocks look as big as softball bats. A paddler friend of mine once saw a particularly large root on the surface, and wasn't sure if he had a young alligator awaiting him.

Water depth, as mentioned at the start of our paddle, plays a major role in the structural characteristics of aquatic plants. In the shallowest zone of water, from the shoreline to about knee-deep, grows the emergent plant community (bulrush, cattail, bur-reed, arrowhead, pickerelweed, and others), a community known for its tremendous productivity. Emergent plants are well adapted to the fluctuating water levels typical of shorelines and shallows. Complexly interlaced roots stabilize their position in the wash of water and sediments, and not incidentally, stabilize the shoreline soil as well. Spongy leaves filled with air spaces provide buoyancy to match the swings in water levels. And they've evolved to utilize a set of flexible reproductive strategies, vegetatively reproducing from spreading roots and rhizomes in high water conditions, and by dropping seeds on newly exposed mud flats during low water.

In deeper water, from around knee-high up to chest-high depths, the floating-leaf community dominates. Water lilies, watershield, and floating-leaf pondweeds rule the roost, their flexible leaf stalks giving with the wave-action and helping the leaves ride out the surface tumult. The circular or elliptical leathery leaves resist

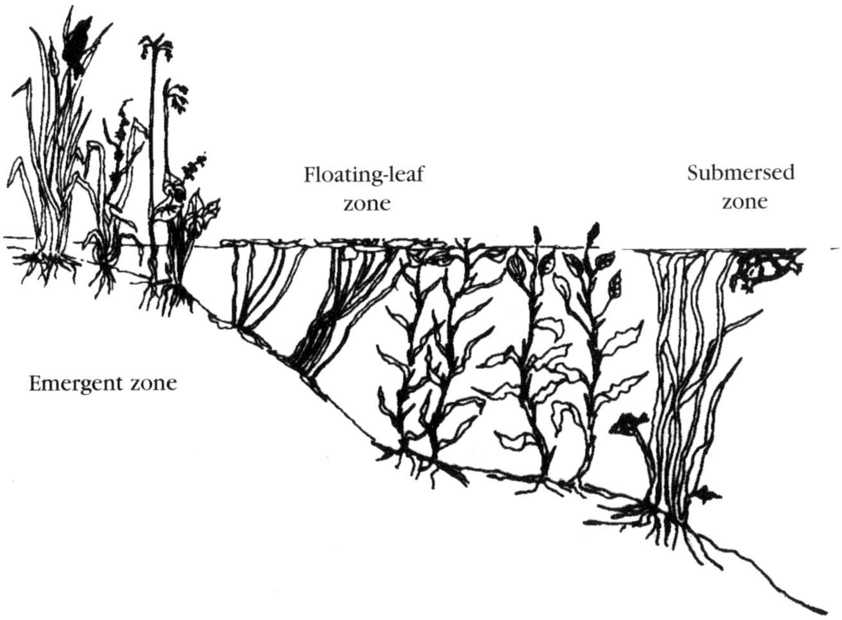

the pounding of waves, while their waxy surface protects the leaves from water loss. Tiny pores called stomata cover the upper surface, allowing air to be exchanged without absorbing water—I guess a bit like how Gore-Tex is supposed to work (but seldom does). To overwinter, floating-leaf plants die back to their rhizomes. In spring, they shoot up quickly to reach light and begin photosynthesis before their rhizomes run out of stored food.

The submersed plant community resides in water from waist-high and deeper, growing under water as deep as available light permits. Stems and leaves flex with the movement of water and contain spongy tissue to create buoyancy. Leaf shapes vary from the wide arching leaves of muskie weed (large-leaf pondweed) to the ribbon-like leaves of wild celery, but all provide essential habitat for invertebrates to rest, feed, and call home. Many submersed plants also produce some floating or aerial leaves, flowers, and fruit. Reproduction primarily occurs through vegetative cloning from rhizomes, but some, like naiads, produce copious seeds in fall.

For an in-depth, easy-to-read, well-illustrated guide to common aquatic plants in the Upper Midwest, try *Through the Looking Glass: A Field Guide to Aquatic Plants* produced by the Wisconsin Lakes Partnership, University of Wisconsin Extension, and the College of Natural Resources at the University of Wisconsin-Stevens Point.

≈ ≈ ≈ ≈ ≈

This stretch of river opens into a wide wetland complex that can be a bit confusing to paddle since the river channel seems to disappear amidst all the plant life in late summer. Have a topo and compass handy in case you aren't sure where to proceed, because following the negligible current may be problematic. An eagle's nest is located on the west shoreline just before the juncture with Johnson Creek.

Johnson Creek enters from the east, the juncture marked by a DNR campsite on the left side of the river. Stay well to the right (west), or you could inadvertently paddle quite a ways up Johnson Creek—the creek, the wetlands, and the Manitowish look much the same here.

If you have the time, a purposeful exploration up Johnson Creek offers a quiet wilderness paddle. The Johnson Creek macrosite has been listed by the Biotic Inventory and Analysis of the Northern Highlands/American Legion State Forest as a "primary site of high significance." On the 220,000 acres NH/AL State Forest, only 36 such sites received this designation, which calls for strong habitat and native community protection. The Wisconsin Bureau of Endangered Resources believes these sites to be of comparable ecological significance to existing state natural areas.

Stands of watershield intermix with acres of water lilies at this juncture. Watershield produces a much more demure flower than the water lilies, and often goes unnoticed for its lack of showmanship. Still, watershield has one unique characteristic that is sure to yield a response, usually in the form of "yyyyuuucckk." Watershield's

Coontail

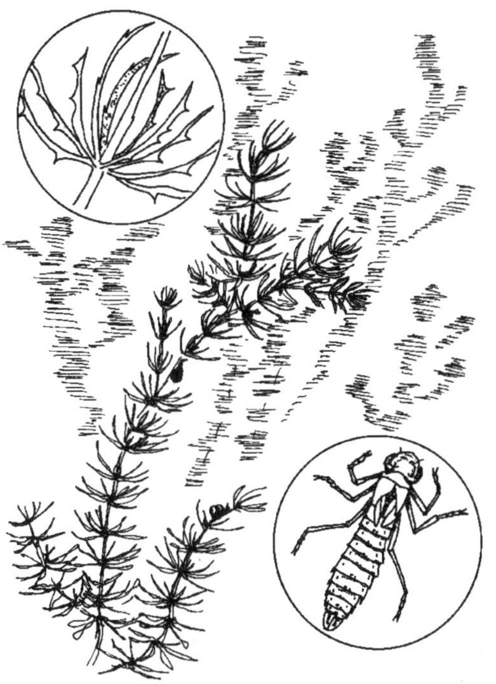

Dragonfly larva utilize coontail

Coontail (*Ceratophyllum demersum*), a common submersed plant in slow streams and ponds, grows in dense beds so thick that the colony sometimes appears as a literal underwater forest. Feathery whorls of leaves showing teeth along one edge of each leaf segment offer a sure identification key—coontail is our only submersed plant with teeth on its leaves. The leaves tend to bush up at or near the end of the stem like a raccoon's tail, hence the name.

Like many submersed plants, coontail has no actual roots, though it may sometimes appear rooted in the mud. Coontail typically floats just beneath the water surface, its small flowers so tiny as to be unnoticeable. In an unusual flowering process, the male stamens mature and detach, floating to the surface where they release their pollen. The pollen then rains down on the female flowers, which produce fruit. But like in most aquatic plants, fruit and seed reproduction is uncommon.

Vegetative reproduction is much more successful. In autumn, the branch tips turn very green and become bunched up. These winter buds detach from the dying branches and sink to the bottom, where they will rise again in spring and produce new branches.

Coontail can grow in poor light, so it can be found as deep as 18 feet in the water. Given its high adaptability to low oxygen content and pH, it grows well—sometimes too well for people's tastes.

Coontail serves as host for some smaller plants, mainly algae, that use the larger plant as a surface to grow on. But its major value is as a haven for aquatic invertebrates like snails and insect larvae. Since it harbors up to 50 percent more small organisms than most aquatic plants, its main attraction to wildlife may be its invertebrates.

Johnson Creek Macrosite

The Wisconsin Bureau of Endangered Resources identified four important macrosites on the Northern Highlands/American Legion State Forest. One of them, the Johnson Creek macrosite, totals 3,376 acres, and contains an unusual concentration of softwater springs, spring-fed headwater streams, and associated wetland communities. Four high-quality, softwater streams emanate from spring ponds within this area. Garland Creek, Goodyear Creek, Salsich Creek, and Siphon Creek all flow for 1 to 3 miles before joining Johnson Creek. Garland Creek feeds into Johnson Lake, a shallow 24-acre undeveloped drainage lake ringed by sedges, open bog, and conifer swamp. Siphon Creek joins these waters just below Johnson Lake, and Salsich and Goodyear creeks come in another half mile south. "Johnson Creek and Pines" is currently designated as a state natural area, as is one of the two ponds at Goodyear Springs.

To the east of Johnson Lake lies the Johnson Lake Barrens State Natural Area, which has been managed with controlled fires to promote an open community of bracken grassland, dominated by bracken ferns, sweet fern, blueberries, and barren strawberry. The 189-acre site contains 76 acres of "pine barrens" or "bracken grasslands" in the uplands. For birders who covet sightings of uncommon northern birds, Connecticut warblers and spruce grouse can be found here.

GOODYEAR SPRINGS—EAST
STATE NATURAL AREA

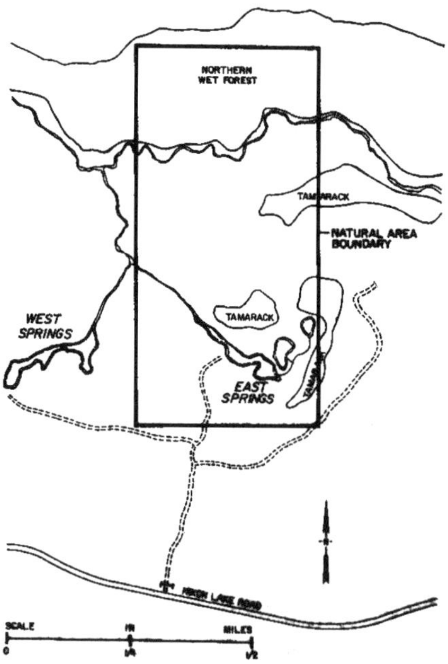

Nearby Goodyear Springs-East State Natural Area covers 180 acres, 135 of which are sedge meadow. A tamarack/black spruce wet forest covers another 40 acres. The 3-acre spring pond is one of the few unaltered spring ponds left in Vilas County, and is part of a larger spring pond complex known as Siphon Springs.

For descriptions and maps of every State Natural Area in Wisconsin, open the DNR Web site at <www.dnr.state.wi.us/org/land/er/snas >.

Watershield

rounded leaf is held up by a long hollow stem that attaches dead center to the underside of the leaf. Pick one of the stems. The stem is completely coated with a clear gelatin-like substance that can only be described as slimy. The evolutionary purpose of slime is what, you may ask? Beats me. Possibly the coating discourages hungry browsers or helps to deflect water currents that could uproot the plant or . . . (add your best guess).

≈ ≈ ≈ ≈ ≈

Wend your way through these vast lily and watershield fields for another mile. I've watched great blue herons in this area stand on top of the mat of dense lily pads while they forage for fish.

Herons are Zen masters, pliers of the trade of total attention, authors of the unwritten book *Zen and the Art of Stalking Fish*. To watch a great blue totally focused over its prey is both a portrait of still life and warp-speed performance art. I've dabbled in yoga for many years, and to stand this still for this long, and then be able to strike with eye-blinking speed, demonstrates the perfect blend of mind and body so seemingly unattainable to yoga bunglers like myself.

Great blues also help give breadth to the word "adaptable." They eat with equal ease in fresh and salt water, in uplands and wetlands, in tidal and nontidal marshes. Opportunists, they typically feed heavily on fish and crustaceans, but when the need arises, they will feed even in urban settings. They've been seen stalking and eating gophers on public soccer fields.

They do have an Achilles' heel—the shrinking number of places suitable for their nesting colonies, or rookeries. One researcher mapped heron colonies over the entire length of the Mississippi River from St. Louis to Minneapolis and found flood plains represented the best refuges. Specifically, the best colony sites occurred in undrained and undiked flood plains protected from wind and humans, with nesting areas within 100 meters of the water. Most colonies were located below dams and/or near river junctions where there were lots of trees and channels. Herons frequently utilize a screen of trees between the colony and water for protection from human disturbance.

Great blues not only have trouble finding good colony sites, but when they do, they often kill the trees they nest in, sometimes by a process called *guanotrophy*. I like this word. It sounds like a trophy for the greatest guano-producer, or maybe it's a trophy made of guano. We all have people in our lives we'd like to give that trophy to.

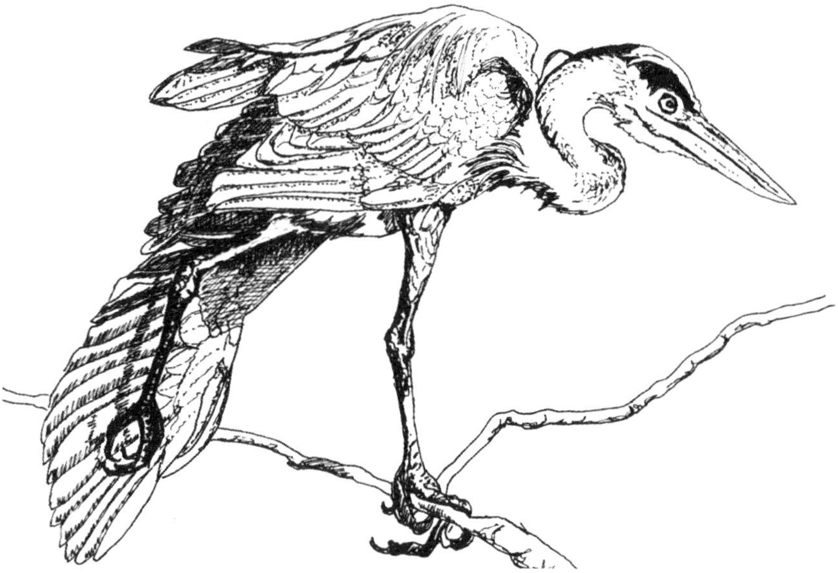

Great blue heron

Anyway, the real definition of guanotrophy is the killing of trees by the guano masses that accumulate beneath them, forcing the herons to move on to another site.

Sometimes colonies are abandoned when a pair of bald eagles decides to nest there. Eagles prey on heron chicks, and sometimes attack even the adults.

Disturbance from any source can cause colony abandonment. When frightened herons fly off their nests, crows that have been waiting in the shadows swoop in and eat the unguarded eggs. Sometimes herons move in response to specific human disturbance, from too much noise to too many ecotourists. And sometimes herons pull out of a colony for no apparent reason at all. So they're often on the move, ancient transients along ever-changing rivers and wetlands.

Great blues most often nest in the tops of mature deciduous forests, but I've seen colonies in old growth white and red pines, too. The colonies can range from 10 nests to several thousand nests.

Protecting feeding sites is just as important as protecting the colony sites. Great blues can fly 30 miles or more from their nest sites to feed, but they still need relatively undisturbed, rich prey sites to fill their large appetites. Chicks hatch at less than two ounces, but weigh four pounds within two months. That level of growth requires a lot of fish and frogs.

Great blue heron numbers appear to be stable or increasing across most of the U.S. and Canada based on 30 years of random volunteer surveys. However, we have no ongoing statewide or regionwide census like the sandhill crane count that gives us long-term population trends. Perhaps it's time a citizen group proposes teaming with the DNR to quantify heron numbers so we know what we have. Otherwise, how will we know if herons are losing ground?

Great blues are a species virtually unchanged from those stalking prehistoric swamps 1.8 million years ago, so observing a large colony's sounds and sights and smells is a free ticket on a time machine. Take the ride.

≈ ≈ ≈ ≈ ≈

The river briefly narrows down in section 14 from its flowage-like spread and now wears the cloak of a river channel again. The first road (Stiloki Lane) visible from the river in several miles leads down to the river

Manitowish

The river loops and coils,
lassoing islands of marsh.
To go three miles,
I paddle eight.
But the fastest way is seldom
the path to any heart.

I anticipate each bend
like a childhood present,
for eagles and painted turtles,
belted kingfishers and otters await.

Feeling five again,
I sneak up on them
soundless, like the river,
expectant, like the child.

Around one corner a beaver dam
backs up a channel of the river.
Primitive grunts roll over the beaver pond.
Several herons land on the tip spires
of black spruce and balsam fir
enclosing the rookery.
They wave in a ballet of wind
like Christmas tree stars.
Others glide and circle,
glide and circle,
probing the air for clues,
ancient voices urging caution, caution.

Several spiral around their nests.
One folds her wings as she lands
wrapping herself in a slate blue shawl.

and brings houses right down to the northern shoreline. One home historically has offered the less-than-pleasing natural perspective of some 15 lawn chairs all lined up along a lawn mowed right down to the water's edge, a rather jarring contrast to the last hour or more of paddling without any human habitation in sight.

Vilas County, as well as every other northern-lakes county, has wrestled with appropriate land-use/shoreline regulations that must somehow balance aesthetics, ecology, and property rights. How far should homes be set back from the water? How much shoreline should remain natural and how much can be developed? What should minimum lot sizes be? How long should piers extend out into the river? On and on it goes.

Are we the Northwoods or the Northlawns?

What we see in lakes depends much on what we bring to the shore—King Arthur's sword or the Loch Ness monster. — Peter Steinhart

The Wisconsin DNR has begun efforts to dampen the anticipated profound impacts of future development around rivers. Called the Northern River Initiative (NRI), its mission statement is to "provide protection options for northern Wisconsin streams and rivers that have high ecological significance, outstanding natural scenic beauty, or special recreational values."

The River Initiative came to life as an offshoot of the DNR's more general Northern Initiative, which asked Wisconsin residents back in 1994-95 what they wanted to happen with the Northwoods. The overwhelming response was "to keep the North the North." People expressed concern, among other things, about development on the banks of previously undeveloped rivers, recognizing that as lakeshore properties dry up, the push for development will shift to rivers. This scenario is already unfolding—you need only look at real estate listings and see the increasing number of river listings and the rising prices.

Continued development appears inevitable, so the question is whether development can be directed responsibly and positively to limit the impacts on a river's natural characteristics. Can we be proactive rather than reactive or remedial?

The WDNR solicited public comments on the Northern River Initiative in 1998, receiving more than 500 responses. People expressed concerns about increased use, crowding and conflict, water pollution, and declining access. They suggested a range of possible remedies including acquisition of land, the promulgation and enforcement of zoning ordinances, providing landowners with financial incentives to forgo development, and education initiatives focused particularly on those who own land along the rivers. Interestingly, few respondents supported the notion that we could rely on landowners to be good stewards. History has taught us, in spades, to believe otherwise.

The NRI teams then developed a river rating scheme to evaluate their cultural, recreational, and natural resource values for the 20 northern counties included in the initiative. They started with 5,500 streams, then

Shoreland Ordinances

Shoreland zoning aims to protect water quality, fish and wildlife habitat, recreation opportunities, and natural beauty. It's a tall task, one beset by resistance of those who fear government regulation of private property. The true fear should not be about the laws that will be imposed, but rather about what will occur if comprehensive laws are not enacted. Wisconsin Dells jumps to mind.

"Resistance is futile" (as the Borg told Picard in "Star Trek"), so let's agree that shoreland ordinances must be done, and done well. Consider all the changes that have taken place on our northern lakes in the last 20 years, and then try to look ahead at the development pressures that will occur over the next 20 years (or 50 or 100 years). Ask your crystal ball to give you the picture with strict ordinances, and then without ordinances. All hyperbole aside, the fate of our lakes hangs in the balance. Only our will to enact strict laws will protect our waters.

Wisconsin's Shoreland Management Program is a partnership between state and local governments that requires the adoption of county shoreland zoning ordinances to regulate development near navigable lakes and streams. The state has written minimum standards, found in ch. NR 115, Wis. Administrative Code. You can print out a copy from the DNR's Shoreland Management Web site: <www.dnr.state.wi.us/org/water/wm/dsfm/shore/title.htm>.

The four major goals of ch. NR 115 are:

- control the density of development
- create a protective buffer of vegetation along public waterways
- minimize disturbances to water resources
- protect wetlands

Individual counties may write additional requirements, as well they should, given that recent studies indicate these minimum standards fail to protect many of the values we associate with our lakes.

Vilas and Iron counties, the two counties through which the Manitowish River flows, have adopted their own county shoreland zoning ordinances, while individual townships within the counties have crafted additional specific regulations to fit their locales. To summarize these regulations here is impossible, but the DNR Shoreland Management Web site referred to above provides links to view each county's shoreland management ordinances.

narrowed the list down significantly, but have found that many rivers are rated inappropriately because of insufficient information. They're currently (as of 2000) working on collecting additional data. Ultimately, every county will end up with its own "top 10 list" of rivers in need of protection. A "top 10 list" for the entire 20 county area will also be generated.

In the meantime, a video, "Ribbons of Life," was produced in 2000 as an education tool designed to provide information on the functions of river systems and practical ways to protect them. It's available free to the public, and should be found in most public libraries.

Land-use controversies will only intensify over the years. I'll address some of them as we float this beautiful river. Unfortunately, we have no choice but to address them.

For more information about NRI, contact the WDNR—Northern Rivers Initiative, Box 220, Park Falls, WI 54552. See the WDNR Web site at <www.dnr.state.wi.us/org/water/fhp/nrifaq99.htm>.

> *Few of us can hope to leave a poem or a work of art to posterity, but working together, we can yet save meadows, marshes, strips of seashore, and stream valleys as a green legacy for the centuries.* — Stewart Udall

≈ ≈ ≈ ≈ ≈

Fishtrap Dam lies a mile ahead, the first of two maintained dams on the Manitowish. Nixon Creek enters on the left, appearing initially as a large flowage area, but narrowing in about a half mile into an easily navigable wilderness creek. An upstream paddle of about two miles will bring you to Nixon Lake, an undeveloped wilderness lake worth exploring—canoe campsites are available on the lake. The presence of an eagle nest and the possible presence of trumpeter swans might further induce you to check this area out.

Nixon Lake was the release site for a pair of trumpeter swans, but the swans have failed to reappear since 1997.

Wisconsin initiated a trumpeter swan recovery program in 1987, with a goal of establishing a migratory and breeding flock of at least 20 nesting pairs by the year 2000. WDNR biologists flew to Alaska for 10 consecutive years to collect surplus trumpeter swan eggs. The eggs were then hatched in incubators at the Milwaukee Zoo and used to establish a Wisconsin population. Over the decade, 383 trumpeters, ranging in age from one week to two years old, were released. By 1998, Wisconsin boasted 18 nests, of which 15 were successful. But the breakthrough years came in 1999 when 32 pairs nested in the state, and in 2000, when 42 pairs nested. More than 100 cygnets fledged from nests in 2000.

Trumpeters hadn't nested in the state since the 1880s. Their original range extended south into central Illinois and Indiana. Only about 3,000 trumpeters currently reside in the lower 48 states, though 12,000 populate the wilds of Alaska. In Wisconsin, trumpeters are listed as an endangered species.

Threats to the restoration effort abound. Seven swans died due to lead

Nixon Creek and Lake

The Wisconsin Bureau of Endangered Resources rates the Nixon Lake Complex "high" in its ecological significance within the NH/AL, and also "high" in its ecoregional significance. Nixon Lake and Nixon Creek represent undeveloped, undisturbed waters surrounded for the most part by equally undisturbed peatlands, which support several rare plant species.

The 110-acre Nixon Lake reaches a maximum depth of 5 feet, providing an excellent site for aquatic plants. An eagle's nest is located on the east side of the lake where the creek comes in. Look for the nest in a red or white pine—the pair has alternated its choice of nest trees since 1973, when the DNR began monitoring nests.

Nixon Lake was once a superb wild rice lake, but has failed to produce rice since the early 1980s. Two attempts by the DNR at reseeding the lake in the late 1980s failed. Rice does grow in the creek channel on either side of Nixon Road, but no longer in the lake itself.

The creek averages 50-feet wide and 4 to 5 feet deep, a depth likely enhanced by water backed up by the Fishtrap Dam. Large expanses of open, tussock-sedge meadow drain into the creek, with sweet gale, bog rosemary, cattail, leatherleaf, tamarack and black spruce dominating the shoreline, and white and yellow water lily and pickerelweed flowering in any slough-like shallows. Many submersed plants, from pondweeds to bladderworts, occupy the underwater world.

Kingbirds, song sparrows, and common yellowthroats should escort you along the way.

Before reaching the Manitowish, the creek enters a large lake/wetland complex. The retreating glacier dropped off a dozen or more large boulders here.

A relatively rare bracken grassland community with scattered aspen, white birch, and balsam fir clothes the north and east shore. Several active beaver lodges protrude along the shores. Wood ducks, mallards, and great blue herons, among others, commonly feed here.

poisoning in 1998, even though Minnesota, Wisconsin, and Michigan have banned the use of lead shot. Illegal shooting continues to take its share of swans—two swans were killed in a senseless shooting on Little Rice Lake in Boulder Junction in 1999. Power lines have been responsible for trumpeter swan losses. And, as always, the loss of wetland habitat could affect the long-term growth and stability of the trumpeter populations.

In the fall of 1999, 10 trumpeters were outfitted with transmitters on neck collars that allowed satellites to track their migration paths and wintering locations. They returned the next spring after a very mild winter season, and were found to have migrated as much as 600 miles and as little as 40 miles, appearing to seek out habitat that was similar to their summer nesting habitat in Wisconsin. For instance, five swans that nested on diked pools and impoundments in central Wisconsin selected reclaimed strip mines now managed for wildlife near Carbondale, Illinois. A swan that nested on a northern Wisconsin flowage migrated 607 miles and selected an impoundment at a state wildlife area in southwestern Illinois. A swan that nested along the Bad River in northern Wisconsin migrated and overwintered mostly on the Lower Wisconsin River.

One swan didn't return, succumbing to lead poisoning from ingesting lead shot after migrating to the Fox River in Yorkville, Ilinois.

The information from the satellite tracking is posted on the Wildtracks Web site at <www.wildtracks.org>. The site also has information about trumpeter swans and Wisconsin's effort to restore the species.

Trumpeter swans are named for their resonant, trumpet-like call and are North America's largest waterfowl. Adults swans are easily identifiable from other non-swans. They all have full white plumage, black bills, a 7-foot wingspan, stand 4 feet tall and weigh between 20 to 30 pounds. However, sub-adult swans are smaller and have grayish plumage. Swans released through the reintroduction program have yellow or green collars around their necks that display identification letters and numbers.

≈ ≈ ≈ ≈ ≈

What you can do: Join the Trumpeter Swan Society, which is dedicated to restoring trumpeter swans to as much of their former range as possible. Write or call the TSS at:

3800 County Road 24
Maple Plain, MN 55359
(612) 476-4663

Or participate in the "Adopt a Swan" program coordinated by the WDNR's Bureau of Endangered Resources where donors can adopt trumpeter swan broods by contributing money to the restoration program. Write the Natural Resources Foundation at:

P.O. Box 129
Madison, WI 53701
(608) 266-1430

At this point, the river is known by locals as "Fishtrap Flowage," in reference to how much water is backed up behind the Fishtrap Dam. In fact, this tiny dam backs up water all the way into High Lake, thus changing the natural character of the river for its entire length behind the dam.

In the last mile before reaching the dam, many, many stumps protrude from the water or lie just under the surface. The stump-fields make a good argument for paddling the river in plastic kayaks. The shallow draft of the kayak and its flexible hull allows passage over stumps that would ordinarily hang up any other craft.

The stumps provide excellent habitat, or structure, for fish. A hot-button issue these days is retention of coarse woody debris. A river or lake without structure is like a forest without an understory. There's little to eat,

The Swan

Across the wide waters
something comes
floating—a slim
and delicate

ship, filled
with white flowers -
and it moves
on its miraculous muscles

as though time didn't exist,
as though bringing such gifts
to the dry shore
was a happiness

almost beyond bearing,
And now it turns its dark eyes,
it rearranges
the clouds of its wings,

it trails
an elaborate webbed foot,
the color of charcoal.
Soon it will be here.

Oh, what shall I do
when the poppy-colored beak
rests in my hand?
said Mrs. Blake of the poet:

I miss my husband's company
he is so often
in paradise.
Of course! The path to heaven

doesn't lie down in flat miles.
It's in the imagination
with which you perceive
this world,

and the gestures
with which you honor it.
Oh, what will I do, what will I say, when those
white wings
touch the shore?

nowhere to rest comfortably or colonize, and nowhere to hide from predators. We spend a lot of time making fish cribs and dropping them into lakes for fish habitat, while simultaneously cleaning our shorelines and shallows of weeds and downed trees, branches, and stumps.

Nothing like working at cross-purposes.

We need to see the world through a fish's eyes if we want to keep recreational fishing strong. We can't have lawns down to the shoreline, remove the emergent and submergent plants in the shallows, take out the woody structure, and still catch lunker fish.

Moderation in all things said the Buddha. Coarse woody debris, or CWD, falls under that adage. A moderate amount of branches and logs fallen into the water provides a host of benefits. Here's the long list:

•As a tree slowly decomposes, it provides a steady rain of organic materials and nutrients.

•Tree debris traps and collects drifting particles and sediments, eventually becoming coated with plant and animal remains that aquatic insects and other invertebrates consume.

•Trees provide shelter and cover to aquatic organisms in the branches that are held under water.

•The debris blunts the impact of waves and ice upon shorelines, deterring erosion.

•Snagged trees provide shade, helping to keep the water cool, and better oxygenated.

•CWD helps protect the river and shallow lake bottoms from being scoured clean by currents and ice, creating better conditions for aquatic plant germination or rooting, and growth.

•Often a stream is slowed by a tree fallen across it, and a pool is formed, allowing wood and fallen leaves to accumulate. Pools become excellent sites for aquatic insects that feed on organic matter, like some

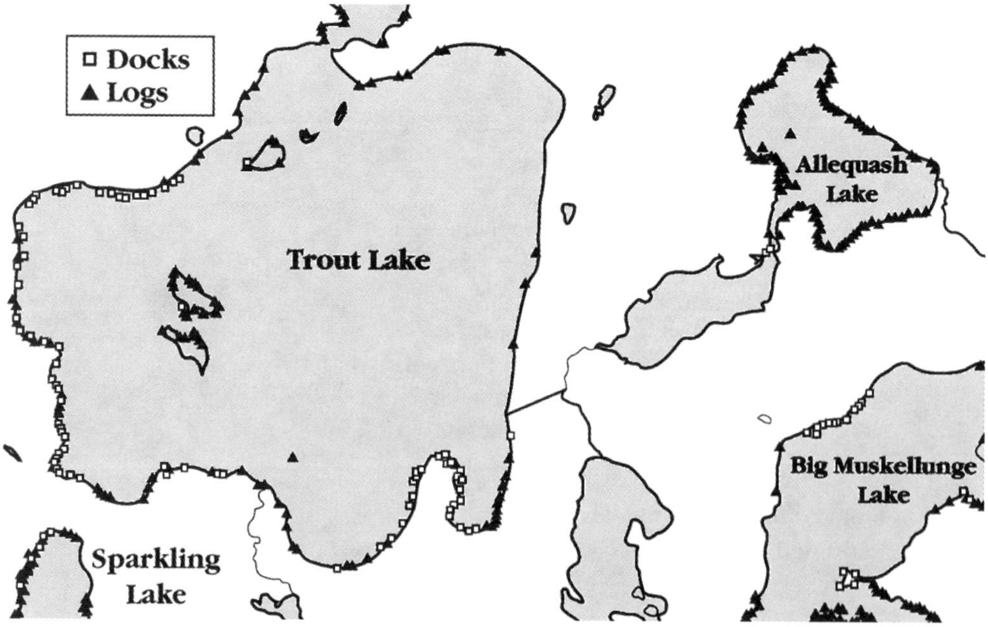

Map courtesy of the University of Wisconsin Trout Lake Station

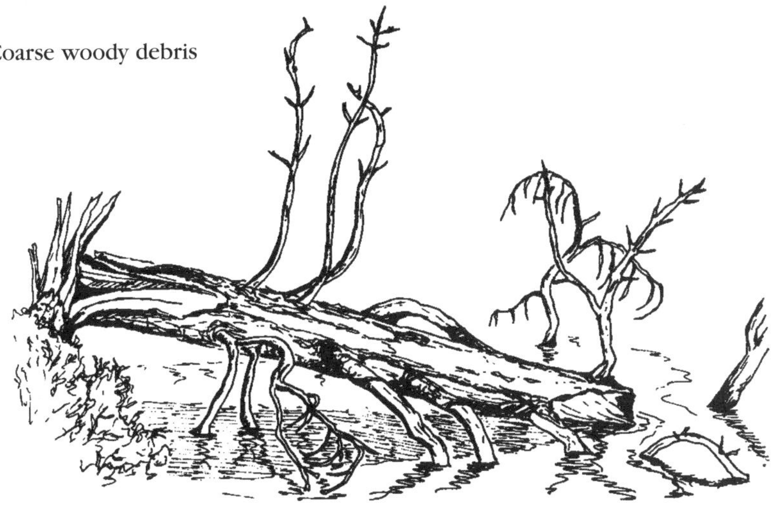

Coarse woody debris

species of stoneflies, mayflies, and caddisflies.

•Fish and other species feed on the plants and animals that have colonized the debris.

•In floodwaters, fallen trees help divert flood surge.

CWD is described as snag habitat. Not only are old, dead trees called snags, but they function as snags, grabbing and holding organic debris flowing downstream.

A fallen tree across a stream often remains for decades. Colonization by boring and tunneling insects and moisture penetrating the bark slowly softens the wood, permitting fungi to take hold and begin decomposing the wood fibers. Decay rates vary with stream temperature, the amount of sedimentation, and the size and type of wood. Thus, warmer streams tend to accelerate decay. But if sediments bury the wood, decay is slowed. Logs decay slower than branches simply because of a size differential. And different species vary in their decay rate —for instance, white cedar will outlast aspen.

The problem with CWD is it looks "messy." Maybe the issue is better restated: The problem is that we perceive woody debris as messy. The "Mr. Clean" ethic we have so taken to heart makes sense within a house, but not in the natural world. Recent lakeshore development has greatly reduced the presence of CWD in streams and lakes through direct removal, as has the cutting of shoreline trees that would eventually fall into the water. A study of cabin sites on 16 lakes in northern Wisconsin and the Upper Peninsula of Michigan found that cabin sites had only 15 percent of the average wood debris density of similar forested lakes in a natural condition. The authors of the study estimated that given how slowly trees grow, replenishing the mean density of CWD found in typical developed lakes could take 200 years!

Remember though, moderation in all things. Too much woody debris can smother fish spawning grounds, remove dissolved oxygen from the water through its decay, and shade the lake or river bottom to such an extent that plant germination and growth is significantly reduced.

So while I would encourage you to leave the dead, I'd discourage you from making your shoreline into a shoulder-to-shoulder debris field.

Balance is always the key.

But I'm onto my standard stump speech here.

≈ ≈ ≈ ≈ ≈

Fishtrap Flowage hides both its stumps and its history well. No one seems to know the exact year the dam was built, but given the logging history of the area, it very likely was constructed within the time period between 1887 and 1892. No pictures exist of it in operation.

Today, the dam is little more than a short spillway with a culvert running around its right shore. It holds back a head of about 6 feet of water. No licensure exists for the dam because it doesn't generate electricity. No study exists to impart light on the impacts of the dam on the river. No reason can be given for its continued presence.

It just is.

Could it be removed? Certainly. Wisconsin currently leads the nation in making rivers free-flowing again. From 1969 to 1999, Wisconsin removed 73 dams from rivers, compared to 47 in California and 39 in Ohio. Across the country, 467 dams were dismantled, all because of the benefits of improving water quality and fish and wildlife habitat by allowing rivers to flow unimpeded.

The litany of ecological problems associated with dams runs long, and is, shall we say, damning. For fish, their migration, spawning, and nursery sites are often dramatically altered. Warmer temperatures created by pooling water in reservoirs often eliminates coldwater species like trout. Siltation buildup behind the dams can render dams completely ineffective, while profoundly altering the normal distribution of sediments downstream. Elimination of the scouring effects of high water levels on spawning beds can destroy spawning opportunities.

While dams have often been built with the best of intentions for the protection of people from floods, and do in fact often accomplish that purpose, a solid argument can be made that the ecological losses outweigh the gains. In some cases, a dam represents an economic trade-off, preventing one economic loss by creating another.

Wisconsin has more than 3,700 dams, 10 percent of which are badly deteriorating or obsolete. The River Alliance of Wisconsin, a nonprofit conservation organization, advocates for selective dam removal on sites where the long-term economy and ecology would be improved. Check out their Web site—<www.riveralliance.org>—for more information on dam removal.

I hasten to add that not all dams are damnable. Removal of dams can cause some major health problems, such as release of toxic sediments downstream. As with virtually everything in the natural world, there are no absolutes. Never say never, never say always.

Every dam must be evaluated individually for its cost/benefit ratio. The problem is, of course, that politics, property values, hysteria, and the lack of a voice for critters like clams and aquatic insects often make the ultimate decisions on a river's future weighted more on human bias than ecological integrity.

Tough decisions await many communities and private companies that own obsolete dams. For decades, or even a century or more, property owners around such dams have come to see the altered landscape as natural. It's not. Each community will need to wrestle with its collective conscience, and hopefully arrive at decisions based on the greatest ecological good rather than short-term economic concerns. Dam removal may cause initial disruption, but in the long run, the greatest ecological good is almost inevitably the greatest economic good.

Two campsites are located just above the dam. An easy, very short, and obvious portage trail runs along the right shoreline from the boat landing. A parking lot that can accommodate seven vehicles or so is just up from the landing.

The total length of the trip from High Lake is about 2.5 hours.

The face of the water, in time, became a wonderful book—a book that was a dead language to the uneducated passenger, but which told its mind to me without reserve, delivering its most cherished secrets as clearly as if it uttered them with a voice. And it was not a book to be read once and thrown aside, for it had a new story to tell every day. Throughout the long twelve hundred miles there was never a page that was void of interest, never one that you could leave unread without loss, never one that you would want to skip, thinking you could find higher enjoyment in some other thing. There never was so wonderful a book written by man; never one whose interest was so absorbing, so unflagging, so sparklingly renewed with every reperusal.

— Mark Twain on the Mississippi River

Neighboring Lakes—A Tale of Altitude

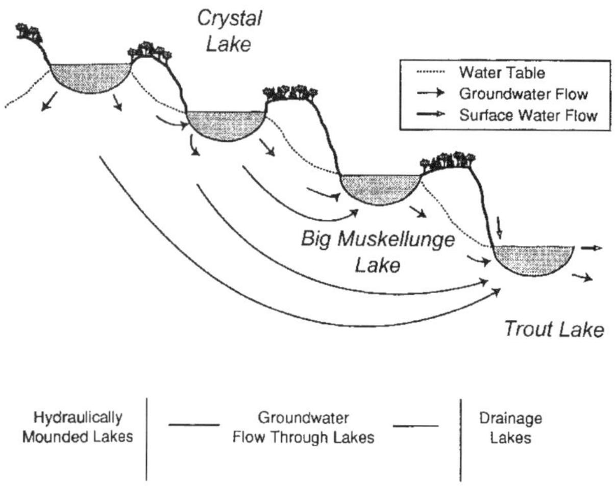

Diagram courtesy of the University of Wisconsin Trout Lake Station

Crystal and Big Muskellunge lakes lie right next to each other, separated by only a few hundred feet of land, and more importantly, by 5 feet of elevation. They are distinctly different lakes, as different as twins can be in the human population. You'd think that neighboring lakes would be the same, that being born from the same parent materials, they would act and look alike.

Ha.

The individual personalities of lakes tend to drive people crazy. All the lakes in the Northern Highland Lake District lie on about 130 feet of non-calcareous sandy tills and outwash. All receive virtually the same weather, yet they differ markedly in a number of characteristics. The most easily understood value is the number of fish species in each lake. Species may vary from zero to 37 in different lakes.

How can one lake be so different from another, particularly when the lakes are within the same geological setting (on the same soils) and experience the same weather? You'd certainly think that the 2,500 lakes in the Northern Highlands Lake District would be similar in biological and chemical properties.

Reasons for these differences may occur due to variations in morphometry (the shape and depth of the lake basin), to internal processes such as trophic dynamics (food webs), to nutrient cycling, or to interactions with the surrounding landscape.

Let's just look at the last variable. A major factor in the interaction of a lake with its surrounding landscape is its position in the landscape (as

Continued

Neighboring Lakes—A Tale of Altitude (Continued)

discussed earlier in Chapter 1). Lake position determines the relative importance of groundwater and precipitation inputs to a lake, with lakes high in the landscape receiving a greater proportion of their water from precipitation than lakes lower in the landscapes. Because of reactions that occur within the soils, groundwater typically has a much higher ionic strength than rain or snow. Therefore, chemical differences will occur between lakes based on the relative inputs of groundwater and precipitation. The greater the groundwater entering a lake, the greater the concentration of minerals in the lake, including calcium, magnesium, potassium, and sodium.

The significance? For instance, calcium concentrations need to be great enough for the development of the shells of crayfish and snails. Many lakes in this area fail to support crayfish and snails due to low levels of calcium.

Lake area and fish species richness tend to be correlated with landscape position. Larger species-rich lakes are typically lower in the landscape; smaller lakes with fewer species tend to be higher in the landscape.

The conceptual framework of the river continuum theory (see Chapter 6) allows for the prediction of how characteristics of streams change as one moves downstream from small, headwater streams to the ocean. Similarly, we can consider the concept of a "lake continuum" in which characteristics of lakes change as one moves from headwater lakes higher in the landscape to lakes lower in the landscape. In the case of both streams and lakes, characteristics of any particular lake or stream are relative to the lakes and streams above or below it.

It's important to note that landscape position is best seen as a constraint, not as a determinant or absolute cause. Lakes high in the landscape are not predetermined to be small or have species-poor fish communities, but they are much more likely to be.

Here's how Crystal and Big Muskie stack up next to each other:

	Crystal Lake	**Big Muskie Lake**
Area (acres)	88	930
Mean Depth (feet)	34	26
Water Temperature	20.8°C	20.9°C
% Groundwater Input	6%	16%
pH	6.0	7.3
Total Phosphorous	8.6 parts per billion (ppb)	22.5
Total Nitrogen	207 ppb	489
Silica	20 parts per million (ppm)	145
Water Clarity (avg.)	7.3 meters	6.7
Fish Species	21	28

Groundwater

Groundwater is one of those mysteries of life that few of us understand, yet all of us rely on. Because we can't see underground, groundwater seems to live a secret subterranean life beyond our understanding. We drill for wells and sometimes the water is there, and sometimes it isn't. Sometimes it can be found at 10 feet and sometimes not until 300 feet. Sometimes it appears in springs, sometimes it flows continuously out of artesian wells, seemingly never-ending and bottomless. What gives? How does it work?

In a nutshell, it works like this (at least most of the time). Soil sponges up water. Some goes only as far as the roots of plants, where it is absorbed. But most passes beyond the root zone, draining downward through pores in the soil until it is stopped by a layer of bedrock. There it accumulates, filling spaces in the rocks and saturating the area between the stones and soil, and becoming groundwater. The top of this saturated zone is called the water table.

Think of it this way. Fill a glass dish half full with sand. Pour water into the dish. The water is absorbed in the sand, and seeps down between the sand grains until it reaches the impermeable layer—in this case the glass bottom of the dish. The sand becomes thoroughly saturated before any water accumulates at the bottom of the dish.

Pour more water in, and the water accumulated at the bottom rises. Wherever the top of the accumulating water occurs is what we call the water table.

Make a v-shaped channel across the surface of the sand deep enough to reach below the free water that is accumulating, and water will appear in the channel, creating a river-cut valley.

Make a round depression with your finger, and water will fill the hole to the level of the water table, creating a lake.

Rivers and underground water are locations, applying to fundamentally the same water. The terms simply tell us where the water is at a given time.

Springs form where the water table intersects the ground surface. During long periods of dry weather when one might expect a river to dry up, the continuous flow of a river comes from groundwater in the saturated zone below the ground surface.

In humid climates like ours, there's usually enough precipitation to raise the water table high enough for even small creeks to carry water year-round. In arid climates like the American Southwest, the streambeds are often dry, and only the deeply cut river channels carry water all year.

(Continued)

Groundwater (Continued)

So, how does water move underground? Water seeks its own level, running downhill. Whether above ground or underground, water flows toward the lowest position of the water table, which is the rock layer. Unless the water slopes, it will not move. Using the glass dish example again, the water in the saturated sand won't move anywhere, because the surface of the free water is flat. If you pour water into the dish just along one edge, the water will flow sideways until it is all distributed uniformly throughout the dish. Likewise, if you tip the dish up on edge, the water will flow downward toward the lowest edge and accumulate there as a "lake."

Water moves slowly through the pores and cracks in the soil, so it takes time for a mound of water created by a heavy rain (or a heavy pouring of water into your dish) to flow down through the soil. Thus, the water table is seldom flat, but is instead undulating, depending on where the water is moving through the soil at the time.

How long does it take water to move through rock? Rates range from a few centimeters per year to a meter or more per day. In one study of groundwater movement, water that was pumped in Milwaukee fell first as rain and snow 25 miles west in Oconomowoc. The water pumped from the well in Milwaukee may have taken hundreds of years to travel from Oconomowoc. But some of the water may have entered the ground nearby and only have taken months or a few years to reach the aquifer.

About 22 percent of the Earth's freshwater exists as groundwater—there's a lot of water down there. Only the glaciers and ice caps hold more water (about three-fourths of all freshwater). Surfacewater in the form of lakes and rivers represents less than 1 percent of our freshwater. We get to see, and paddle on, a miniscule amount of the Earth's freshwater as flowing water.

The numbers get much smaller yet if you look at the entire Earth's supply of water—freshwater and saltwater combined. The oceans hold 97.3 percent, the glaciers and ice caps 2.2 percent, groundwater 0.5 percent, soil moisture 0.005 percent, atmospheric water vapor 0.001 percent, and finally inland lakes 0.018 percent and rivers 0.000096 percent.

Gradient

The Wisconsin section of the South Fork of the Ontonagon River begins at an elevation of about 1,695 feet, 55 feet or so higher than the Manitowish, and empties into Lake Superior at the city of Ontonagon in Michigan. Lake Superior is listed at a mean lake elevation of 602 feet. Thus, the Ontonagon drops nearly 1,100 feet from its headwaters in northern Wisconsin, producing some fine whitewater along the way.

The Manitowish follows a much shorter and milder course, beginning at an elevation of 1,640 feet and meeting the Bear River some 44 miles southwest of High Lake to form the North Fork of the Flambeau. The confluence of the Bear and Manitowish occurs at an elevation of about 1,574 feet, a drop for the Manitowish of only about 66 feet from birth to merger. The gradient of the Manitowish, a number determined by simply dividing the elevational drop by the length of the river, is around 1.5 feet per mile, barely enough to create a riffle now and again. Quiet-water paddlers love this kind of river.

The rate of descent is another term used in relation to gradient. Look at topo maps to determine the rate of descent of any river. Measure the difference in elevation from your put-in to your pull-out by counting the difference in feet between contour lines, and then dividing by the length in miles of your trip. For example, if your paddle will descend 100 feet over the course of 10 miles, the average rate of descent will be 10 feet per mile.

Average rates of descent offer you one picture of how turbulent the river may be. Gradients of 5 to 10 feet per mile typically indicate the presence of some riffles and rapids, while gradients of over 10 feet per mile usually predict serious whitewater.

Average rates of descent can be deceptive, however. Long stretches of flatwater can dampen the average descent rate and lull a paddler into thinking the river is a piece of cake, when in fact there may be a few difficult, short drops with intense whitewater along the route. Likewise, a 20-foot-high waterfall may account for nearly the total descent over a 10-mile stretch of river—a piddling 2 feet per mile average descent, but a potential killer.

If you have any doubts about a river's gradient, consult any of a number of excellent canoeing/kayaking books or ask experienced paddlers. Remember though that "easy riffles" or "a short distance" may mean very different things to different paddlers.

Chapter 2
Our Course: From Fishtrap Dam to County H and K

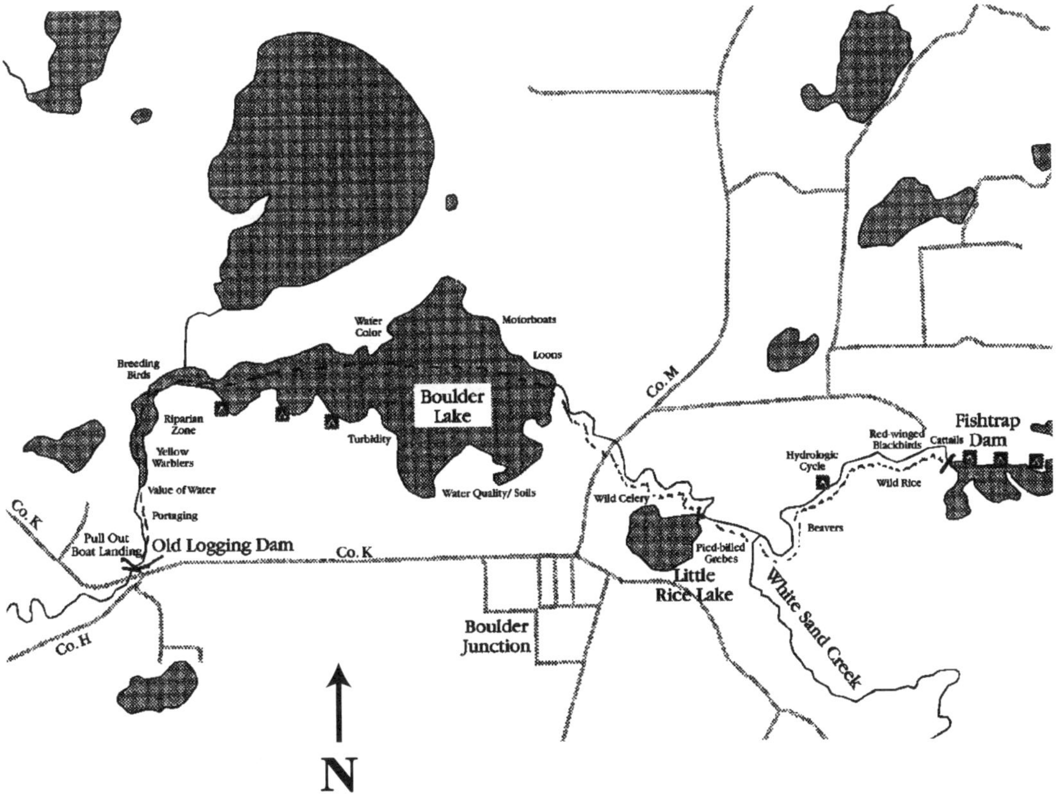

I wish all roads were rivers.
—Celine

Chapter 2

Of Wild Rice, Beaver, Motorboats, Grebes, and Values

If nature is fond of us, it says so with water.
—Peter Steinhart

Below the Fishtrap Dam stands an extensive colony of tall cattails, an atypical sight along the Manitowish. Wetland communities along the river tend to be either an alder/willow shrub complex or sedge meadow community. Cattails sprinkle their way down the river, but rarely dominate for acres as they do here.

Does the dam alter the water and nutrient flow here in a manner that encourages cattails? It's difficult to say. Cattails typically require fairly mineral-rich water, a constant flow of water, and reasonably stable water levels. Many stretches of the Manitowish provide such conditions, but only here do cattails grow in such a luxurious stand.

Most of us know cattail from its long sausage-like seed spike that, when ripe and broken open, instantly expands into 200,000 silky seeds or more. Children are delighted by how the seeds unfold in their hands into a downy pillow. Each seed comes equipped with a parasail that lifts the seed in the slightest breeze, often carrying it many horizons away. If a seed lands on a good site and germinates, it can produce hundreds of shoots in just one season from its network of new rhizomes.

Cattail's astonishing seed numbers, its wide distribution, and its extraordinarily fast growth make this plant seem unfair competition for other marsh dwellers, but its vegetative cloning ability is every bit as remarkable. Cattail rhizomes creep under the wet soils, sending up tiny clonal shoots that will rise in spring, extending the colony farther into the marsh. One study of colony expansion showed the clonal spread moving out at 17 feet per year. Employing what is known as a "phalanx strategy," cattails must appear to other marsh dwellers just like a formation of close-ranked infantry invading a village.

The good news is this invading army of cattails provides an inordinate array of ecological services to a marsh, from food to cover to building materials. Painted turtles eat the seeds and stems. Muskrat, geese, and beaver forage on the shoots and rhizomes. The submersed stalks provide spawning habitat for sunfish and cover for other fish. Many wetland birds weave the stalks into nests. Marsh wrens, swamp sparrows, and red-winged and

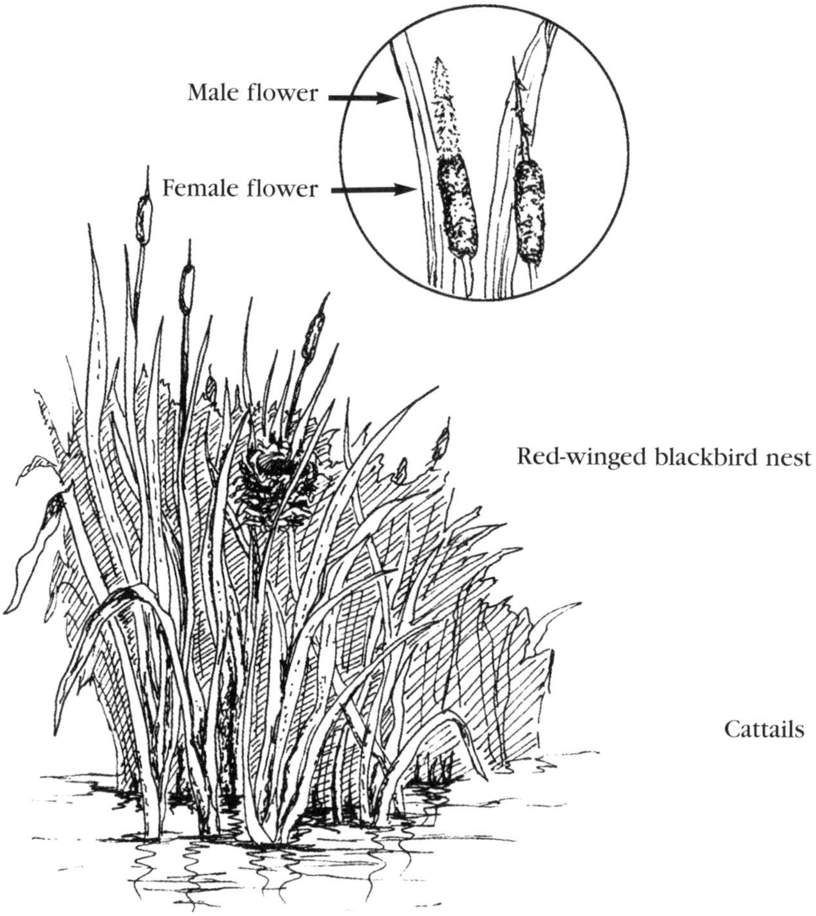

yellow-headed blackbirds suspend their nests amid the dense leaves. Other wetland birds build nests on cattail platforms, while still others construct nests from cattails right on the water. Look for nesting least bitterns, grebes, black terns, Virginia and sora rails, American coots, blue-winged teal, and trumpeter swans in and around northern cattail marshes. The downy seeds make an excellent nest liner—American goldfinches are known to use this natural mattress material in their nests.

But as I've said before, too much of a good thing can be a bad thing. Cattails vegetatively clone themselves so efficiently and extensively that they can become a monotype in a marsh. While humans try to engineer machines with the simplest and fewest parts, nature tends to do best with a diverse array of characters in its plays.

Here muskrats hopefully enter the picture. Muskrats forage on the starchy rhizomes, eat the grassy stems, and build their huts with the leaves, providing an essential control mechanism on cattail stands. They do so well in cattail stands that a new problem soon arises—muskrats often grow too high in number, and "eat-out" the marsh, again proving that too much of a good thing is a bad thing. Mink now enter as the last piece of the puzzle, providing predatorial control of the muskrat colony. The balance of these three species tips and tumbles over the years, but in undisturbed marshes, usually maintains a rough, dynamic equilibrium.

Cattails appeal to human tastes, too. On trips I lead, I frequently offer the inner core of the bottom four inches of stalk to everyone in the group, and most are quite surprised to find that it tastes pretty good. It's often one thing to say a plant is edible, and quite another thing to say it actually tastes good. The other good thing about eating cattails is that you don't have to feel guilty about removing a wild plant—there's often so much cattail in a marsh that you're doing the marsh a favor by opening it up a little.

Humans have recognized cattail's utility for centuries. In 1616, the French explorer Samuel de Champlain wrote of the Huron Indians, "During the day they bind the child to a piece of wood and wrap him in furs or skins ... under the child they spread the silk of a special kind of reed—the one we call hare's foot [cattail]—which is soft for it to lie on and helps to keep it clean."

Over a century later in 1749, Swedish explorer Peter Kalm described the use of cattails by his countrymen, "Formerly, the Swedes employed the down which surrounds its seeds and put it into their beds instead of feathers; but as it coalesces into lumps after the beds have been used for some time, they left off making use of them."

Ethnologist Frances Densmore, who lived with the Ojibwa in Minnesota, wrote in 1926, "The outer covering of cattail rushes was formed into toys representing human beings and ducks. The latter were usually made in groups of five. They were placed on the surface of smooth water, and the child agitated the water by blowing across it, which caused the ducks to move in a lifelike manner."

Shortly thereafter in 1932, ethnobotanist Huron Smith added these uses for cattails: "The Flambeau Ojibwa used the fuzz of the fruit for a war medicine. They claim the fuzz thrown into an enemy's face will blind him ... The Flambeau Ojibwa women use the cattail leaves to make mats to be placed on the sides of the medicine lodge or any temporary wigwam ... The fuzz or seed is used to make mattresses and sleeping bags."

≈ ≈ ≈ ≈ ≈

Several hundred yards farther downriver, wild rice begins to emerge, replacing the cattail as the dominant emergent plant. A September paddle through these many acres of wild rice should kick up flocks of waterfowl, particularly mallards, wood ducks, and coots. In fact, almost every species of North American duck and goose relishes the nutritional grains of wild rice. Toss in swans, rails, bobolinks, red-winged blackbirds, and song sparrows, and you'll understand why wild rice marshes resonate with bird life.

Throughout the growing season, probably the most common bird in a wild rice marsh, and certainly the easiest to recognize, is the red-winged blackbird. The male's scarlet epaulets announce its identity with a splash.

A badged uniform such as this is worn for a reason by a bird, just as uniforms with badges clearly play a role in human society. One Christmas when I was a young boy, I received a "Dick

Tracy" badge that, as far I was concerned, bestowed great status upon me. I would flash it around the house to indicate that I was an important guy, one not to be trifled with. My older brother thought otherwise.

Likewise, epaulets clearly convey military rank to soldiers, visually designating authority without the bearer having to constantly prove his or her physical power.

Along marshy river shorelines in spring, male red-wings also use badges to prove their status, apparently ranking one another by the size and brilliance of their scarlet shoulder patches, or epaulets. Just what function do the badges play? Researchers in two studies altered the color of the epaulets by dyeing them black, and found that over 60 percent of the dyed males lost their territories, while only 10 percent of the undyed control males lost theirs.

Another study placed mounts of male red-wings with variations of colored epaulets within the territories of other male red-wings. Their epaulets were totally blackened, half-blackened, kept normal, or made twice the size of normal. The territorial males attacked the mounts in proportion to their epaulette color and size. The double epaulette mounts received the highest level of aggression.

The same researchers experimented with removing territorial males to see how intruding males would respond. Initially, the intruders kept their epaulets covered, but within minutes after the removal of the territorial male, the intruders began incrementally exposing their scarlet patches. A half hour later they were fully displaying as if they owned the territory.

These studies have lead to the "coverable badge hypothesis," which says that birds with badges can signal their intentions to fight or be submissive by the extent to which they cover or uncover their colors. By covering his badge, a male can feed in another male's territory or search for a new territory with limited fear of attack.

When paddling in the spring or early summer, you'll know if you're in a territorial male's province by a number of cues. If a male leans forward from a branch, spreads his tail, droops his wings, fluffs his feathers, points his head downward, raises his scarlet epaulets, and sings "konk-kee-ree," he's performing his "song-spread," telling you that you're in his territory. The male may also display territorial song-flight, performing a stalling flight between perches as he flares his epaulets. Or as you approach his territory, the male may call "cheee-er" from his perch, and utter harsh "check" notes. If you get too close to the nest, the male may hover just behind or above you. Rarely, one may even strike at the back of your head.

Male red-winged blackbird

River Nations

As dawn nears,
the moon on the black riverwater fades,
its ivory polish overwhelmed
by the sun's prismatic entrance.
A color guard begins to rise
illuminating the emerging flags of the river country,
Cattail, Wild Rice, Bulrush, Sweet Flag.

Then the musicians tune up,
a white-throated sparrow's silver tenor flute
a pair of sandhill cranes' primitive bugles
a mallard quartet's derisive saxophones
an ensemble of red-winged blackbirds on three-note piccolos.
The symphony emerges by degrees,
Unconducted, but delivered with wild unity.
The birds understand
these territorial anthems
these matrimonial ballads.

I'm thankful to be in the midst of the flurry.
The instruments and their bearers
are fluent beyond anything
I might interpret.

I try to join one of the choirs,
and sometimes the curious wander over
to glimpse with chagrin
my mimicry of
their discourse that was honed
over a million dawns like this.
I suppose they look at me as we would
at someone yelling gibberish from the audience
at a Broadway musical.
What have I to offer them they ask?

Still, I will return other dawns to look
for breakthroughs in the code,
pledging to maintain its secrecy,
if only once they would reveal it,
just once.

Look for red-wing nests typically no more than 3 feet off the ground. Red-wings weave a basket of plant materials, and line it with soft, fine rushes and sedges.

Many people look unfavorably upon red-wings, because they've earned a reputation as pests that eat farm crops, particularly during early spring and fall harvest. But during their breeding season, red-wings eat mostly insects, species like forest tent caterpillars, grasshoppers, grubs, and cankerworms. So, they can be highly beneficial as well.

In the North Country, red-wings offer the first sign of spring to many of us long weary of snow and cold. Along the Manitowish River, we expect the first male red-wings to arrive near the spring equinox, around March 21. The females typically arrive three weeks later. Females seem to choose the desirable male territory more than the specific male—sort of like marrying someone for his house. Brown and heavily streaked with a light line over each eye, the females often perform a song-spread display in early spring too, but without the ability to flash a colorful badge.

While most paddlers think they're just floating down a river, they're really passing through a number of species' living rooms, trespassing on hard-won territory, and lucky that birds do little more aggressive protection of their estates than singing and displaying colors.

In May or June, there's more action and wild noise to be found in a marsh than just about anywhere in the North Country. And more mystery, too. I've heard more baffling sounds in a marsh than any other habitat.

≈ ≈ ≈ ≈ ≈

Birds abound in a wild rice marsh, but the real story begins and ends with the star of the show, the wild rice itself. Maple sugar and wild rice were the two most important indigenous foods used by American Indians in the Upper Midwest. Wild rice dominated the diet of the native people in the Upper Great Lakes region, accounting for nearly one-quarter of their caloric intake. Some 30,000 Native Americans were likely supported by the extensive stands of rice in what is called by botanists today the "wild rice district" of northern Minnesota and Wisconsin. Archaeological evidence suggests wild rice has been an important food for native people here for over 2,500 years.

Wild rice marsh

While this emergent aquatic grass grows from Maine west to southern Manitoba and Saskatchewan, south to Central Florida, and along the Gulf Coast to Texas, its greatest densities occur by far in northern Minnesota and Wisconsin—some 20,000 acres in Minnesota, and 6,000 acres in Wisconsin. The myriad shallow kettle lakes and marshes provide ideal habitat for rice to thrive.

So, you're in wild rice country, though what we see today is but a small fraction of what was once so regionally dominant. A side trip to Aurora Lake, a Wisconsin State Natural Area about 10 miles southeast of here, offers a feel for how much of this region may have looked. The water in Aurora Lake is little more than a shallow sheet, no deeper than 4 feet, and spilling over 94 acres. By August, a canoe trip on the lake is more like being in a tallgrass prairie; rice literally covers the lakebed from one end to the other.

The Menomonee Indians derived their name from this native grass, *menominee* meaning "wild rice people."

As extensive as wild rice can become, it doesn't come with a guarantee of growth every year. Wild rice is an annual, so must reseed itself every year. It's finicky in its germination and ultimate growth, fluctuating with too much water or too little water; too much current or too little current; too mucky a bottom or too little muck; water temperatures too warm or too cold. All these variables influence growth in any given year.

Read more about wild rice on the next page.

≈ ≈ ≈ ≈ ≈

The fluctuating water levels, which largely account for the presence or absence of rice, occur because of nuances in the hydrologic cycle. Now before you give up reading this, I too remember reading about the hydrologic cycle in junior high school—the rain comes down, it goes back up—wake me up when the bell rings. Well, there's a little more to it than that, and that makes a world of difference for rice and a host of other plants and animals.

First off, there's a LOT of water continuously cycling from the atmosphere to the earth. If the precipitation that falls on the Earth's surface every year was all pooled together, it would form a layer of water on the continents 29 inches deep.

When water falls to the Earth, lots of things can happen to it. A substantial portion of the precipitation that falls never reaches lakes or rivers. Some evaporates from the surface of vegetation immediately. Some is absorbed directly by plants. Some of the water that does reach the ground surface evaporates directly back into the atmosphere. Some water is lost during the exchange of gases necessary for photosynthesis, through transpiration. Some water may freeze into glaciers; some may directly enter the oceans.

Given all of these possibilities, the hydrologic cycle can be viewed as series of storage places and transfer processes, with rivers acting to both store water and transfer it to oceans.

The whole process is powered by solar energy, which drives evaporation and transpiration, the processes that transfer water from the land surface, plant tissues, and oceans back into the

Wild Rice—*Zizania palustris*

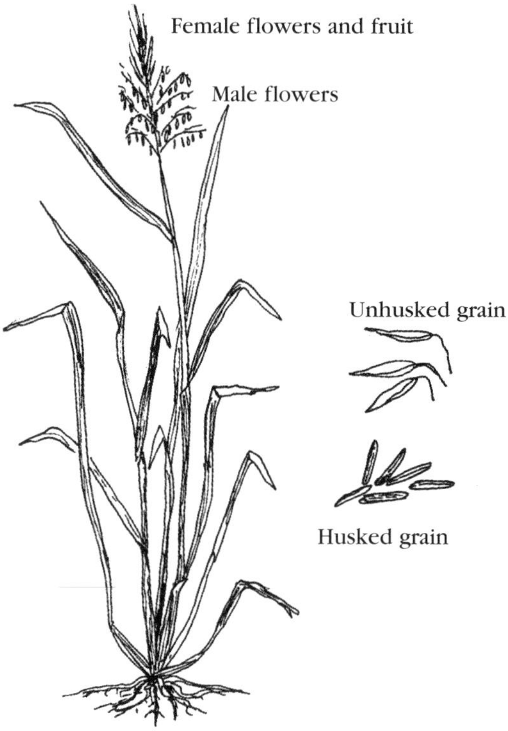

While European wheat is now the staff of life for most Midwestern Americans, wild rice is our native staff of life, the only cereal grain native to North America. A member of the grass family like corn, wheat, oats, barley, and millet, wild rice stores proteins, fats, and carbohydrates in the seed, providing superior nutrition.

Wild rice differed historically in one important way from the other grass crop plants. Out of the water, the seed could only be stored for a few months before it became nonviable, making it difficult to use as a domesticated agricultural crop. It couldn't be sown the next year like corn or wheat until relatively recent genetic strains were developed that are now planted in extensive commercial beds in California and Minnesota. Commercial "paddy" rice can now be mechanically harvested very efficiently, which has led to rice prices dropping below $2 per pound (compared to $7 a pound for the real hand-harvested wild rice).

Methods of wild harvest today remain relatively similar to historical methods practiced and described in 17th century records. Nonmechanical harvesting minimizes damage to the stands and disturbance to local wildlife. A wild rice marsh comes ripe over the course of several weeks, requiring several return visits to harvest as much as possible of the crop. Even then, hand-harvesting typically only removes 5 to 20 percent of the total annual seed production, leaving plenty for wildlife and the necessary reseeding. Rice grains can remain viable in the bottom muds of lakes and rivers for several years, so if a crop fails one year, the stand can still come back the next.

Most small stands of wild rice aren't even harvested today, so the natural genetic diversity of wild rice is considered to be protected from

Continued

Wild Rice—*Zizania palustris* (Continued)

harvesting impacts. Rice is allowed to coevolve with pests and pathogens in local stands, making the rice in one stream a bit different than the rice in the lake a few miles away.

Hand-harvesting wild rice is really the only practical method that can be used—small johnboats and canoes are the only craft capable of reaching wild rice in the few feet of water where it grows.

Harvest is remarkably easy. The rice grains "shatter" and drop from the stalk simply by tapping the seed head. However, there is an art to doing it well. The poler stands in the back, moving the canoe through the rice bed, while the ricer sits in front using two ricing sticks about 3 feet long and an inch or two in diameter. With one stick, the ricer gently bends the rice stalks over the canoe, and with the other stick, taps the heads of the plants to dislodge the grain. The sweep and tap is repeated for hours until the canoe is full. If the rice is ripe, the bed dense, and your skill ample, a good day can yield up to 200 pounds of unfinished rice.

Unfortunately, harvesting the "green" rice is only the first step—the rice still needs processing. Finishing rice involves reducing the moisture content through parching, and removing the sheath that covers the seed. Traditional processing meant roasting the rice, then "dancing" it to loosen the hulls, and finally winnowing the rice in the wind. The best stands of rice are typically found in soft, organic muck, but rice can grow in a wide variety of soils. Water depth and fluctuation are particularly important. Wild rice grows best in water depths between 1/2 to 3 feet, and when depths are stable or gradually receding during the growing season. Rapid rises in water level may drown or uproot plants, so if fluctuations occur, they need to be small and regular. The rapid and extensive ups and downs of waters regulated by dams are often the death knell for rice. Slowly moving water appears to be optimal, because the current brings a continual flow of nutrients and causes enough disturbance to favor a colonizing annual like wild rice.

Rice goes through several distinct growth phases. By early June the plant leaves are in the submerged leaf stage. By late June, the ribbon-like leaves reach the water surface where they lay flat in the floating leaf stage. Soon the aerial shoots begin to develop, and the plant rises from the water, reaching a height of 2 to 8 feet above the water. Flowering begins at the end of July, the showier male flower blossoming below the diminutive female flower.

Rice, like most wild fruits and nuts, has good and bad years. A typical four-year cycle often means one bumper-crop year, two fair years, and one bust year.

atmosphere. Hydrologist Luna Leopold writes that the Earth's atmosphere should be viewed "as a vast heat engine powered by the sun."

As rain falls, the wetted surfaces of leaves, stems, soil, and rocks immediately begin losing water to evaporation. In heavily vegetated areas, a large quantity of the rain is intercepted by branches and stems of plants, and falls or runs down slowly to the soil surface. In a deciduous forest and in mature croplands, 10 to 30 percent of the annual rainfall evaporates at the point of interception.

Transpiration, however, has a larger role in the land/water cycle than evaporation. Transpiration is the means by which all plants expel water from their leaves during photosynthesis. Water is taken up by roots from the soil, moves up the trunks as sap, and is transpired through the stomata—thousands of tiny holes on the underside of every leaf. The leaves of a large oak tree give off nearly 150,000 liters of water per year.

The most important source of moisture in the air is evaporation from the oceans, particularly those in warm areas. Much of the rain that falls on the central U.S. is composed of water that was evaporated from the ocean near the equator or from the Gulf of Mexico. Only a small part was evaporated or transpired from waterbodies and plants in the area of the rainfall.

The combination of evaporation and transpiration returns 70 percent of all precipitation that falls on the land. Only 30 percent "survives" to become surface streamflow.

While that's still a lot of water, it's held in storage by a variety of ways and means that are highly unequal in volume. In the atmosphere this very second, only 0.001 percent of the Earth's water is stored. Rivers hold a tiny, tiny amount of the Earth's water, and even then, that volume is disproportionately distributed. The world's 10 largest rivers account for nearly 40 percent of the total runoff volume—the Amazon contributes 15 percent alone. In the U.S., the Mississippi River contributes some 40 percent of the total discharge.

Aldo Leopold wrote an essay entitled "The Round River" to illustrate the biotic continuum, the perpetual flow of energy from soil to plant to animal and back to soil. He used the analogy of the fabled Round River, a river that flowed into itself in a never-ending circle. Leopold wrote that "Wisconsin not only had a Round River, Wisconsin is one. The current is the stream of energy which flows out of the soil into plants, thence into animals, thence back into the soil in a never ending circuit of life. 'Dust unto dust' is a desiccated version of the Round River concept."

Water unto water might then be the eulogy for those buried at sea, but it's also the synthesis of the hydrologic cycle. In his book *The Closing Circle*, ecologist Barry Commoner wrote that there is no such place as "away." This is one of the most basic ecological principles. All the water flowing under your canoe right now has been here since the earth awakened. Hence, on the Manitowish, as on all rivers, you are floating on the very oldest of museums.

≈ ≈ ≈ ≈ ≈

Several large boulders eventually appear to mark the end of the rice stand and the return of the pickerel-

Transition

At dusk the snow
settles in the March woods,
snow once as fine and light
as the clouds it was orchestrated in.

Heavy with age now,
the snow collapses
in a soft crescendo
compressing
into water.

Water that will soon percolate
through the sandy loam
under the white pines
into the Manitowish River.

Water that will flow
on its spring migration
to the Flambeau, to the Chippewa,
to the Mississippi, to the Gulf.

Water that will flow
through walleye gills,
through primordial pincers of crayfish,
over and under the slick bodies
of otter and muskrat.

Now, a smooth path crosses ours,
the snow dirty with the muddy
belly of a beaver,
her love for popple evident
in the strewn bodies of softwood
just visible under the darkening ice.

In the old homestead opening
on the ragged nest in the dying white pine
a motionless white head stares
across the river.

The river sags and breaks.
Water stands on the ice.

Then,
from the deflating snow
evening fogs slowly lift
like emerging butterflies.
Winter hovers between
life and death.

Dark settles in.
A snowshoe hare, mottled white,
ambles from the deepening balsam
into the purple alder
a final winter frost
illuminated,
molting into spring.

weed and water lily that typify the emergent plants in the Manitowish.

A beaver lodge emerges shortly before the boulders, though there's no guarantee the lodge will still be there in any given year. Lodges are typically used for many years, and some lodges have housed beaver clans for 50 or more years. Even if the lodge is still active, you're unlikely to see any activity unless you're paddling by during the early morning or evening. Later in the fall, you're more likely to see beavers during the day since beavers often work around the clock to make their winter storage cache in front of the lodge.

Beavers have learned their noctur-

Beaver eating the cambium layer

Beaver cutting down a tree

nal ways from contact with humans. In areas with limited human contact, beavers commonly do their work during the day.

Beavers, through their feeding and dam-building activities, act as a "keystone" species to affect ecosystem structure and dynamics far beyond their immediate requirements for food and space. By stopping up free-flowing streams with their dams and creating backwater ponds, beavers alter entire stream communities, from plant species to aquatic insects to fish to birds. One wildlife biologist estimates that 30 to 50 percent of all the streams he has studied in his lifetime were directly affected by beaver. A keystone species has pervasive influence over an ecosystem, and beavers fit that definition.

A beaver is an all-purpose logging machine, a one-man-band in the woods, working as feller, limber, skidder, and debarker. Trees smaller than 6 inches in diameter can be felled in 15 minutes or less, depending on how soft the wood is. Once the tree is felled, beavers limb the smaller branches from the top of the tree to get to the leaves and buds that are the youngest and most succulent. Then they skid the branches along well-worn paths between the water and cutting area. Once in the water, the beavers swim the branches over to their lodge or dam where the wood provides food and/or building materials.

Beavers are relatively selective in what they eat. While they may utilize hundreds of species of plants, one study listed beaver using 53 species of trees and non-woody plants. Another study in Massachusetts found that only 16 tree species were cut by beaver over a two year period, and only six species accounted for over 90 percent of all trees cut. Aspens, alder, and willow are the most prefered species in northern Wisconsin.

Beavers savor the thin smooth bark of seedlings and saplings. Older bark contains repellant chemicals and is tougher to chew, so it's far less desirable. Aspen, alder, and willow head the beaver menu, and provide the added benefit of responding to cutting by resprouting from their roots. Aspen bark is particularly nutritious, because photosynthesis occurs in the bark and the leaves.

Hardwoods like maple, oak, and birch are taken occasionally, but if you've ever tried to drive a nail through one of these species, it's easy to imagine how hard it would be to bite through them.

Conifers like red pine also show up in a beaver's diet. However, the amount of pitch exuded by a conifer must make eating one like chewing on the stickiest taffy humankind ever invented.

It's very common in the Northwoods to see lakes where all the aspens have been removed, leaving only pines and birch along the shoreline, and often only pine. The journals of the first French explorers in northern Wisconsin often describe the forests as all pine. Their myopia stemmed from the fact they seldom got off the water and saw much beyond the pine-clad lakes that were greatly altered by beaver activity. If they had looked inland a hundred yards, they typically would have found a variety of forest communities replete with sugar and red maple, red oak, yellow and white birch, aspens, ironwood, and ash, as well as a variety of other conifers.

Beaver take big trees as well as little ones. There are reports of felled trees over 3 feet in diameter, but a beaver probably has to work several days to drop one of these monsters. As an energy-efficient exercise, cutting small trees makes much more sense than big ones.

Just to kill a myth once and for all, beavers can't control the direction a tree will fall. Most trees on a shoreline fall into the water because they're leaning that way to get the most sun, not because a beaver knows geometry. So, where a tree falls depends on its lean and the slope of the ground, not on beaver wisdom. I've seen photos of a beaver flattened beneath a fallen tree as further extreme evidence of their lack of directional engineering.

How much does an average beaver need to eat daily? Captive beavers have maintained their weight on 1 1/2 to 2 pounds of aspen bark a day, but a noncaptive animal weighing 50 pounds or more in a cold climate needs lots of fuel, so figure maybe 3 pounds a day as a rule.

Beavers look pretty chubby for good reason. Wood has high amounts of cellulose, a substance difficult to digest and from which to extract calories. So beavers, like porcupines, must eat large volumes of food to extract a minimum of nutrients.

Digesting cellulose requires some specialized adaptations. Bacteria in a beaver's cecum—a large pouched digestive organ—allow a beaver to break down about one-third of the cellulose in a meal. The rest stays in the digestive tract for several days. Beavers also re-ingest some of their feces to give their digestive system another crack at breaking down the cellulose.

Re-ingestion is fairly common among small mammals, like rabbits, that eat lots of vegetation.

Beavers don't just eat woody species. In summer they feast on herbaceous species like cattail, pondweed, arrowhead, smartweed, milfoil, pond lily, and a variety of sedges. They don't just eat the leaves either. They like cattail and pond lily roots as well as the roots of some sedges and reeds.

≈ ≈ ≈ ≈ ≈

To spend childhood days along creek banks is to be drawn into the wider world. A creek reaches upward into the hills and mountains, where clouds brood and God's bluster. It reaches down to the lowlands and the fat old rivers, sad and murky with the silt of experience. A creek teaches one the curve of the Earth, the youthful swell of mountains, the age of the seas. Above all, a creek offers the mind a chance to penetrate the alien world of water and think like a tadpole or a trout. That is one of the great experiences of otherness, one of the leaps of perception that makes us human and allows us to live with dream and obscurity. What drifts in creek water is the possibility of other worlds inside and above our own. — Peter Steinhart

Red maples line much of the shoreline through this section, the leaves coming to life in late September in a blaze of red. But colorful fall palettes are a bit limited on the Manitowish. Red and white pine, white birch, red oak, balsam fir, and black and white spruce clothe most of the shorelines along the Manitowish. Red maples are a member of this dry soil forest community, but tend to do better in soils that have some more moisture. The first half of this section of the Manitowish from Fishtrap to Island Lake offers one of the best red maple stands along the Manitowish, if you're looking for vibrant autumn color displays.

The extensive wetland communities along the river also offer exceptional tamarack displays of gold in October. You know your life is good when you come upon a wetland stand of soft gold tamarack backlit by the brilliant scarlet of red maple on the upland.

The river makes a big swing south here and then turns north in section 16, flowing through a large wetland complex. White Sand Creek enters here, a small stand of wild rice acting as the entry marker. The creek emanates from White Sand Lake, a large (734 acre), deep (71 foot maximum depth), and partially developed lake. The creek is generally navigable in high water, lies almost entirely within state lands, and is very much worth an exploration. One pole building on the uplands to the west mars this beautiful wetland complex that is otherwise devoid of development. Sedges, leatherleaf, marsh cinquefoil, tamarack, and pickerelweed are all quite common through here.

Not too many folks appreciate plants like sedges and leatherleaf. They don't offer colorful flowers, sweet smells, dramatic size, or a clear economic benefit. We all have plant biases, like the biases we have for virtually every other category of living or nonliving things. We tend to value plants more highly if they're one of the following:

•rare (we like underdogs)

•profitable (we like money)

•valuable to wildlife for cover, food, or nesting (but then we have biases relative to wildlife—we want hummingbirds and deer, but not starlings or raccoons)

•native (we don't like those Europeans)

•utilitarian (we like firewood, carving wood, and other useful products)

•sensory (we like the sound of wind in pines or the smell of balsam or basswood flowers, and lilac flowers)

•tasty (apple trees, juneberry, thimbleberry)

•beautiful (we are awed by the height of white pine, love the scarlet leaves of red maple or the gold of sugar maple)

•fast-growing (we like shade, like to erase the bulldozer scars, like the erosion control)

•thick growing (we like solitude and reduced noise)

•soil improvers (farmers like to use alfalfa to fix nitrogen, but we can use maple to improve soil through the decomposition value of their leaves)

•easy to grow (we like no-brainers that resist disease, drought, and insects)

•convenient (we don't like "messy" cottonwoods or aspens that clog screens)

Plant Values

Rank the following plants according to how greatly you value each one. Use a "1" for very low value, and up to a "10" for very high value.

- Oak tree
- Dandelions
- Black cherry tree
- Cattails
- Sphagnum moss
- Algae
- Wild rose
- Hazelnut
- Milkweed
- High bush cranberry
- Poison ivy
- Pondweeds
- Tag alder
- Wintergreen
- Bracken fern
- White pine tree
- Blackberries
- Lichens
- Silver maple tree
- Kentucky bluegrass
- Water lilies
- Willows
- Wild rice
- Mushrooms
- Aspen trees
- Bulrush

Now go back and rank them again according to how valuable you think each plant is to the ecosystem it's found in. Compare your rankings, and analyze why there are differences.

Make up your own list specific to your area if these species don't trip your trigger. The point is to see how much our personal likes and dislikes reflect the needs of the ecosystem we live in.

• traditional (grandpa and grandma liked it)

• historical (we like cherry trees in part because George Washington lopped one off)

• colorful (we like color all year round—in fall, red maple; in winter, highbush cranberry; in spring, marsh marigolds; in summer, blue flag iris)

• companionable (we like plants that help us—marigolds inhibit insects in our garden)

The list can go on. The values issue gets sticky when it involves those plants and animals that don't fit easily into our personal value system of needs and desires. It's hard to love something that's difficult to identify, hard to get to, small, plain, and otherwise "just there." But it's necessary nonetheless. Loving the megafauna—the eagles, osprey, bears, and wolves—is easy. Loving the megaflora—white pines, sugar maples in autumn, and showy orchids—is also easy. But loving, or at least appreciating, sedges or algae or decomposers takes an additional leap (or two or three).

The meek shall inherit the earth, we're told. In the case of most plants and animals, the meek make up most of the earth. They are the inheritance. Without them, there would be no earth, at least as we know it.

≈ ≈ ≈ ≈ ≈

Just a quarter mile downstream from the entry of White Sand Creek lies Little Rice Lake, a shallow, 59-acre lake located nearly in the middle of Boulder Junction, but with only one house on it. Stands of wild rice, wild celery, and other desirable wetland plants attract flocks of waterfowl in the autumn. Look here for trumpeter swan, coot, scaup, Canada goose, mallard, teal, pied-billed grebe, and others. Northern pike can keep anglers busy here, too.

Little Rice Lake supports a rich plant community, both above and below the water. Unfortunately, the emergents that flower so brightly in July and August get all the "oooh's" and "aaah's." While they deserve the praise, submergent plants merit the positive press, too. The problem is we tend to see only "weeds" under the water.

Wildlife, though, know the value of underwater plants, and one of their favorites is wild celery (no relation to the celery you find in grocery stores). Duck hunters know wild celery, too, because it imparts a celery-like flavor to the meat of ducks that forage heavily on it. Canvasback ducks take their appreciation of wild celery to a higher level. They utilize wild celery so heavily that they're known to change their migration patterns to find it. In fact, canvasbacks are so tied to wild celery (*Vallisneria americana*) that they were given the similar Latin name of *Aythya valisneria*.

Diving ducks seem most attracted to the growing tips of the rootstocks, while other species are content with feeding on the leaves and seeds. Mallards, ruddy ducks, greater scaup, coots, black ducks, ring-necked ducks, wood ducks, goldeneye, and buffleheads feed less extensively on wild celery, but still include it in their diet, making wild celery one of the most valuable waterfowl foods in the Upper Midwest.

Wild celery's former species name was *spiralis*, due to the odd spiral-coiled stalk that it produces to lift its female flowers to the surface. The tiny

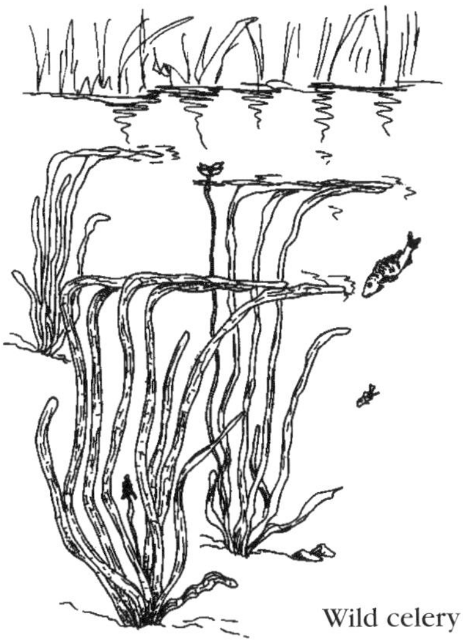

Wild celery

male flowers blossom on separate plants, clustered in a head that develops underwater. Individual male flowers eventually are released and float to the surface where they open, hopefully to contact and pollinate the female flowers. The female flowers cooperate by forming a slight dimple or depression in the water that helps gravity draw the male flowers toward them, a pollination system unique among American aquatic plants. Once fertilized, the female stalk contracts, withdraws under the water, and eventually develops a 2 to 4 inch-long fruit capsule.

Most folks may know wild celery by its cellophane-like ribbony leaves, which emerge from the base of the plant. You can see right through the margins of the transparent leaves. The submersed leaves typically reach to the surface of the water. Look for a prominent central vein to help distinguish wild celery from the similar submersed leaves of bur-reed, ribbon-leaved pondweed, and arrowhead.

Wild celery's thin "skin" allows the absorption of gases and nutrients directly through its cell walls, an adaptation shared by many submersed plants. Since submersed plants have no need of supporting tissue, they typically collapse into a soggy mass when removed from the water. When submersed, the leaves sway and billow in the currents like a prairie grass.

Wild celery belongs to the Frog's-bit family (*Hydrocharitaceae*). I note this only because how can you not like a plant that belongs to the Frog's-bit family?

In his excellent *Book of Swamp and Bog*, John Eastman writes that wild celery is a fine edible plant, either raw or cooked, though he cautions that it should be picked from clean waters and then washed. With a little salt and butter, I'll bet it tastes just as good as just about everything else cooked with salt and butter.

≈ ≈ ≈ ≈ ≈

Open-water marshes like parts of Little Rice Lake provide habitat for 19 bird species in Wisconsin. In the Northwoods, those species that nest over water in marshes include pied-billed grebe, American bittern, least bittern, Canada goose, northern harrier, sora rail, Virginia rail, black tern, marsh wren, red-winged blackbird, and yellow-headed blackbird. Other open-water marsh nesters typically found in southern Wisconsin include black-crowned night heron, redhead, ruddy duck, king rail, common moorhen, American coot, red-necked grebe, and Forster's tern.

Several species of ducks also raise their broods in marshes, while other

Pied-Billed Grebe

Pied-billed grebe

At only 13 inches long, and subdued, if not plain, in its coloration, the pied-billed grebe seldom attracts "oooh's" and "aaah's " from observers. Its blunt, unduck-like bill, neglible tail, white undertail rump, black throatpatch, and dark-ringed bill give it visual identity, but little pizazz. However, the pied-bill makes up for its unassuming appearance with a resounding and utterly distinctive call, phonetically penned as "cuck, cuck, cuck, cow-cow-cow - cowp-cowp - cow-ah, cow-ah." But while all those cows makes it a good Wisconsin bird, they don't begin to give the flavor of the call, which accelerates rapidly and then descends in a lingering wail that almost sounds painful. The call, once heard, cannot easily be forgotten.

The other characteristic that brings distinction to the pied-bill is its diving. Pied-bills are the Mike Nelson of the divers (older readers will remember television's "Sea Hunt" from their youth). Rather than flying from danger, pied-bills usually escape by diving, often sinking like a submarine. They accomplish this by expelling air from their body and feathers

Continued

Pied-Billed Grebe (Continued)

to reduce buoyancy, and then simply slipping below the surface like some rubber toy that someone let the air out of. This ability has earned them some wonderful nicknames, like helldiver, dabchick, dipper, or water-witch.

This same knack for sinking permits them to swim with only their head above water. Their large, broadly lobed and webbed toes, dense bones, and the placement of their feet far back on their body propels them quickly enough under water to capture aquatic insects, which make up nearly half of their diet. Crayfish, small fish, snails, and frogs make up the rest of the dinner menu.

The upside of evolving the ability to swim and dive so well had an adaptive compromise—not being able to walk well on land. Pied-bills also must run across the water for a long distance before becoming airborne. Their slow takeoff makes them the last bird off the runway. It's the price a fish-eating bird pays to outswim the fish.

In flight, they hunch up, dip their outstretched neck and tiny tail, and extend their feet.

Pied-bills are pond ducks. They prefer the more densely vegetated shallow ponds. The more rushes and reeds, the better, though they want some open water for foraging too. Don't look for pied-bills on large lakes—they're not usually a big open-water bird.

They're shy and solitary. A pied-billed seldom associates with other waterbirds or even its own species, living by the maxim that it's better to be heard than seen. In fact, they often carry out territorial displays through calling. The males face off across their boundaries and call loud and long to try to intimidate one another.

Small territories are the rule—only about 30 yards on a side. They build their nests right on the water on floating platforms of dead vegetation anchored partially to cattail or bulrush or reed. Nests aren't real homey. They're usually sodden and "cooking" with a wet mass of vegetation that's decomposing—definitely not the Ritz.

Pied-bills exhibit one last remarkable characteristic—they eat their own feathers. Up to 50 percent of their stomach contents may be feathers! It's unknown why, but some researchers speculate that the feathers may line the stomach walls and protect against fish bones.

birds use marshes to feed their young at distant nest sites. These species include herons, swallows, belted kingfisher, Eastern kingbird, common grackle, and song sparrow.

Birds are typically distributed throughout a marsh, partitioning the available resources to minimize the stresses of competition. For instance, least bitterns prefer the deeper marsh and open-water edge, while American bitterns are a shoreline wader. Marsh wrens utilize cattails and bulrushes, while sedge wrens utilize sedge meadows. Yellow-headed blackbirds need cattails, bulrushes, reed grasses, or shrubs surrounded by or next to open water. Black terns like deeper marshes, nesting on floating vegetation mats. Red-winged blackbirds are simply ubiquitous.

Dramatic differences in birdlife occur between similar marshes. Even within a marsh, there are yearly differences. These modulations occur in large part due to the changes in vegetation that frequently take place in the normal hydrological cycle of a marsh. In dry periods, plants that can only germinate on aerated soil may succeed, while in wet years, plants that can survive long-term inundation are those most likely to succeed.

Wetlands gradually change over time. Cattail marsh can fill in, begin to dry up, and change to a sedge meadow. The sedge meadow may eventually fill in and change to an alder/willow complex, which, of course, can fill in and become a wet forest complex of tamarack, black spruce, white cedar, and red maple. Or succession can be reversed by a fire or flood that destroys the existing vegetation cover. The marsh then starts from scratch again.

Botanical plant communities aren't fenced-in, distinct entities. Wetlands, like forests, exist often as a continuum, so lines become hard to draw. Where does the bog leave off and the marsh begin? Wetland communities often intermix, so variations, indistinctions, and blends are the norm. Natural communities seldom fit into neat boxes.

≈ ≈ ≈ ≈ ≈

A river is superior to a lake in its liberating influence. It has motion and indefinite length . . . With its rapid current it is a slightly fluttering wing. River towns are winged towns.

— Henry David Thoreau

Back on the river, a tamarack bog clothes the shorelands for a quarter mile until homes appear and the County M bridge crosses the river, just north of the town of Boulder Junction.

The Manitowish now flows about a half mile through private lands until it empties into Boulder Lake. Many boulders line this section of river. On the lake, boulders also appear consistently along the northwestern shoreline.

Boulder Lake, a 524-acre, relatively shallow lake has a mean depth of 12 feet. It is historically known for its excellent walleye and muskie fishery. The shorelines are completely developed on the lake's eastern end, but undeveloped on its western end. The vast majority of the undeveloped shoreline is owned by two private corporations, Camp Manitowish and Dairyman's Country Club. Both organizations are well-known for their conservation efforts; thus, the shoreline

LoonWatch Survey 1985-2000

Year Surveyed	# of Lakes	# Adults	# Chicks
1985	185	2,358	516
1990	207	2,420	608
1995	191	3,017	678
2000	151	3,131	462

stands a strong chance of remaining undeveloped.

Common loons utilize the lake, and are consistently sighted. The LoonWatch program of the Sigurd Olson Environmental Institute in Ashland, Wisconsin, coordinates an annual monitoring program on specific lakes across Wisconsin, including Boulder Lake. Citizen volunteers, called "Loon Rangers," keep track of the loon numbers on each of these lakes.

To estimate the statewide loon population (both adults and young-of-the-year), LoonWatch and the Wisconsin Department of Natural Resources devised an additional survey in 1985. This survey is conducted every five years on randomly selected lakes during a single day in mid-July. Surveys take place on lakes covering 25 to 49 acres, 50 to 149 acres, 150 to 499 acres, and 500 acres or more. The results are listed in the LoonWatch Survey table on this page.

The increase in loon adults and chicks between 1985 and 1995 was statistically significant, indicating minimally a stable and healthy loon population. However, while the average number of adult loons per lake was highest on 50 to 150-acre lakes, the greatest proportion of chicks (44 percent) occurred on lakes 25 to 50 acres. Of the state's 12,400 lakes situated north of Highway 29, approximately 66 percent are less than 10 acres, and 90 percent are less than 50 acres. Shoreline development rates are highest in Wisconsin on lakes smaller than 50 acres, thus placing these encouraging population trends at some risk.

Loon territories average about 50 acres in size, but loons have been known to nest on lakes as small as 12 acres. Loons also prefer clear-water lakes so they can search out prey before diving for the catch.

Lake size and water quality are only part of the equation though. Lake shape may be even more critical. Lakes with islands appear to provide the best odds of hatching success. Lakes that also contain quiet bays that can be used as "nurseries" for the chicks provide optimal conditions for raising young. Round lakes without islands or bays may have great water clarity and fish populations, but typically support fewer successful pairs than lakes of equal size with more shoreline character.

Loons epitomize the feeling of wild lakes. It has always been so. While camped in July of 1820 with the Cass/Schoolcraft expedition on Lake Winnibigoshish in Minnesota, Captain David Douglass wrote:

"The stars presently shown out with increased lustre, and the sweet moon shot her mild rays through the foliage of the old oaks which overshadow our camp, while

Loon platform

her image came reflected from the surface of the still smooth lake. No sound disturbed the silence of the scene except the deep solemn, yet melodious night cry of the loon, which came from the lake with a peculiar and indescribable effect. What a season in this remote region for indulging the imagination. Then images of my home and of dear friends, how strongly did they recur to my mind . . ."

≈ ≈ ≈ ≈ ≈

Whenever the topic of loon populations comes up, inevitably so does the issue of motorboats. Just about everyone has a horror story about a motorized watercraft running down a loon, or of a boat wake swamping a nest, or of incessant motor activity harassing loons into abandoning a lake. However, the loon-versus-motorboat issue begs the larger question—what are the impacts of motors on the overall aquatic ecosystem?

The question has large implications, because the number of registered boats in the Upper Midwest has skyrocketed. In Wisconsin, boat registration has increased by 87 percent since the late 1960s, from 303,000 in 1968 to 69 to 567,000 in 1997 to 98. The size of boats has also increased: from 18 percent of registered boats between 16- and 39-feet long in 1968-69 to over 40 percent in 1997 to 98. Horsepower has also profoundly increased, doubling on new boats registered in Minnesota between 1981 and 1999. Add in the explosion in personal watercraft—6,500 registered in Wisconsin in 1991 compared to 28,900 in 1998—and there are considerable quantitative and qualitative impacts.

The impacts must also be seen in light of the $250 million spent per year on boating equipment in Wisconsin, and the additional $200 million spent on boating trips per year in Wisconsin. Restrictions on boating won't, and don't, take place in an economic vacuum.

So, what impacts do motorized watercraft have? Motorized watercraft can affect aquatic ecosystems through emissions and exhaust polluting waters, propeller contact ripping apart plants, turbulence from propulsion systems scouring bottom sediments, waves eroding shorelines, and noise and movement of the boats affecting animal behaviors.

A 2000 study by Tim Asplund of the Wisconsin Department of Natural Resources summarized the effects of motorized watercraft on aquatic ecosystems. He looked at six issues—water clarity (turbidity, nutrients, and algae), water quality (metals, hydrocarbons, and other pollutants), shoreline erosion, plant communities, fish, and aquatic wildlife. Here's the short version of his findings just on the issues of water clarity, water quality, and shoreline erosion.

1- Water Clarity. The clarity of water affects the ability of animals to find food, the depth to which plants can grow, the dissolved oxygen content, and the temperature of the water. Complicating the matter, however, is the fact that water clarity fluctuates normally due to cycles in food webs, storms, seasonal changes, wind and water mixing, algal growth, and other factors.

Still, the impacts of motorized watercraft on water clarity can typically be sorted out. Boat props often disturb the bottom sediments, and boat wakes can contribute to shoreline erosion. The erosion clouds water and stirs up nutrients from the sediments, which increases algal growth.

A U.S. Army Corps study on the impact of motorboats on the Fox River Chain of lakes in Illinois found the rate of re-suspension of sediments varied with water depth and sediment type. In silt, highest re-suspension occurred in depths of 3 feet, half as much at 6 feet, and none at 8 feet.

The conclusion of this study, and many others, is that shallow lakes and shallow rivers are most susceptible to the impacts of motorboats on water clarity.

It's important to note that the speed and number of passing boats, as well as the boat shape and motor size, also determine how significant wake damage may be. Long flat-hulled boats cause greater wakes than keeled hulls, while small inboard motors cause smaller wakes than large outboard motors. Long propeller shafts will shred and uproot more aquatic plants than short propeller shafts.

2- Water Quality. Two-stroke engines make up the vast majority of motors in use today. They are very inefficient, expelling on average about 25 to 30 percent of their fuel into the water column, depending on engine speed, oil mix, horsepower, and tuning.

You would think that gas and oil in the water would produce significant toxic effects. However, most studies have shown minimal toxic effects because the amount of pollution is small compared to the water volume of most lakes. Equally important, most hydrocarbons are volatile and disperse quickly.

However, in areas with high concentrations of motor usage such as marinas, a build-up of compounds in the sediments has been documented.

This is a cause for concern given the potentially detrimental effects on bottom-dwelling organisms.

Four-stroke engines are a far cleaner technology, with the added benefit of being much quieter. The expense of replacing a two-stroke with a four-stroke remains a barrier, but eventually four-strokes will be the rule rather than the exception.

As for personal watercraft (PWCs or jet-skis), the California Air Resources Board estimates that two to three gallons of unburned fuel per hour are released into the water from jet-skis operating at high speeds. PWCs can operate in waters less than 12 inches deep, and are highly maneuverable, presenting impact possibilities that far exceed those presented by most motorboats. The even more contentious issues of intense, continuous sound levels and irresponsible usage have led to near-universal hatred of these craft by everyone except those who use them.

The EPA has ruled that all marine engines will reduce their emissions by 75 percent by 2025. In the meantime, the good news is that sales for PWCs are declining nationwide, perhaps in response to the efforts of local and statewide ordinances that restrict their time of use and place of use.

3- Shoreline Erosion. Two main factors determine the impact of waves on shoreline erosion:

•the characteristics of the bank material itself. For instance, sand erodes more easily than clay. And bare shorelines erode more rapidly than vegetated shorelines.

•the intensity of the eroding force. Stronger waves, higher water levels, and more powerful water currents will all create more erosive force.

Obviously, motorized watercraft produce a wake, but the height of the wave varies according to the speed, draft, and length of the boat, as well as the distance to the shoreline from the point the waves were produced. As a general rule, the deeper the draft and the longer the craft, the bigger the waves. Conversely, the faster the speed and the greater the distance away from the wave source, the smaller the waves.

A three-year study on two comparable stretches of the Mississippi River looked at shoreline erosion. In a channel with heavy boating activity, the shoreline receded up to 14 feet, compared to less than 3 feet along another channel with light boating activity. Eight years later, a resurvey of these channels showed the shoreline had receded a total of 28 feet in the intensely active channel, compared to a total 4 foot recession in the lightly used channel.

Another study along the Lower St. Croix National Scenic Riverway investigated 14 sites for shoreline erosion over four years. Nine sites showed net erosion, two sites had net deposition, and three sites had no net change. As one would expect, the sites with no boat waves and no foot-traffic had no net change or sediment deposition. Interestingly, little net change was seen on the sites with boat waves only, while shoreline erosion was greatest at all sites that combined boat waves with shoreline trampling.

The study also found that wave heights below 0.4 feet did not erode the banks and re-suspend sediments. However, the more boat waves greater than 0.4 feet, the more erosion.

Obviously, narrower rivers and lakes typically are most influenced by boat wakes, since boats must operate closer to shore. Those shorelines that have steeper grades and looser soils are also most impacted.

Motorboats are almost certainly here to stay. To me, the question then is can boat owners become aware of how they impact lakes, and will they care enough to alter their recreational practices to favor the lake community?

Read more about motorboats at the end of the chapter.

≈ ≈ ≈ ≈ ≈

The water quality on Boulder Lake looks good, but how can one tell without fancy instruments? Many of us may want to know what the water quality of a river or lake is without performing a bank of tests. Is there a way to "eyeball," or to "read," the health of a waterbody? I think so, but it's a long-term process. We need to direct our study both landward and into the water. First, look toward the watershed and the overall structure of the shoreline, and ask yourself these questions:

• Are the banks of the river stable? Are they eroding?

• Is more than 90 percent of the streambank covered by vegetation?

• Is the width of the riparian vegetation (the transition zone between the river and the uplands) more than 50 feet with no evidence of human disturbance?

• Is the channel natural, or has it been altered by dredging, channelizing, placement of artificial structures like dams, or other activities?

• Are snag areas, tree roots, and overhanging branches present along the shore, or have they been "cleaned up"?

• Are there lawns or fields right down to the river edge that would allow fertilizers and pesticides and herbicides to be washed more readily into the water?

• Are houses closer than 75 feet to the water, indicating older construction that likely still has original, and thus poor, septic systems?

• Are there roads/sidewalks/paving of any kind leading right down to the river where sediments can wash directly into the water during a storm?

• Are there obvious sources of direct "point" pollution like factories and sewage treatment plants?

• Is the flood plain intact and able to do its job?

Now look at the river itself, and ask:

• Does the water flow fill the channel, or are the banks exposed? Is the water unnaturally shallow?

• Have sediments accumulated on the bottom of the river, covering up the natural substrate?

• Scoop up a handful of water. Are there suspended particles in the water that make the water murky?

• Does the river have structure? Are there branches, stones, boulders, logs in the water, or have these been "cleaned up"?

• Does the water flow naturally, or has it been altered or impounded by some artificial structures?

• Do concrete walls line the stream to supposedly protect landowners from bank erosion, but which alter the stream/land interface?

• Are aquatic plants growing in the

Measuring Water Quality

Water quality is truly a fluid term, usually defined in reference to human goals like:

- Will the water system support a particular sport fishery?
- Will the water be swimmable or drinkable?
- Will the water be deep enough for skiing?

Sometimes we refer to the negative possibilities:

- Will the water be a habitat for noxious or pathogenic organisms?
- Will the water be able to decompose organic wastes we introduce?

Rather than look at human health or recreational pursuits to measure water quality, we need to consider the requirements of survival, growth, and reproduction of aquatic life in a self-perpetuating ecosystem where the structure and function of the waterbody are "normal." The problem with normal is that it tends to escape easy definition since normal depends on a host of variables, which do what variables do—they vary. One year water clarity on a lake may be 22 feet, and the next year it may be only 12 feet. Did the lake lose its "water quality" in the interim due to some human action(s)? Or does water clarity, as just one measure of water quality, typically fluctuate from year to year?

Answers to these questions are often very difficult to tease out. To do so requires an integrated, holistic approach that looks at the physical and chemical measures of water quality, the populations of plants and animals that live in the water, the land use around the river, and the human impacts on the river. Just like we need a constellation of characteristics to describe human health, we need a battery of measures to get the big picture on water quality.

One model, the Water Quality Index (WQI), can be used with minimal scientific training to measure different sections of a particular river over time. The results can then be weighted and totaled to provide an overall picture of the health of the river.

In the WQI, nine tests are performed, including measurements of dissolved oxygen, fecal coliform, pH, biochemical oxygen demand, temperature, total phosphate, nitrates, turbidity, and total solids. Procedures for sampling are clearly outlined, as is the purpose of each test.

The process for sampling benthic macroinvertebrates (bottom-dwelling insects visible to the eye) and tabulating and analyzing the results is also included. The presence of certain aquatic insects can be good indicators of water quality.

All aspects and processes of the WQI are thoroughly discussed in the book *Field Manual for Water Quality Monitoring: An Environmental Education Program for Schools*—see the bibliography.

water in what seems like an adequate number—not too many, not too few?

• Have recreational pursuits changed the waterbody from an ecosystem into a motorized playground?

This list could go on, from describing the obvious—"Are there cows standing in the water (this is bad, even in Wisconsin)?"—to the less obvious—"What aquatic insects are living in the water?" The point is to see the whole through the interplay of the parts, each of which alters the health diagnosis perceptibly. It's also absolutely necessary to apply these same observations over the course of the seasons, and over many years, in order to understand how each river defines "quality" for itself. Then, and only then, can one "read" the true health of a river.

≈ ≈ ≈ ≈ ≈

> *To paint water in all its perfection is as impossible as to paint the soul.*
>
> — John Ruskin

Whenever I paddle across beautiful blue lakes like Boulder Lake, I wonder why the lake water varies in color, given that all the water turns clear when I hold the same water in my hand.

At first glance, it's simple physics. Because sunlight is broken up into the color spectrum as it enters water, some long wavelength colors like red are absorbed rapidly within the first 30 feet, while the short wavelengths at the blue-green end of the spectrum are transmitted and absorbed at the greatest depth. All colors are ultimately absorbed except blue, which is scattered and reflected throughout the water.

The clearer the water, the deeper the light penetrates and the more pure its blue appears. However, if the water contains many sediments or microorganisms, light penetration is reduced, and the blue may not be perceived.

Bottom sediments may impart color, too, as can surface reflections of the sky or trees overhead. Water is a chameleon. It can mimic the bottom, from green plants to brown sands to black silt. Or organic matter such as algae can give water a greenish color, while bogs and swamps typically impart a tea-like brown. Or the surface can appear golden on a small bog lake on a late afternoon in October, reflecting the shoreline ring of tamarack trees.

Water color gives you a summary of the whole scene that surrounds you: the chemistry of the water, the composition of the bottom, the quality of the shoreline, and the type of day. Watch a river or lake over the course of a day, and the water will likely change color throughout the day. If water changes from lighter blue to darker blue, it usually indicates the bottom drops abruptly from shallow to deep.

Every body of water has two kinds of color—true color and apparent color. True color is the real color, the result of the chemistry of the water and the materials suspended in it. Apparent color is color "on loan" at that moment due to reflections on the surface or from the separation of light in the water's depths. This allows you to understand why a glass of northern lake water and a glass of water from the tap may be equally clear. The tap water has had all the suspended matter filtered out. Both sources of water have no apparent color because in small quantities, water can't gather enough light to reflect the sky or shoreline. If you were bored and long-lived enough to take buckets of tap water back outside to fill a lake basin, it would gather and reflect light to produce a shade of blue.

Reflections on the water occur much more readily when the sun is low in the sky. At sunset and sunrise, nearly 100 percent of the light is reflected, so we receive the blessings of orange/red/purple on the water as well as in the sky. When the sun is directly overhead, only about 2 percent of the light is reflected off the surface. The rest penetrates, so the lake will appear its "bluest."

I'm told the Red Sea is indeed red due to the dense blooms of cyanbacteria, which contain red pigments. And the Yellow River in China is indeed yellow because it averages 34 percent sediment, earning it the title of the muddiest of all rivers in the world as it carries its heavy load of loess to the sea. Polar waters, though "pure," are most often olive green, due to diatoms blooming by the billions. (Read about turbidity at the end of the chapter.)

≈ ≈ ≈ ≈ ≈

> *Lakes do to light and sound what sleep does to thought.*
> — Peter Steinhart

Three campsites are sited relatively close together on the far southwestern shore of Boulder Lake, just before the lake narrows back into a river. The river turns south here and flows rather directly toward the County K bridge, the site of a historic logging dam now long gone, though remnants remain.

It would be interesting to paddle back and forth between shorelines to compare which birds are singing from the undeveloped shore and what birds are singing from the developed shore. Fortunately, the Wisconsin DNR already did this study for us back in 1997. The researchers studied 34 lakes in Vilas, Oneida, and Forest counties, conducting breeding bird surveys on 17 undeveloped lakes and 17 lakes with a range of shoreline development. Each undeveloped lake was paired with a developed lake that had similar acreage, shoreline length, alkalinity, and pH.

The researchers found that a large number of breeding species nested on developed lakes, an equally large number of different species utilized undeveloped lakes, and an equally large number had no apparent preference for either lake type (see tables). While the quantity of birds was unaffected by development, the type of species was. Ground-nesting birds such as thrushes and many neotropical migrants were found to be less numerous on developed lakes, while birds often associated with suburban residential areas were more common.

Total number of species and total number of individuals was remarkably similar for both developed and undeveloped lakes, so the average lakeshore owner unfamiliar with specific bird species would not be likely to notice any difference. However, to those who take the time to identify birds, habitat alteration along our northern lakes clearly triggers a very significant shift in avian communities, and current zoning regulations have proven insufficient to prevent this species change.

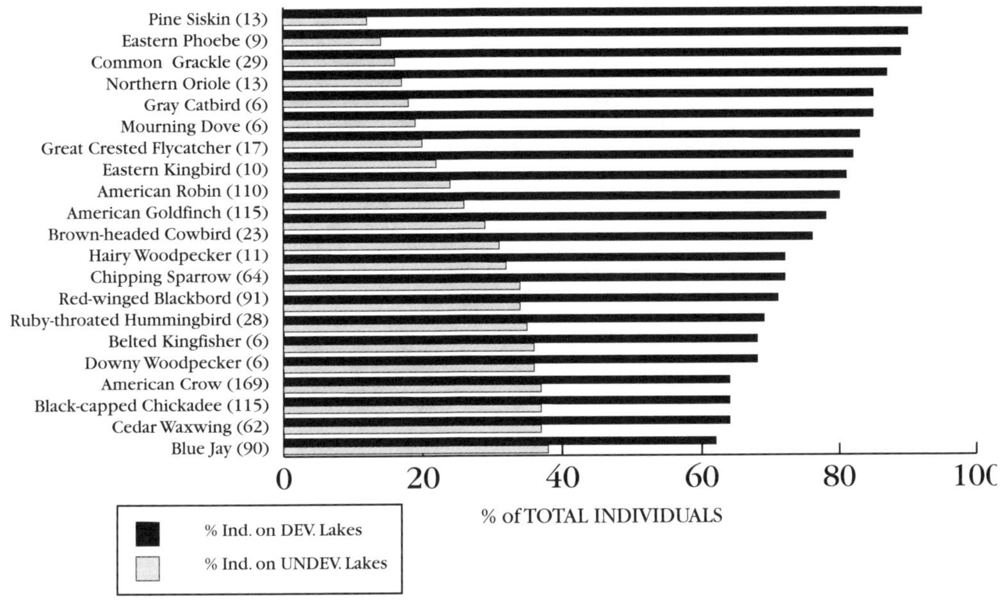

Avian species who were found primarily on developed lakes during 1996 and 1997 point count surveys in Vilas and Oneida counties. Number in parentheses reflects total individuals observed in both years. From Meyer, Woodford, & Gillum, WDNR (1997).

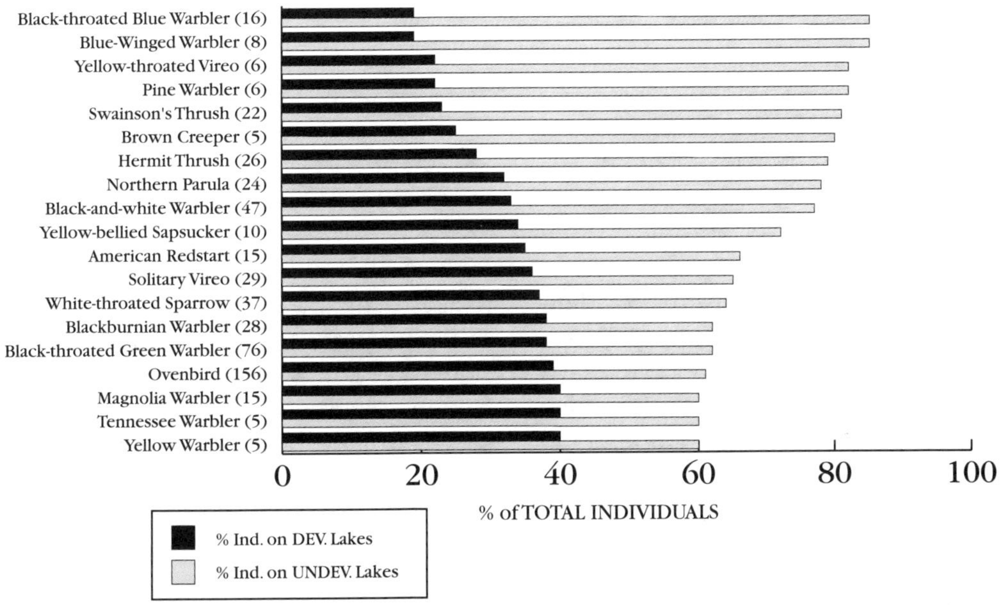

Avian species who were found primarily on undeveloped lakes during 1996 and 1997 point count surveys in Vilas and Oneida counties. Number in parentheses reflects total individuals observed in both years. From Meyer, Woodford, & Gillum, WDNR (1997).

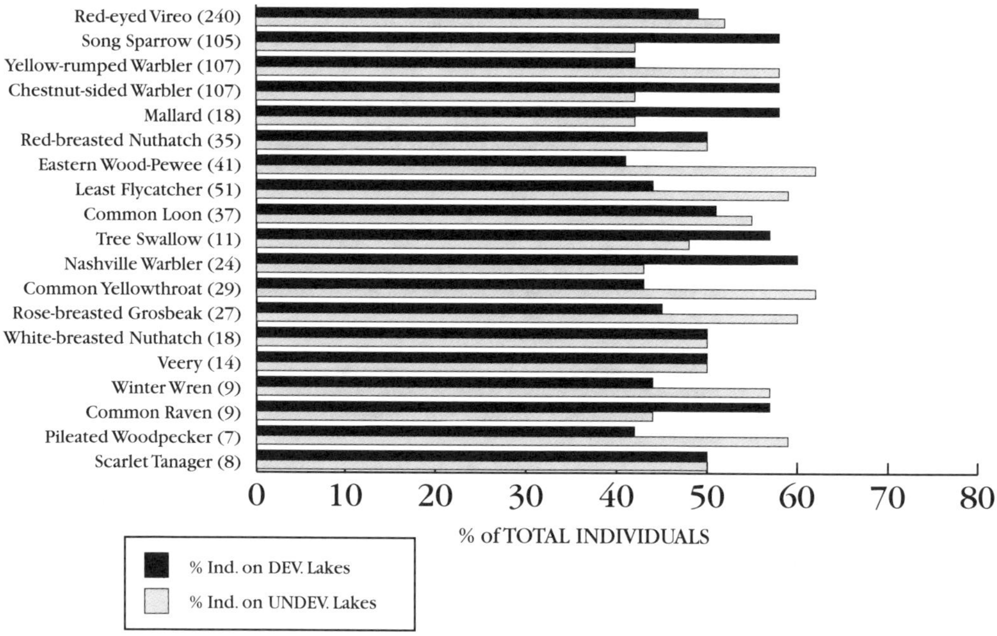

Avian species whose distribution was not associated with lake type during 1996 and 1997 point count surveys in Vilas and Oneida counties. Number in parentheses reflects total individuals observed in both years. From Meyer, Woodford, & Gillum, WDNR (1997).

Many people would not attach any importance to this shift in community members; in fact, some bird observers might prefer the "suburban" constellation of species. That's unfortunate, because the larger issue is that the species that are displaced from developed lakes are also thought to be "source/core" species. This means their core, or main source, of population lives in northern Wisconsin, providing a surplus of young that forays into less desirable habitats to maintain regional populations. By diminishing their number through shoreline development in northern Wisconsin, we diminish their numbers in less favorable habitat elsewhere. Thrushes, ovenbirds, black and white warblers, black-throated green warblers, and brown creepers are all highly productive in northern Wisconsin and are considered "source/core" species.

≈ ≈ ≈ ≈ ≈

You never get so close to the birds as when you are wading quietly down a little river. — Henry Van Dyke

Shoreline areas are often referred to as the "riparian" area—the transition area where land and water meet. These ribbons of land where the stream and land interact are of tremendous biological importance because they often blend the biota from several ecosystems, and thus are quite diverse and productive. Ecologists call these areas "ecotones"—places where one community adjoins another, or grades into one another.

But riparian zones are more than just the banks that define the channel or the area where land-based species have to tolerate high waters. They may be a narrow thread of vegetation in a constricted canyon, or a broad zone encompassing the flood plain as is typical on much of the Manitowish River. Riparian zones defy easy definition, because they comprise a complex system of forest edges, side channels, gravel bars, and islands. They're also biological hot spots that provide optimal conditions for plants and animals.

Riparian areas are often structurally diverse because sunlight can reach the vegetation along the lake or river, as can disturbances like wind and flooding. The more complex the riparian zone, the better. In one study, complex riparian zones had twice the number of plant species as upland areas, and streams that flowed through diverse areas had twice as many trout as simplified riparian zones.

Animals use the riparian ribbon as a migration corridor. Aquatic mammals like mink, otter, beaver, muskrat, and raccoon live in and along this corridor. Deer frequently winter along these edges where water remains open. Even adult insects will fly upstream to lay eggs, a migration that compensates for the inevitable downstream drift of larvae.

In riparian areas, big trees, whether as nesting platforms or as nest cavities, are essential for a number of bird species. Specifically, eagles, osprey and five hole-nesting ducks—hooded and common mergansers, wood ducks, buffleheads, and goldeneyes—need old trees preserved in the riparian zone.

Colonial waterbirds place their nests in trees or shrubs near or in water. Islands or beaver ponds in the Northwoods are commonly used by great blue herons and green herons, while more southern areas see great egrets, black-crowned night-herons, and yellow-crowned night-herons.

Riparian lowland conifers like cedar and black spruce also provide needed habitat for some conifer-dependent species like boreal chickadee and bay-breasted warbler.

Unfortunately, riparian zones are optimal sites for humans, too, and are typically the first ecotone we disturb with our human development. Our attraction to shorelines seems genetic. We have historically lived where lands and waters meet. Where better to find water to drink, to bathe, to cook, to dilute our wastes, to protect us from enemies, to provide transportation, to offer a mild microclimate for farming and living, and to be amidst beauty?

We humans see riparian areas from a host of viewpoints: lakeshore property owner, forest manager, fisheries biologist, highway engineer, pollution control engineer, realtor, recreational user, environmentalist, et al. The viewpoints are all right, and all wrong, because there's no way to simplify the complexity of riparian areas into one viewpoint. Which means we have to do a lot of talking to understand one another, something we don't always do very well as a species.

≈ ≈ ≈ ≈ ≈

It is not hard to imagine that, when we stop to look into the sea or listen to a mountain creek, the attraction we feel is the water inside calling to the water outside, two ponds, perhaps, stopping by the road of time to trade the news. — Peter Steinhart

Riparian Areas

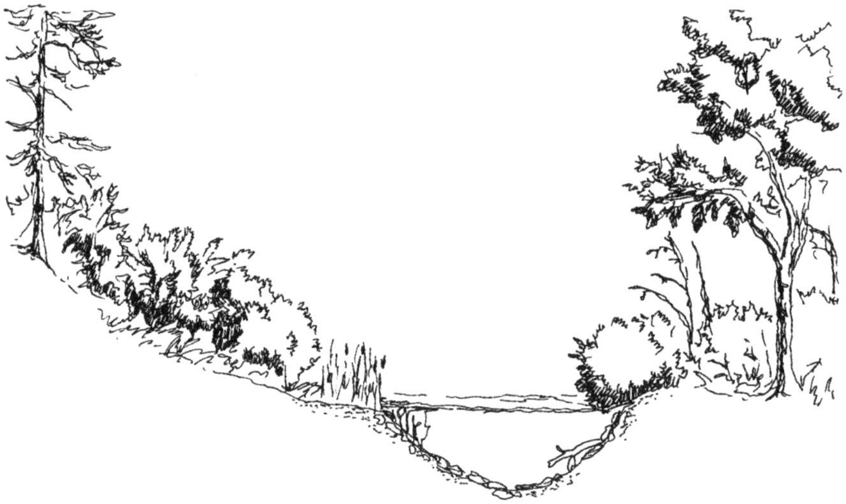

One of the major ecological tenets associated with riparian zones is that streamside vegetation exerts the primary control over the biotic associations in a stream. This "control" occurs through four means:

- The introduction of coarse woody debris into the water
- Bank stabilization by the shoreline network of intertwined plant roots
- Shading, and thus cooling, of water by overhanging trees
- Nutrient inputs from shoreland and in-stream plant growth

Streamside forests control how much sunlight reaches the water, stabilize the banks during floods, filter groundwater entering the stream, provide leaves and wood to the food web, and are a source of large logs. However, the shoreland doesn't hold all the cards. The stream also interacts with and alters the riparian zone, shaping it by such actions as cutting new channels, connecting sloughs, disconnecting sloughs, and taking out trees.

These variables all combine to determine in large part what animals, from aquatic insects to fish, live in a stream or lake.

Human activities in and near the riparian zone greatly influence the water quality. Clearcutting, fertilizing, road-building, development, and pesticide applications all threaten aquatic ecosystems by contributing sediments, nutrients, organic material, and chemicals.

The riparian area serves in part to help filter out these materials and diminish their impact on the water. But how much riparian area is needed to protect water quality? How wide a buffer zone do we need?

One study looked at 50-, 150-, and 250-foot-wide buffer zones, measuring the rate of processing of organic material by aquatic organisms.

Continued

Riparian Areas (Continued)

The study found no significant difference between the processing rates in each buffer zone—a 50-foot buffer was good enough to protect water quality, at least as measured by insect ability to decompose leaves fallen into the water.

The study looked additionally at the same sample of buffer zones, and in hardwood areas, the researchers found mammal use was greatest close to the river. Use decreased out to about 400 feet, and then leveled off. In conifer areas, mammal use was fairly high near the river, but actually higher at 400 feet away. For conifer forests then, buffers should be planned at least 400 feet wide. While in hardwood forest, buffer zones of 100 to 400 feet wide would optimize mammal use.

For birds, many species use the riparian area for breeding and feeding. Researchers found higher numbers of species and individuals near the river before they cut trees and shrubs in the buffer zones. After the cuts, breeding numbers dropped from 24 to 17, particularly in the 50-foot zone. So it would appear that a 50-foot buffer fails to protect many breeding bird species, even though it's adequate to protect water quality.

If you are a riparian landowner, and you don't feel you can maintain large areas, you should try to protect microhabitats. A good rule of thumb is to maintain structural diversity of the vegetation, both vertically (layers up) and horizontally (patchiness). Some microhabitats may be permanent, like cliffs, rock outcrops, seeps, and springs. Be sure to leave vegetative buffers around them and maintain the canopy species. Some microhabitats may house temporary occupants, like an otter den, bird nests, and turtle egg holes in the sand. Be aware of these sites and provide them a buffer. Leave bog areas for orchids and carnivorous plants, sedge openings for yellow rails, dead trees for cavity nesters, downed logs for insects/amphibians/lichens/and others. Even look at disturbed microhabitats as potential habitats, like old gravel pits, which may be good for nesting kingfishers and swallows.

By all means, avoid the desire to make the area into a park. Parks are very poor habitat for wildlife, unless you're a squirrel.

If you can think of the potential of everything you have as a habitat, you will be well ahead of the game.

Even if after all this discussion you're still having trouble remembering the term "riparian" (ry-pair-e-n), think of it as an area "ripe" for development, or ripe for birds and mammals to live

Why are we so attracted to water? I think that in part it's because humans are akin to a leather water pouch. We're nearly 70 percent water. We begin as babies at about 90 percent water, and gradually dehydrate. About 10 gallons of water course through us, constantly circulating, the heart acting as the aquarium pump. To breathe, we dissolve oxygen into our blood through the moisture of our lungs. To eat, we dissolve our food in the acidy waters of our digestive system before it can be absorbed. To remove wastes, we use internal waters to carry them away.

We flow like rivers, our minds as well as our bodies doing the work of a river system. Sometimes we emotionally flow at an even rate within our banks, and sometimes we flood. Sometimes our intellect feels in drought, and other times, it cascades with ideas.

I like the fact that a potato is 80 percent water, an apple 85 percent, a tomato 95 percent, a watermelon 97 percent. When we eat, we often really drink.

How is it that we see and use water every day, that no compound on earth is more abundant, but we will spend hours looking at it, and we will pay astronomical sums to own 100 feet of shoreline along it? We humans, ordinarily so prone to quick boredom, seldom fail to be entranced by water.

Yet water is usually free, unlike any other substance we take from the earth.

What an interesting conundrum!

Wisconsin means "gathering of waters." Our 15,000 lakes, 5.3 million acres of wetlands, and 43,000 miles of rivers and streams dulls us into believing water is so plentiful that it should be free. The problem is how do we assign a monetary value to a river, a lake, a wetland? (For that matter, how do we assign a value to the call of a loon or the choruses of spring peepers in the spring?) In Wisconsin, no one is charged for withdrawing water from a lake or stream, and no one pays to draw water from a well. In urban areas, residents pay to have their water treated to make it consumable and then pay to have the used water disposed of, but the water itself remains free.

What value then should we place upon water? Buying a gallon of bottled water costs more than gasoline; even a liter of spring water costs as much or more than a liter of milk. We've become accustomed to such prices just in the last decade. In my youth, I would have laughed at the very notion of someone charging $1 for 16 ounces of water, but now we pay it without blinking.

But that's "processed" water. We will pay for it if it comes in a bottle because we understand processing costs. What about "wild" water, water that doesn't come from processing bottlers or city treatment plants? What about a lake? What's it worth? One way to look at it is to sell the lake by the bottle. Let's see—Lake Superior . . . The dollars end up in the high trillions, or more.

How about figuring how much it costs to build a lake? You'd have to buy the land, build a dam or dikes, hire in a crew of bulldozers, haul in sand and gravel, plant aquatic vegetation, bring in fish/amphibians/aquatic insects, line the lake bottom with natural materials so the water wouldn't simply drain out, potentially pump in millions of gallons of "free" groundwater if your source of water wasn't sufficient to maintain water levels . . .

Now consider building a river, and the complexity and cost staggers the imagination (except possibly for a few engineers in the U.S. Army Corps of Engineers). In Richard Brautigan's *Trout Fishing in America*, the main character visits a Cleveland wrecking yard and sees a sign that says, "Used Trout Stream for Sale. Must Be Seen to Be Appreciated." He asks how much the stream is selling for. He's told, "Six dollars and fifty cents a foot. That's for the first hundred feet. After that it's five dollars a foot." He asks how much for the birds. "Thirty-five cents apiece. But of course they're used. Waterfalls are available for 19 a foot. Insects are selling for 80 cents a square foot. The flowers are sold by the dozen, but the trout come free with the stream."

Those were 1967 fictional prices. Imagine the prices a writer could dream up now.

Ten thousand years of post-glacial natural forces have conspired to create the landscapes and waterscapes we now enjoy. Their value is literally impossible to calculate. More important is the value each of us places on these gifts.

For more information on some of the concepts used to determine the economic value of water, you can order a series of fact sheets from the University of Wisconsin Extension—Water Issues in Wisconsin, #G3698.

≈ ≈ ≈ ≈ ≈

A public boat landing sits on the eastern shore just above County K. Parking areas are also available on the other side of County K. A public wayside was built at the southeastern corner of County K and County H—one can also put-in from just below this wayside on County H.

One way or another, unless the water is very high, you will have to portage around this section, or drag your canoe through the very shallow, rocky riffles here.

There's not much you can do about portaging—it's water over the dam, as the cliché goes.

Water seems to bring out the best, or worst, in us relative to clichés. We've all used watery clichés to describe everything from the state of our present bad luck to our moral character.

- I'll return come hell or high water.
- She took to it like a duck on water.
- Don't throw out the baby with the bath water.
- He has one oar out of the water.
- That idea's dead in the water.
- That's water under the bridge.
- That's water over the dam.
- Blood is thicker than water.
- The newcomer was like fish out of water.
- He's in hot water.
- Keep your head above water.
- Quit muddying the water.
- You can lead a horse to water, but you can't make him drink.
- Cry me a river.
- We're sending you up the river.
- That politician's as crooked as a bend in a river.
- Don't piss off the alligator until you've crossed the river.

Portaging

No matter how hard you think your portage is, it's a cake-walk in the park compared to the portages made by people 150 years ago.

In 1820, David Douglass wrote of the 9-mile-long Grand Portage of the Fond du Lac:

"There is one, a half Frenchman, a tall elegantly made man to whom I have given the name Lord Byron . . . This man would shoulder at a load two kegs of bacon weighing not less than 125 pounds each and a bag of flour or corn nearly a hundred pounds more, and this he would walk or rather trot off in style which would have fatigued me had I been perfectly unloaded . . . But if I was astonished by the loads borne by the men, I was even more so by the scarcely less burdens carried by the women, for all joined in the labour of the portage. Lord Byron's mother carried at one load her birch canoe of sufficient size to carry a whole Indian family with all their baggage. . . "

Remember too, most voyageurs were relatively small men. Space for trade goods was at a premium and big guys took up too much room in the canoe. They weren't wearing the latest high-tech hiking shoes, the DNR or the Forest Service didn't keep the portage trails clear for them, Power Bars came in the form of moldy pemmican, DEET and 40-pound Kevlar canoes that didn't require continual sealing with pitch were the stuff of dreams . . . well, enough said.

How tough were the trails to cross? Doty wrote of one portage:

"We had not proceeded a mile on this course before we were led into swamps and morasses which one would think impenetrable were he not led on by an Indian. We seldom found footing before we had sunk to our knees in mud, and frequently to our hips . . . Early in the morning we rose, shouldered our packs, and commenced our route which the whole of this day laid over wind-falls and through cranberry and tamerack swamps. It is impossible to describe the fatigues of this days march."

So, if you've portaged canoes before and whined like a dog as you dragged yourself onward, buck up! The path you're on likely has the footprints of thousands of men and woman over a thousand years who passed before you. Honor their memory by bearing your load with some dignity—or at least learn to swear in French or Ojibwa.

• That's a drop in the bucket.

Perhaps Nietzsche deserves the last say—"I'm tired of the poets of the day who muddy their waters that they might seem deep."

It was like being in the belly of a wave, below waterline in a surf traced by points of foam. Perhaps a trout sees a rapids like this, points of dancing foam, while he holds in an eddy waiting. — Diana Kappel-Smith

Soils

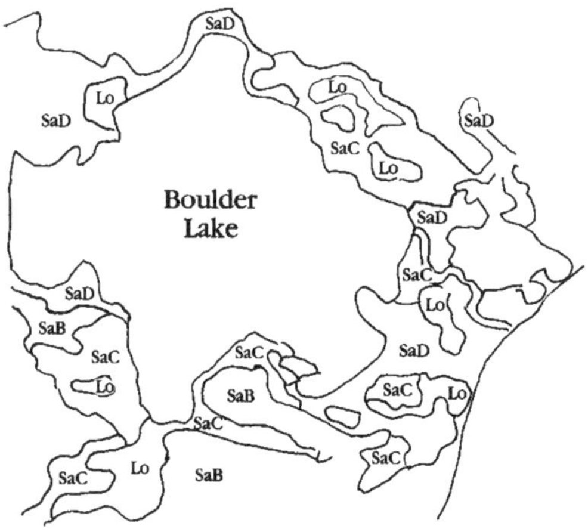

For the most part, four basic interacting factors determine the nature of a body of water: the watershed and soils around and in the lake, the basin shape of the lake (its "morphology"), the elevation of the lake, and local climatic effects. As discussed earlier, lakes quite close to one another can be remarkably different in their water quality, and in their plant and animal life.

The health of a waterbody mirrors its watershed. Still, intact watersheds don't guarantee crystal-clear water. Some waterbodies are naturally nutrient-rich, or "eutrophic," despite a relatively undisturbed watershed all around them. The bay of Green Bay on Lake Michigan comes to mind. The historic adjective—"green"—tells you the waters are naturally highly productive. The soils around a lake may be nutrient-rich or nutrient-poor, as may be the plants all around the edge of the lake that contribute their leaves every autumn. A large watershed that funnels down nutrients from a rich countryside will likely produce lots of plants in the lakes at the end of those funnels.

But in this area of northern Wisconsin, the soil is sandy. Forty-two percent of Vilas County soils are an association of "Rubicon-Sayner-Karlin" soils, which broadly could be said to be little more than an upland beach. Rubicon and Sayner soils are described as "excessively drained," while Karlin soils are "somewhat excessively drained." A typical Rubicon soil has 2 inches of partly decomposed forest litter at the surface, while the surface layer itself amounts to 1 inch of dark reddish gray sand with 23 inches of dark reddish brown subsoil sand laying beneath it.

It's fair to say this ain't farming country. In fact, agriculture in Vilas County is less important than in any other county in the state. On the other hand, we grow some of the best white and red pines in the country.

Around Boulder Lake, the soil is classified mostly as "SaB", "SaC" or "SaD," each describing a Sayner-Rubicon complex defined as "excessively drained soil formed in sandy glacial outwash." "Lo" refers to Loxley and Dawson Peats, or what we usually call peat bog.

Turbidity

The distance that light can penetrate through water is one way to define turbidity. The more free-floating suspended particles or microscopic plants that grow in the water, the greater the turbidity.

When water becomes murky, light gets absorbed quickly in the water column. Without light, plants can't photosynthesize at greater depths. When photosynthesis is limited, the production of oxygen declines.

Turbid waters are also warmer waters. When water is warmer, it holds less dissolved oxygen than colder water—a negative scenario for many organisms.

Sediments suspended in the water can also clog fish gills, depriving them of oxygen. If the sediments settle onto the stream bottom, they can smother fish and insect eggs, as well as insect larvae. Or the sediments can fill in gravelly spawning beds, preventing successful egg hatching.

Many aquatic species have evolved and adapted to turbidity over time. Catfish and carp are well-adapted to low oxygen and light levels, whereas trout have little tolerance for even low levels of turbidity.

Turbidity also affects visibility. Walleye and loons must see their prey in order to chase them down. For visual predators, turbidity dramatically affects their ability to eat.

Time may be the key element in determining the impact that turbidity has on aquatic life. The longer water remains turbid, the more stressed many species of fish and other aquatic life become.

As discussed earlier, some lakes and rivers are naturally more turbid than others. Turbidity becomes an issue when cultural sources artificially decrease water clarity through excessive algal growth, eroding stream banks, erosion from fields and construction sites, and urban runoff from storms.

Water transparency can be measured with a Secchi disc (pronounced seh-kee), named for an Italian astronomer and Jesuit priest who invented it in 1865. The disc is alternately white and black, 20 centimeters in diameter and weighted to make it sink. It's lowered into the water on a line marked with depth measurements. When the disc disappears from view, the depth is noted as the "Secchi disc transparency" of the water. This point occurs where only about 5 percent of the sunlight penetrates the water column. In water over 4 feet deep, a Secchi disc reading of less than 1 foot is considered "turbid" water. If the disc remains visible at 4 feet or more, the water is considered "clear."

The Sweetest of Warblers—The Yellow Warbler

"Sweet, sweet, sweet, I'm-so-sweet!" So goes the most common version of the yellow warbler's song repertoire. Learn this song, because no bird is likely to be more common along the Manitowish than the yellow warbler. Low, scrubby vegetation near water, like the dominant willows and alder that typically line shallow, warmwater rivers, makes for ideal breeding habitat for yellows.

Individual territories can be as small as two-fifths of an acre and less than 100 feet apart. Where shrubs occupy the shoreline, a paddler is seldom deprived of the yellow warbler's song. The yellow seems to sing most of the day as well, not saving his voice for dawn and dusk.

Visual identification of a yellow requires little gnashing of teeth and arduous thumbing of manuals. The male's bright yellow face, greenish-yellow topside and rusty-streaked underside, though somewhat variable, have no imitators. Naturalist Frank Chapman called yellow warblers "a bit of sunshine"—no other warbler exhibits such extensive yellow. The female, more demure and plain as female birds are wont to be, may create more identification stress.

Continued

The Sweetest of Warblers—The Yellow Warbler (Continued)

No other wood warbler breeds more extensively throughout North America than the yellow. This yellow bundle likes shrubby habitat, preferring to nest in wet deciduous thickets, but also nesting in old orchards, overgrown pastures, regenerating clearcuts, and even reclaimed strip mines. Certainly not a bird of old growth stands, the yellow is drawn to disturbed and early successional habitats thick with new vegetation.

The willow thickets along the Manitowish offer dense shelter and forked stems aplenty for construction of a cup-shaped nest usually 2 to 8 feet off the ground. The usually monogamous yellows mate in May. The female lays 3 to 6 whitish eggs, and incubates them for 11 to 12 days, during which time the male often feeds her on the nest.

Unfortunately, yellow warblers frequently host brown-headed cowbird eggs. The good news is that yellow warblers have had a long association with brown-headed cowbirds, and many have learned to recognize the much larger eggs. Because the eggs are too large to eject from the nest, the adults often simply bury all the eggs under another nest layer. If parasitized again by cowbird eggs, the pair will build another floor—the record is six tiers high! Cowbird parasitism on yellow warbler nests ranges as high as 30 to 40 percent in some regions.

Yellow warbler chicks fledge on average in 10 days, and if the season permits, the adult pair will get to work on raising another brood. The summer goes fast for yellows, which are one of the earliest fall migrants, often heading south simultaneouly with the blooming of goldenrods. Wintering grounds range from Central America through the northern half of South America.

In their brief window of time in the Northwoods, the yellow warblers' diet consists almost entirely of insects, like spiders, gleaned from outer foliage. Look for their tail pumping up and down from their singing perch, and spend time learning the local songs. Yellows have two song types: accented ending songs and unaccented ending songs, and several variations of each. They can fool you if you don't pay attention.

Yellow warblers also show the strongest geographical variation of all wood warblers; 43 subspecies occur. Expect some variation in yellow warblers and enjoy their evolutionary diversity. They're guaranteed good company along the Manitowish throughout the summer.

Motorboats—Effects on Fish, Wildlife, and Plants

Fish populations are affected by water quality, turbidity, pollutants, and altered aquatic plant habitats. Surprisingly, very few studies have documented the direct impacts of boat activity on fish mortality and behavior. We know that boat activity can disturb fish from their nests and increase re-suspension of sediments, which can cover eggs and alter aquatic plant composition, quantity, and quality. But how much boat activity is necessary to set these consequences into motion isn't known.

The effect of motors on aquatic wildlife like eagles, waterfowl, otters, muskrat, turtles, and frogs is even more difficult to answer. Some species are sensitive to human presence, while others seem adaptable. Some species are more sensitive during particular times—for instance, during nesting or raising young—than other times.

Two recent Midwestern studies tried to determine motor impacts by directly measuring the flushing response of 16 waterbird species to five different human activities—walking, ATV, motorboat, canoe, and automobile. As anticipated, the birds varied considerably in their flushing response to human activities. The more habituated birds, like gulls, flushed much less rapidly and over a shorter distance. Interestingly, walking and canoeing tended to flush birds at a greater distance than motorized activities, just the opposite of what one might expect. The authors speculated that perhaps the slower speeds of approaching canoeists and walkers gave the birds more time to respond.

The bottom line is that all the activities flushed birds, leading the authors to recommend minimum buffer zones of 100 meters from all activity to protect most bird species.

The impacts of motorized watercraft on small mammals, shorebirds, amphibians, and others are less studied. The easily recognized heron rookeries, eagle nests, and loon nesting islands need to be protected. But how do we protect frog habitat, turtle loafing sites, waterfowl shoreline nests, migratorial shorebird needs, otter dens, and other sites from motorboat impacts?

For plants, the negative impacts of motorboats are easier to measure. A 1997 study on Lake Ripley in Jefferson County, Wisconsin, showed that excluding motorboats from enclosed experimental plots significantly increased the amount of growth of aquatic plants, their overall coverage, and their shoot height.

Restrictive ordinances such as no-wake zones, or total motor restriction, in plant beds and shallow waters would protect plants and fish. They would also offer the added benefit of separating anglers from high-speed boaters, a source of continual conflict.

Breeding Birds and Shoreline Structure

Nature abhors a vacuum.This maxim of ecology applies as much to birds as it does to everything else. Birds have evolved over time to fill nearly every niche that the natural world can present. So, in a 1997 study comparing breeding birds on developed and undeveloped lakes, it's not surprising that 87 species of birds were observed on undeveloped lakes, while 82 species were observed on developed lakes. Or that 1,093 individuals were counted on undeveloped lakes (13.0 per point count site), while 1,260 individuals were counted on developed lakes (13.4 birds per point count site).Visit a clearcut forest and you might be surprised by the number of birds that nest and feed there.That's not a ringing endorsement for clearcutting, but a simple acknowledgment that if a habitat provides food, cover, and breeding and nesting opportunities, animals will find a way to utilize it over time.

So, numbers aren't an issue.What is the issue is how the different habitats alter the species composition, and even more specifically, what the "guild composition" of birds was at each lake type.A guild is a group of species in a community that exploits the same set of resources by using different means.A nesting guild, for instance, includes categories such as ground-nesters, bank-nesters, cliff-nesters, cavity- or snag-nesters, shrub-nesters, floating-mat nesters, and others.A feeding guild includes major categories such as omnivores (eats whatever is available), nectivores (feeds primarily on nectar), frugivores (eats primarily fruits), insectivores (eats primarily insects), seed eaters, and others.

In this study, foraging, diet, and nesting guilds were compared on developed and undeveloped lakes. In the nesting guild, ground-nesters declined at developed sites, an expected result since ground-nesters are less likely to nest where lawns replace forested groundcover.

In the diet guild, insectivores were less common at developed sites, another expected result given that fewer insects are typically found on less vegetated sites. However, the proportion of birds that were omnivores, nectivores (nectar eaters), frugivores (fruit eaters), or seed eaters were much higher at developed lakes.

Most important may be the fact that on developed lakes, no birds were observed that had a preference for undeveloped lakes—all the birds present showed a preference for developed lakes.Thus, on a developed lake, birds that prefer to breed, nest, and feed in undeveloped habitats simply are completely displaced.

Many of these displaced species are "edge-sensitive," including Swainson's thrush and black-throated blue, black-throated green, and

Continued

Breeding Birds and Shoreline Structure (Continued)

black and white warblers. This means they don't do well where two habitats come together and form an "edge"—for instance, a woodlot and a lawn.

Other displaced species have special habitat needs (northern parula, pine warbler, and brown creeper) that apparently can't be met on developed sites.

The bottom line is that habitat alteration changes the composition of bird species in a community, and that our current Wisconsin shoreline regulations permit these alterations.

Chapter 3
Our Course: County H and K to Island Lake

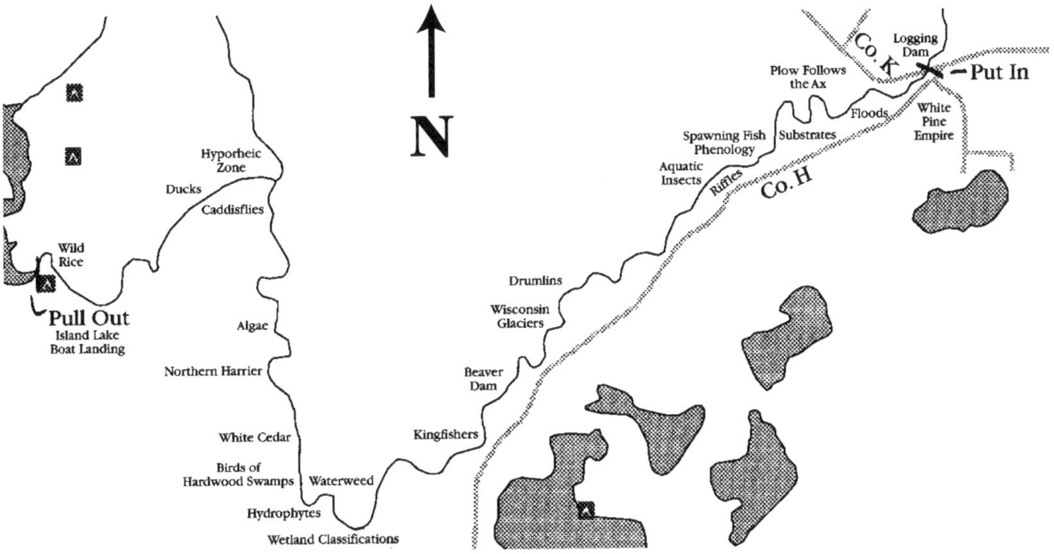

Rivers course through my dreams, rivers cold and fast, rivers well-known and rivers nameless, rivers that seem like ribbons of blue water twisting through wide valleys, narrow rivers folded in layers of darkening shadows, rivers that have eroded down deep into a mountain's belly, sculpted the land, peeled back the planet's history exposing the texture of time itself.

—Harry Middleton

Chapter 3

Of Logging, Floods, Glaciers, Kingfishers, and Algae

If there is one tree which can be used to exemplify the northern forests, it is the white pine . . . When the vacation-bound traveler from the hot and steamy cities of the south sees his first white pine, he knows that he is entering the Northwoods. — John Curtis

Put-in at the boat landing below Highway H and K. This section will take about three hours to paddle, or more if curiosity takes the lead.

The logging dam that was once here was built around 1890, in the same time frame as the Rest Lake and Fishtrap Lake dams. Many of the pilings and structures from the dam can still be seen in the riverbed.

In the spring, each of these three logging dams would hold back a massive head of water, and then release the water in one pulse to carry the logs downstream. Given how shallow the Manitowish is today, you can imagine how much water would have been needed in order to successfully shoot logs downstream. The water from behind the Boulder Dam is said to have backed all the way up into White Sand Lake, 3 1/2 miles away, as well as into North Creek Springs, which is nearly 2 miles to the southeast.

Estimates of Wisconsin's white pine empire suggest the state once contained over 100 billion board feet of white pine. White pine had the greatest commercial value of all tree species in the state. Its ramrod-straight growth, clear grain, huge diameter, great height, often limbless trunk, soft but strong wood, and its buoyancy to make the float downriver to sawmills, all characterized a wood that would eventually build not only most of Wisconsin's towns, but prairie towns desperate for lumber like Chicago and Kansas City.

The best pineries produced 10 to 12 million board feet per section (640 acres). Individual trees could be 5 to 7 feet in diameter, and scale 6,000 to 7,000 board feet of lumber. These giants dated back to the 1400s.

The lumber era peaked in 1899, generating 3.4 billion board feet of lumber that year, and then gradually dissipated, until in 1930, "only" 0.3 billion board feet left the woods for southern sawmills.

Log driving conditions influenced the pattern of urban development. The first sawmills were all built on the banks of rivers, typically at the foot of a rapids or a waterfall where fast water could be directed to power the saws. The Wisconsin River was difficult to drive because of its many rapids, so sawmills were built as close to the logging camps as possible and at the base of rapids in cities like Wausau, Stevens Point, Wisconsin Rapids, Merrill, Mosinee, and Rhinelander. On slower and easier rivers to drive, sawmilling could be more centralized, giving rise to bigger towns like Oshkosh, Chippewa Falls, Eau Claire, and La Crosse. Bigger towns with better transportation and trade connections also made it more profitable to invest the necessary money in steam-powered mills.

When railroads appeared on the scene, water transport mills began to disappear, unable to compete with all-weather routes to market. By the 1890s, most Wisconsin mills were located along railroad tracks, particularly in the north where there were still forests to cut.

The little Manitowish River was a bit player in this massive siege, but still a player. The Northwoods was the source of the logs, but not the best place for mills or easy marketing. As a result, relatively few large mill towns arose in this area during the logging era, logs being cheaper than milled boards to float or train to market.

We were fortunate. Sawmills produced prodigious amounts of waste in the form of sawdust, bark, edgings, and slabs. The companies typically disposed of mill refuse by dumping it in the nearby stream, filling in lowlands, or paving streets with sawdust. At Peshtigo Harbor north of Green Bay, the Peshtigo Lumber Company created 10 acres of new ground on the west side of the river by dumping slabs and sawdust. In 1880, sawdust as deep as 8 feet from mills on the Menominee River was burying whitefish spawning grounds, inducing the Wisconsin legislature to pass its first law prohibiting dumping mill refuse into streams. Oshkosh was called "Sawdust City" because it filled in many acres of wetland along the Fox River with sawdust. Entire blocks of Oshkosh now stand on a foundation of slabs and sawdust.

In Wisconsin, lumbering left in its wake nearly 75 ghost towns. We have no ghost towns along the Manitowish River, though we have river towns much smaller and less "alive" than they once were. Whether that's good or bad depends on your point of view.

The major players in and around the Manitowish River logging drama were the Chippewa Lumber and Boom Company, Salsich and Wilson, Patterson Brothers, Yawkey Bissell, Wright Lumber Company, and the Vilas County Lumber Company. Logging began in 1892 in this area, and by 1911, 90 percent of the land was cut over. The poor sandy soils of this local region didn't tend to produce the truly big pines like the better soils did in Presque Isle and Winchester, but white pines over 3 feet in diameter were common.

Logging camps tended to be run by small independent outfits who sold to the bigger companies. To identify whose log was whose once the logs reached a mill, log hammers were used to stamp the ends of the logs. Some 6,000 log marks were registered in the Chippewa District from the Mississippi River to High Lake, representing each one of these companies, from one-man operations to major companies.

Today, some of the pine forests of

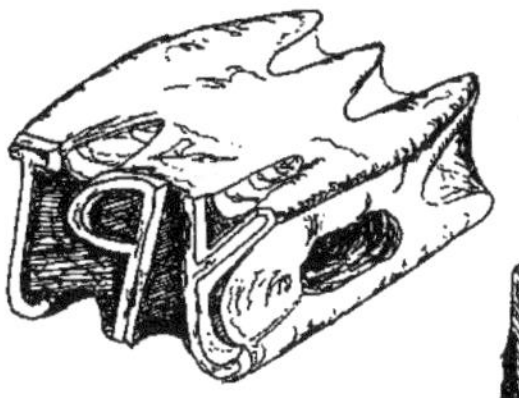

YPJ: Jim Plunkett's personal logmark

FLY: Chippewa Logging Co.

HK: Chippewa Lumber & Boom

Vilas and Oneida counties are resurrecting themselves after their amputation, and their subsequent takeover by the pioneering forests of white birch and aspen. When forest succession is left to age gracefully, the aspen/white birch communities on sandy soils typically give way to white pine. The most recent inventory of white pine in Wisconsin done in 1983 showed a 48 percent increase in white pine compared to a similar inventory taken in 1968. In Michigan, the story is even more heartening. White pine increased 66 percent from 1966 to 1980.

Under the patchy shade of thinning birch and aspen canopies, white pine seedlings and saplings have come back strong.

In his book *Listening Point*, Sigurd Olson wrote:

"The pine stands were thought inexhaustible, and no one could have imagined a day might come when trees had other values than lumber.

"But the pioneer attitude survives and there are some today who still look at a tree as having just so many board feet and no other values. I visited with a cruiser of the old school not long ago while he was estimating the timber in a small stand on a neighboring lake. We stood in the shade of a big pine . . . 'take this one,' said the cruiser as though reading my thoughts, 'this is an old one and overripe, should be cut to make room for the young stuff coming underneath. Even the seeds aren't as good as they should be, and with the decay inside it's a nest of fungus and beetles. That tree is dangerous, ought to come out, and that's true of most of the big stuff left. People don't get any more pleasure from such a relic than they do from a healthy young tree.'

"As I stood there, I could hear the soft moaning of the wind in the high dark tops and feel the permanence and agelessness of the primeval. In among those tall swaying trees was more than beauty, more than great boles reaching toward the sky. Silence was there and a sense of finality and benediction that comes only when nature has completed a cycle and reached the crowning achievement of a climax, when all of the inter-relationships of the centuries have come at last to a final glory."

We have some people today who still see aging trees as "waste, danger, and bad economics." As with everything, it depends on what you value. They're right that by cutting and opening up the canopy, that more trees will germinate and grow. They're right that by cutting a tree at the moment its growth curve begins to decline that they will optimize the board foot output of the forest. But they're wrong when they call mature trees a "waste." By my value system, the benefits of old trees far exceed the loss incurred through the narrowing of growth rings.

Describing white pine stumps, Michigan essayist John Eastman said,

"They are time warps in the landscape, relics of the golden age of timber, amputees of an America I never knew . . . The big forest is gone, but the ancient languages can still be heard."

Balance, it's always all about balance. We need wood products—I heat with wood and build with wood—but we need equally the "products" of beauty, history, inspiration, and reverence. I vote for the Menominee Indian model—sustainable old-growth forestry. Tourism and timber and ecology, in balance.

≈ ≈ ≈ ≈ ≈

Thoughts from the Old Growth—Pine Voices

I look across the marsh
to the river behind it
where the old pines rise
like long feathers
scattered in random fall winds.

'Cross river, the camp road,
with blackberries narrowing its scar,
leads me to the base
of an ancient white pine.
Here years of needle duff have settled
two feet above the forest floor.
I pat the plates of bark and ask
how and why it escaped,
and thank you, and
how many owls have clutched
your open branches in the moonlight?

The wind rises
and finds voice in this pine,
a rushing like a hare racing
over powder snow,
like a mother
comforting a waking baby, *ssshhh*,
ssshhh, like gentle waves
falling back along
a fine sand beach.

The logging companies were given free reins to do what was needed to move logs downriver. Just how free? The following language is from the 1911 statutory authorization, section 182.71(3)b, for the Chippewa and Flambeau Improvement Company: "The company may construct, acquire, maintain and operate dams, booms and other structures in, along, or across any of these portions of the rivers and their tributaries to accomplish the purposes of this section. The company may clean out, straighten, deepen, or otherwise improve any of these rivers and tributaries to improve navigation or to prevent injury to property bordering on the rivers."

Meanwhile, Wisconsin developers were heavily promoting the myth that "the plow follows the ax." Jobs, cheap land, and farming opportunities were the selling points. Despite the obvious reality of poor soil, short growing seasons, and lack of markets, people came because of excessive rhetoric like the following: "With farms supplanting the forest, northern Wisconsin will not revert to a wilderness with the passing of the lumber industry, but will be occupied by a thrifty class of farmers whose well-directed, intelligent efforts bring substantial, satisfactory returns from fields, flocks, and herds."

Or so maintained Dean William Henry of the University of Wisconsin College of Agriculture who wrote the above in 1896 in a fit of chamber of commerce oratory.

Read more about early logging at the end of the chapter.

The effect on the riverbed and shoreline due to the release of these sudden "floods" carrying thousands of careening logs downstream was profound. The shorelines were gouged much wider than their natural width from the scouring effect of the water and the scraping effect of all the logs. How much wider? Unfortunately, it's nearly impossible to say. Accounts of early traders and explorers who paddled through this area provide some insight, but they offer few specific details regarding the original depth and width of the river channel. Nevertheless, common sense tells us that the Manitowish's river channel must have been narrower prior to the artificial scouring created by the floods from the logging dams.

This is not to say that floods never occurred naturally on the Manitowish. The Manitowish often floods naturally in the spring. The lower reaches of the river have extensive flood plains that absorb the impact of heavy spring snowmelts. But natural floods on smaller northern rivers like the Manitowish typically build over a few days and are released over time. These floodwaters typically flow with far less velocity and scouring effects than the instant massive releases from the logging dams.

It's important to note that in general natural floods, like natural fires, are ecologically beneficial. A flood moves sediments down the river channel, sweeps away encroaching plants, shifts the river's banks, and moves the riverbed—from sands to boulders. This allows for the formation of transient gravel bars that provide spawning habitat; otherwise fine sediment particles can fill up the bed and destroy the spawning. Recent studies on rivers have concluded that a river needs to mobilize its channel bed every other year on average in order to provide ideal spawning habitat.

Floods—Beneficial and Necessary Forces

Most floods need to be seen as normal and essential disturbances. Like fires, windstorms, icepush, intense cold, insect outbreaks, disease—the whole laundry list of possible disturbances in the Northwoods—floods are part of the ecological scene. Our river communities, plant and animal, have had to adapt to this disturbance, or fail to exist.

The question then is what occurs to a river during a flood? A healthy river will usually absorb all but the most catastrophic floods. The riparian plantlife holds the shoreline soil in place with its network of roots. The leaves of shrubs and trees further protect the groundlayer from the pounding and erosive forces of heavy rainfall.

Still, sometimes a flood scrubs an area clean of vegetation, but most vegetation has adapted to regenerate quickly. Silver maple and willow flower and go to seed immediately in spring, so the seeds are ready to drop on exposed river banks when the traditional spring floodwaters recede.

Floods often improve a river's complexity. Logjams serve to slow and redirect surging currents, causing the river to dig out pools and flow into side channels and sloughs. Floods carry wood and rock debris that drops out of the flow along the way and provides new structure within the river. Since different fish need different habitat structure for spawning, feeding, and resting, floods are actually key events in shaping and maintaining high-quality fish habitat.

During floods, rivers cut new channels, resculpt older ones, clean silt out of spawning gravels, and flush accumulated leaves and woody debris into the water from the flood plain. Flood plains work simultaneously to soak up the high water, and then release it slowly, keeping the river flowing in drier months.

Floods carry older organic materials like dead plants from the flood plain into the river, and replace them with newer materials. The floodwater also scours the long strings of filamentous algae, as well as other plants, off the riverbed. Afterwards, plant populations often explode due to the nutrient-rich organic debris, fueling a growth spurt all the way up the food chain.

Though some fish and other animals die in floods, most survive by resting in side channels, sloughs, and logjams. When the waters have receded, they multiply on the abundance of food and new spawning habitat.

Rivers, and the life within them, survive through resilience. A healthy river is self-sustaining and self-healing. Destruction of one life leads to new life in a dynamic ebb, flow, and flood.

> *I know of no solitude as secure as one guarded by a spring flood; nor do the geese, who have seen more kinds and degrees of aloneness than I have.*
> — Aldo Leopold

The logging and damming were considered necessary precursors to the arrival and settlement of homesteaders. Most Europeans understood farming or urban life, but making a life in the wilderness woods was well beyond their realm of experience. Logging provided winter income, and removed trees so that the land could be cleared. From the viewpoint of most rural Europeans, civilization required subduing the forest. Damming rivers made a great deal of sense to most homesteaders. They needed to transport logs, limit flooding, and tame rivers enough for safer boat travel. To acquire the benefits of civilization required the free movement of trade, though the hardships of northern winters remained extreme even when homesteaders had the ability to purchase goods.

For the American Indians and European settlers, trade and agriculture set up opposite approaches to living in the Northwoods. Although engaged to a significant degree in the fur trade, American Indian people still were dependent on the seasonal cycles of the natural world to sustain their lives. Thanks to their much more extensive trading systems, Europeans could become significantly disconnected from seasonal cycles and still survive. Thus, the American Indians are classified by some anthropologists as Ecosystem People, while Europeans are referred to as Biosphere People.

These two opposing land-use perspectives appeared to be the only choices for most of those who wanted to live in the wildlands of northern Wisconsin. You either appreciated nature as bountiful and as a provider, or feared it as a wilderness that had to be quashed, altered, and settled. The first perspective required use of local resources, but it also mandated protection. Direct life-threatening consequences could be suffered by the Ojibwa if the use of a species exceeded its sustainability.

The American Indians in this region used over 200 species of plants for food, medicines, building materials, hunting and trapping, clothing, transportation, household goods, art and music, games—the entire gamut of life. Prior to the fur trade, commerce between indigenous peoples occurred, but to a limited degree, given the difficulties of transportation, storage of goods, and time constraints. Replacing a native species with a nonnative trade item was seldom an option.

By contrast, European settlers brought a set of skills and understandings that didn't fit the ecosystem they wished to live in. As settlement grew, they were able to maintain their European agricultural mind-set, because they brought with them a strong trade, communication, and transportation network that gave them the ability to destroy species without consequence, in fact often without even noticing.

Put any of us in their place, even knowing what we know today, and, for better or worse, nearly all of us would have done the same. The settlers implemented what they knew, and their initial survival depended on making that implementation happen quickly. As a result, native flora and fauna were often being perceived as pests, because they didn't fit into a European agricultural vision. However, nonnative plants and animals did fit their cultural understandings. Europeans had the tools and knowledge to alter the ecosystem to fit their needs, and so they did.

Today, it's astounding how extensively that European influence has changed the landscape. It's very likely that nothing you ate today, last week, or even this month is native to North America.

The only native grain we commonly eat today is wild rice, and that is more a delicacy than a main component of our diet. Our northern native agricultural crops in use today consist only of blueberries, sunflowers, wild rice, cranberries, and maple syrup. The origins of virtually all of our major crops today are European or Asian. Consider the following crops and their origins:

Potato - Peru
Corn - Central Mexico
Wheat - Near East (Iraq, Iran, Turkey)
Barley - Near East
Apple and Apricot - Southcentral Asia
Rice - China/Indonesia
Sugar cane - China/Indonesia
Soybean - China
Millet - China
Oats - Europe
Rye - Southwestern Asia
Coffee - Ethiopia

The list could go on. Though many of these crops aren't grown in the Northwoods, their processed versions are eaten by northerners every day. An argument can certainly be made that the introduction of these nonnative farm crops has improved life for most people. That may well be true. The point is not to condemn current foods, but rather to demonstrate that what we typically eat today is quite alien to what was eaten here by American Indians just 150 years ago. Does the future "global community/economy" bode even worse for native plants? It's fair to expect that our connections to our local ecosystem will continue to wither away.

≈ ≈ ≈ ≈ ≈

Below the remains of the logging dam, the river flows through a series of pebbly shallows. This gravel/cobble bottom is uncommon on the Manitowish. Sand bottom appears far more regularly, as is expected in low-gradient streams like the Manitowish. Steeper gradient, rocky bottoms usually indicate fast-flowing streams, while gently sloping, meandering, sandy bottoms usually characterize slow-moving streams.

The type of stream bottom, or substrate, plays a major role in which organisms will live and reproduce in a stream. Some organisms require one type of substrate for reproduction, but require another substrate for feeding and safe resting. Lake sturgeon, for instance, must migrate to spawn on shallow, rocky shorelines, but as bottom-feeders, will nose through a variety of substrates in their efforts to vacuum up aquatic insects and other food. Many aquatic insects and other invertebrates hide within cobble and gravel in order to stay in place, a preferable alternative to being swept downstream with the river current.

Substrates look like this to a fish or an invertebrate:

Silt/clay/mud - Fine, often sticky particles with lots of water between them, making the texture into an ooze. Not a preferred substrate for most aquatic invertebrates because the sediments bury plants and reduce available oxygen and light both in the water and along the streambed.

Sand (particles up to 0.1 inch) - Tiny, gritty rock particles smaller than gravel but coarser than silt; smaller than a grain of rice. Sand scours "clean" in strong currents or during pulses of water from storm surges or spring snowmelt. There's not much to hang on to in a sand substrate, nor are there places to hide or rest. The tight packing of the sand grains fails to trap much organic detritus as it floats by and can limit the availability of oxygen. So, from a general aquatic invertebrate's viewpoint, sands are less attractive substrates than stones.

Gravel (0.1 to 2 inches) - Stones ranging from quarter-inch pebbles (pea/rice size) to rocks about 2 inches (ping-pong ball in size). Clean and porous gravel provides habitat for spawning, egg incubation, and home sites for a variety of aquatic invertebrates. If sediments bury the gravel, water circulation, and therefore oxygen circulation, will decline, resulting in the failure of eggs and/or embryonic fish to survive. Varying size, depth, and volume of gravels are required by different fish for spawning.

Cobble (2 to 10 inches) - Rocks the size of a ping-pong ball to a basketball. Cobble provides excellent habitat for aquatic insects. The spaces between the rocks provide protection from predators and the power of the current. The rocks themselves are colonized by a host of insects that have adapted to living on the rock surface. Cobbles typically produce the most insects, and therefore, the most fish life.

Boulder (10 inches and bigger) - Rocks from basketball to car size. Behind these boulders lie "eddies," or quiet pools of water where fish and insects and plants can get out of the flow, conserve energy, and find food.

Bedrock - Solid rock on the stream bottom. Bedrock, much like smooth sand, provides little habitat.

As important as the type of substrate is, embeddedness, or the extent to which rocks are covered by silt, sand, or mud on the stream bottom, can also radically alter life on the bottom. When rocks become embedded, there's less surface on and between the rocks for aquatic macroinvertebrate habitat and fish spawning. Runoff that carries soil from uplands will tend to embed rocky substrate unless the rocks are scoured clean periodically by fast water or periodic floods.

Still, the mineral substrate and its embeddedness are only part of the in-stream life equation. The presence of organic detritus like logs or woody debris along the bottom provides important fish and invertebrate habitat. The additional presence of rooted plants, filamentous algae, and other plants and animals also strongly influences life along the streambed.

Categorizing an overall substrate is difficult, because how does one average observations of, say, woody debris, mosses, sand, and rocks that all occur in one short stretch of river? In looking at stones, is the size of the stone most important? Or is the surface area and texture of the stone more critical, because the diversity and abundance of invertebrates is greater on rough, irregular surfaces than on smooth, regular surfaces? How about the spaces between the stones? Stream substrates are similar in a way to many forest communities, exhibiting a patchiness both along the streambed and into the streamwater. What's more, the patches change over time, sometimes rapidly, depending on fluctuations in stream flow.

One generalization that usually holds true is that particle size along a streambed typically decreases as one proceeds downstream. Headwater streams tend to have large stones and boulders, while higher order, larger rivers tend to have sand and silts in their river bottoms.

Organic materials collecting on the substrate can act as food or as habitat from which to gather food. Very small organic particles usually serve as food, but large plant stems, branches, and submerged logs typically provide perches from which to capture food. Logs and branches often manage to accumulate food that is floating by, or they provide the surfaces for plant growth, so in effect, they provide food, too. Leaves shed in fall and that bunch together in leaf packs often support the greatest diversity and abundance of invertebrates.

What all this means is obvious—vegetated substrates usually support a much greater fish and invertebrate population than bare substrates. The plants may not actually provide food directly, but instead may offer refuge and act as a trap for organic matter

Fish Spawning Phenology on the Manitowish River

Species	Timeframe	Substrate-Habitat
Burbot	Jan-March	gravel
Northern pike	early April	silt/detritus/flooded vegetation
Walleye	early-mid-April	gravel/cobble
Yellow Perch	early May	silt/detritus/flooded vegetation
Muskie	early May	silt/detritus/flooded vegetation
Suckers (white)	mid-May	gravel/cobble
Greater Redhorse	mid-May	gravel/cobble
Sturgeon	mid-May	gravel/cobble
Black Crappie	late May	silt/detritus/flooded vegetation
Largemouth Bass	late May	sand/gravel/vegetated areas
Smallmouth Bass	late May	sand/gravel/vegetated areas
Rock Bass	late May	sand/gravel/vegetated areas
Bluegill	late May	sand/gravel/vegetated areas
Yellow Bullhead	early June	sand/gravel/vegetated areas
Pumpkinseed	early June	sand/gravel/vegetated areas

Northern lakes and rivers have a very compressed spawning period compared to the southern part of the state.

suspended in the water. Or, as in the case of pondweeds, plants may support large populations of invertebrates that graze on the algae that grow on their leaves. In one study of two species of pondweeds, 18 species of *Chironomidae* (midges), two members of *Simulidae* (black fly family), three *Trichopterans* (caddisflies), and one *Ephemeropteran* (mayfly family) lived on their leaves. (I realize the two species of black flies would not be missed by most of us.)

Most stream-dwelling invertebrates have general substrate preferences but are able to live in smaller numbers across a spectrum of substrates. Fish seldom use the substrate as directly and extensively as a bottom-dwelling invertebrates, but when spawning, many fishes of running waters require particular substrates. Most freshwater fish utilize gravel or large stones for reproduction. Stones offer the advantage of being sculptable into nests, and provide protection from the current, which is always at work trying to sweep away the eggs and sperm. Water can also flow between coarse stones providing oxygen to buried eggs. In general, diversity and abundance of aquatic invertebrates and fish increase from substrate sands to gravel and cobble.

A study of substrates along small woodland streams yielded 120 species of invertebrate animals, with leaves supporting the most species and sand the least. In lowland larger rivers, submerged wood often provided the only stable substrate; it held a disproportionately high number of species compared to the sandy/silty river bottom.

≈ ≈ ≈ ≈ ≈

The majority of aquatic insects live on the bottom of a stream or on a substrate like a log, plant stem, or rock in a stream. Swimming against the current 24 hours a day would be self-defeating, so these *benthic* organisms have evolved remarkable adaptations to their physical habitat. One way of classifying aquatic insects is according to the means by which they move about and maintain their place in the stream.

For instance, insects like water penny beetles cling to rocks or other

Adult water penny beetle

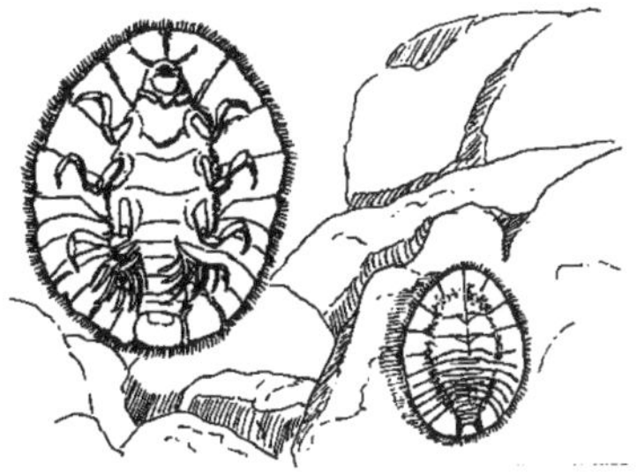

Water penny beetle larva

substrates in fast-moving water, and not surprisingly, are called *clingers* (not to be confused with Corporal Klinger of MASH fame). The water penny's entire body is essentially an attaching disc-like structure.

A clinger more commonly known and reviled is the larva of the black fly (*Simulidae*), the mortal enemy of anglers and paddlers. These devils possess attachment discs at the end of their abdomen to hold them in place on rocks in turbulent water.

Other benthic categories include sprawlers, climbers, and burrowers—see the sidebar for their descriptions.

Read more about benthic categories at the end of the chapter.

Why should you care about benthic organisms? There are many reasons, but the simplest one is that they feed fish and birds. Most river-dwelling benthic macroinvertebrates (*macro* here meaning big enough to be seen) are the larvae or nymphs of adult insects that many of us are familiar with—dragonflies, damselflies, mosquitoes, and others. They're small enough to usually go unnoticed, or weird enough that many folks don't want to notice them, but they're absolutely essential parts of a river ecosystem. As underwater larvae and nymphs, they're fodder for a host of fish. When hatched as flying adults, they're the lifeblood of songbirds from tiny flycatchers to kingbirds and larger raptors like kestrels.

So, while rocks are certainly not the only substrate in which aquatic insects live, they provide desired diversity on a soft-bottom river like the Manitowish.

Pick up a few of the rocks that you're scraping your kayak or canoe over and examine them. A number of clingers and crawlers and sprawlers will probably be scuttling about on the underside. Using a hand lens or microscope can do wonders for bringing these organisms into greater focus and detail.

≈ ≈ ≈ ≈ ≈

Before we leave the subject of rocks in a streambed, let's briefly examine why these rocks are here in the first place. A river often alternates between shallow riffles with a higher current velocity and mixed gravel-cobble substrates on one hand, and deeper pooled areas of slower velocity and finer substrates on the other.

The riffle is a topographical hill in the streambed, and the pool is a topographical depression. Formed by the deposition of gravel beds, riffles typically appear at intervals of every 5 to 7 channel-widths in a gravel-bed river, and characteristically alternate from one side of the channel to the other. In slow, sandy-bottom streams like the Manitowish, this regular riffle-pool alternation is usually not seen.

Why do rocks moved by water in a river tend to end up grouped together in beds? The hypothesis of many hydrologists is that rocks interact with one another, in a manner of speaking, so that the closer they are together, the more force is needed to move them. Objects moving in the same path typically will congregate together, bunching up in groups called platoons, and in the case of rivers, creating riffles. Hydrologists use the analogy of how drivers on a freeway tend to clump together, leaving long stretches of open road between them. Because the interaction of cars leads to braking and going slower (except on Chicago freeways), cars bunch together. Likewise, rocks bump into one another and tend to clump together. But why the pools and riffles are spaced uniformly at distances of 5 to 7 channel-widths apart remains a mystery.

Riffles typically have the highest concentration of organisms, while pools collect the remains of dead plants and animals washed downstream. In smaller, fast rivers, major decomposition occurs in the pools,

Sampling Aquatic Insects

While you may need a sophisticated Ph.D. to identify some aquatic insects down to their species level, the process of sampling aquatic insects is very unsophisticated and plain fun. For gear, all you need are some hip boots (or don't bother and just use some old running shoes or water shoes), a stout rectangular or D-frame net, a couple of white plastic dishpans, and several plastic ice cube trays.

To sample a streambed site, simply hold your net downstream, kick around in the stream bottom to dislodge the insects from the rocks, plant remains, and the like, and see what drifts into the net. In shoreline areas, sweep the net through the vegetation.

Procedurally, samples should be taken in a variety of habitats, like rocky riffles, quiet pools, sandy runs, undercut banks, and vegetation in the stream. Each habitat has its own array of residents that are adapted to living in the particular conditions created by the interplay of substrate, available food, current velocity, water depth, and available oxygen.

Take the net up onto shore and dump the contents into one of your white dishpans. Poke through the debris to find the insects, and then sort them out into the water-filled cups of your ice cube trays. Return them to the water when you're done, of course.

To identify what you have, use one of a host of sources. "The Citizen Monitoring Biotic Index" produced for Water Action Volunteers (WAV) helps you to classify insects into one of four groups based on their sensitivity to pollutants. The Izaak Walton League "Save Our Streams" program groups stream insects into three taxa also based on their sensitivity to pollutants. "WaterWatch," a program created by the Dane County Water Education Resource Center, has produced several excellent booklets on streams that are highly recommended. And the University of Wisconsin Extension has hired "stream basin educators" to provide information to individuals and groups who wish to pursue stream quality studies—see their Web site at <clean-water.uwex.edu.>. "The Water Quality Index" described in the book *Field Manual for Water Quality Monitoring* also thoroughly looks at benthic invertebrates.

while major plant production tends to occur on the rocks of the riffles, particularly in the sunny areas.

Living in a riffle area can require a great energy expense unless an organism has evolved behavioral and physical adaptations. Pools provide important resting and feeding areas for fish. Hydrologists speak of three distinct areas of a pool: the head, body, and tailout. Turbulent water enters the head, bringing high dissolved oxygen contents and food from upstream. Water then slows down in the body of the pool, allowing organic materials to settle out to the bottom of the pool.

Here, bacterial decomposition produces carbon dioxide, reducing the dissolved oxygen content near the bottom, but recycling nutrients. Fish hang out here waiting for drifting insects and plants. The pool *tailout* then leads back into new riffles. Here gravel often collects, serving as spawning beds for many fish.

≈ ≈ ≈ ≈ ≈

The Manitowish now flows southwest through several miles of uplands before turning north and re-entering extensive wetlands. An eagle's nest was located on the south side of the river less than a mile or so down from County K, but the nest blew out in 1999. Perhaps a new nest will be constructed in this vicinity. County H parallels the river through this section, providing a minor backdrop of traffic noise.

These uplands wall-in the river, a topographical feature that is highly unusual in the 44-mile trek of the Manitowish. If you're carrying a topo map, look just north of the river in sections 22, 23, and 27 to see the clear footprint of drumlins left behind by the last glacial lobe. Drumlins are long, narrow, streamlined hills that were formed when the ice molded already deposited soil and rock. Some drumlins can be 2 miles long and over 100 feet high, though in Vilas County, most drumlins are less than a mile long. They appear a bit like giant oblong gumdrops on the landscape.

Drumlins are prominent landscape features in two parts of Vilas County—around Trout Lake in west-central Vilas and throughout much of eastern Vilas. Note that these drumlins run south-southwest to north-northeast. All of the drumlins clearly show the same orientation, providing strong evidence that the retreat of glacial ice some 10,000 years ago occurred in a north-northeast direction.

Twenty thousand years ago at the height of the Wisconsin glacial period, about one-third of the Earth's landmass was covered with ice. Lakes far larger than the modern Great Lakes covered the Midwest. Massive icebergs floated on their waters, calved from tall cliffs of ice. Up to 3 miles thick, the ice cloaked what is now Canada and the northern third of the U.S. Ice dams collapsed at times, releasing huge floods and moving enormous boulders. Swift rivers flowed along the ice margin, shifting course as the ice melted.

That glaciers moved at all seems counterintuitive. After all, that's an incredible mass to move around. Consider this: Ice sheets grow when they are "fed" and shrink when they're "starved." For a glacier to flow, it must be fed a constant supply of fresh snow. The snowfalls of successive seasons must steadily accumulate, compressing the snow at the bottom and turning it to ice. A glacial dome builds up at its center and squeezes the underlying ice, making it flow outward. The friction of ice against the land and the heat of the unfrozen earth itself keeps the bottom of the glaciers partially melted, allowing the glacial ice to slide very, very slowly across the landscape, grinding, gnashing, and rounding off everything in its path.

Snow build-up in the far north near Hudson Bay added to the weight of the last glacier, causing it to flow outward at rates as slow as a few inch-

Drumlins

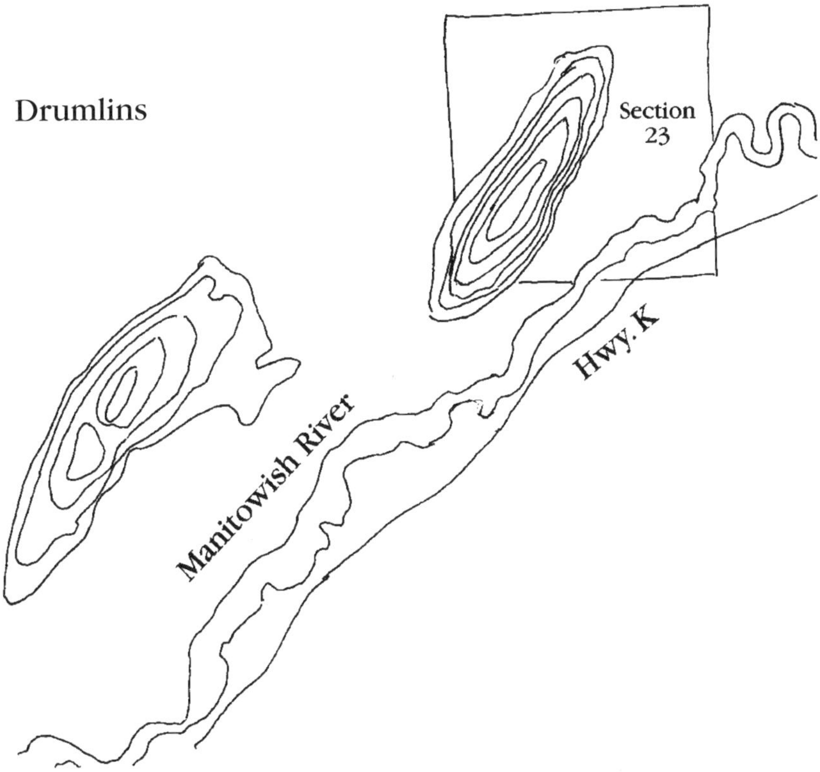

es per year to as fast as a few miles per year. Hence, the adjective "glacially" slow refers to someone or something that seems to take forever to get going or get done. Eventually, the margin of the ice sheet reached regions where the summers were warm enough to melt the ice as fast as it arrived, and a state of dynamic equilibrium was created.

A big ice sheet develops its own climate, and contributes to its own starvation. As the ice sheet increases in size, the precipitation at the center of the sheet decreases because of the enormous amount of cold air that is generated. Cold air can't hold much moisture and greatly reduces evaporation rates, making less moisture available to the atmosphere. The glacier, lacking new snow, starves, and retreats as the melting along its edge gets the upper hand.

At the height of the last glacial period, so much of the world's freshwater was locked up in continental ice sheets that the worldwide sea level was minimally 270 feet below its present level. Huge areas of what is now submerged ocean shelf was then dry land.

The unimaginable weight of the ice pressed down on the land surface, compressing it so profoundly that 10,000 years later, the land is still rebounding. This is the case in the Hudson Bay region, where the land continues to rise at a rate of 15 millimeters (a little more than 1/2 inch) per year. Not much you say. Multiply that rate by a thousand years, and you realize that Hudson Bay will completely drain away in time, unless new ice sheets halt the process.

The Wisconsin glacial period was

Glacial Lobes during the Wisconsin glaciation

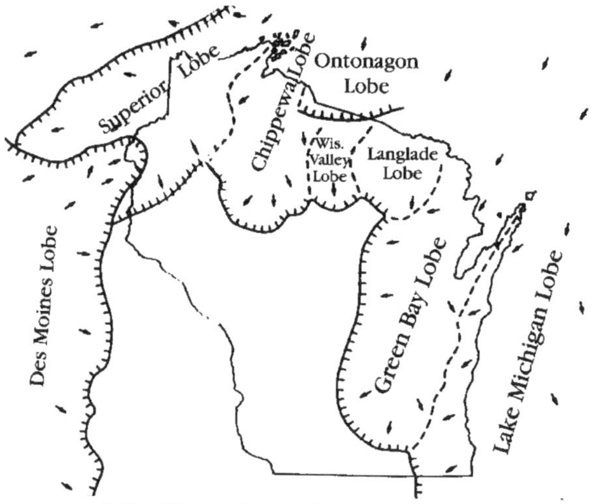

Map courtesy of The Wisconsin Geological and Natural History Survey

the last of the glacial advances, beginning about 25,000 years ago, advancing to its farthest edge about 15,000 years ago, and retreating from our region about 10,000 years ago. The advancing ice ground its way forward in six lobes, each lobe shaped by the local climate and topography, each advancing different distances and in different directions. Previous glaciers advanced further into Wisconsin, but never as far as southwestern Wisconsin, a region known as the Driftless Area because it contains little glacial drift, or debris.

Left in the glacier's wake was a barren, rocky landscape strewn with glacial castoffs. In the case of north-central Wisconsin, we were given the gift of mostly pitted outwash pockmarked with kettle holes and striped with eskers.

Pitted outwash? Eskers? If you're like me, you've always had trouble visualizing glaciers and their various deposits. Maybe I was just an unimaginatively poor student, but this King Kong of bulldozers moving at a rate of a few inches to a few feet a day was just too extreme, too radical, too darn big to get my arms around and bring into my experiential understanding.

Well, it's still too big for me to really fathom, but I've gotten better at imagining it. Like the blade of a bulldozer, the ice picked up and carried with it the surface of the landscape, freezing much of it into the bottom layer of ice. And when it got through playing with these landscape toys—the boulders, stones, and soil—it dropped them and left them behind as it went back to its home in northern Canada. What materials it dropped, how it dropped them, and where it dropped them defines the landscape today.

Glacial outwash simply refers to debris including sand, gravel, and stones that was carried from the glacier by torrents of melting water, the debris "washed out" of the ice far in front of the glacier's edge. But understand, that's an extraordinary amount of water given the height of the glacier, and it carried huge deposits of sand and gravel.

Outwash plains can be just level sandy flats, or if buried ice was left behind under this blanket, the landscape ultimately collapsed into a series of hills and valleys.

"Pitted" outwash occurs where huge ice blocks broke off from the retreating edge of the glacier and were left behind amidst all the other glacial junk. Eventually these giant ice cubes melted, leaving depressions, or pits, in this new blanket of ground. As discussed in Chapter 1, the deeper pits filled with water to form "kettle" lakes, while smaller ice blocks left shallower pits that filled with water and created the array of wetlands that clothes this area.

In describing northern Wisconsin in the late 1800s, the chancellor of the new State University of Wisconsin wrote, "Among the ten thousand undulations there is scarcely one which lifts its crown above its fellows . . ."

That says it well. We lack in extremes. The countryside is pretty rather than sublime or spectacular. The glacier scalped us, but left behind lots of holes and sand. We're lake country, and we're pine country. The holes filled to become lakes, and those sands and gravels that were dropped off in the meltwater now hold onto few nutrients and even less water for plant growth. Pines, being the cacti of the north, are best suited to survive such droughty, impoverished conditions.

We northern Wisconsinites don't apologize for this lack of rugged topography. Unlike much of the southwestern U.S., which is blessed with mountains but has limited seasonal changes in its lowlands, we have nature's theater in four passionate acts. Monotony is in the eye of the beholder. I'll take the scarlet reds and smoky golds of autumn, the deep snows of winter in the heavy boughs of balsams, the ephemeral wildflowers, and the surge of spring migration to the Southwest's lack of seasonal undulations. We do undulate, but we just undulate differently.

Glaciers currently cover about 10 percent of Earth's land surface. We're now in an interglacial period, one that is likely more than half over, having peaked 10,000 years ago. Someday the ice will return.

Rivers give the illusion of outrunning time, at least the kind we're all swimming against, which is no small part of their appeal. Or maybe it's just that, being linear, they tend to connect things. Drifting down a familiar stretch of water, we're reminded of other trips and other rivers until memory and the present merge into one all-encompassing river that cuts through time and space and connects everything. — John Hildebrand

≈ ≈ ≈ ≈ ≈

Several beaver dams are likely to impede your progress along this section, though you could also say they improve your progress by raising water levels in areas that might otherwise be quite shallow.

Beaver dams are a fact of life in paddling most northern rivers. Daniel Greysolon Sieur Du Luth's description of paddling the Brule River in June of 1680 gives a feeling for just how dominant beaver dams were historically: "I entered into a river which has its mouth eight leagues from the extremity of Lake Superior on the south side, where after having cut down some trees and broken through about one hundred beaver dams, I went up the said river, and then made a carry of one-half league to reach a lake, which emptied into a fine river, which brought me to the Mississippi."

I doubt Du Luth would have been a good choice to teach a class on the pleasures of writing in-depth journals. A league is roughly 3 miles, and paddling the turbulent Brule River upstream would be a hard test for modern-day paddlers in high-tech boats. But his bare-bones description at least details the number of times he had to either drag his loaded birch-bark canoe over a dam, or more likely, had to portage around it. He writes that he "broke through" the dams, but the effort of pulling apart one of these engineering marvels would seem to me to greatly exceed that of just going around them. But then again, the French voyageurs were as tough as the landscape they explored, so I wouldn't put it past him. Du Luth is, of course, the man for whom present-day Duluth, Minnesota is named.

Beavers dominated the pre-settlement landscape, their numbers in North America estimated to be somewhere between 60 and 400 million. Despite that enormous population base, and impossible as it may be to imagine, by 1900, beaver became almost extinct in North America. Overharvest was the main culprit, but since 1834, some 120,000 square miles of beaver habitat have been converted to dry land in the U.S., contributing to the precipitous decline in beaver numbers.

Today, the beaver population has rebounded, and is thought to be 6 to 12 million. Given that the present beaver population represents only a small fraction of earlier numbers, rather than cursing the damn dams you encounter, you might praise your luck to see them at all.

The beavers' ability to build complex structures, establish travel networks, and manipulate water levels is unparalleled in the animal world (except by humans, of course). They build dams to raise water levels, so more food is accessible without leaving the water. An average fat, low-slung beaver will win few races on land. They are excellent swimmers though, so remaining in the water makes great beaver sense.

Beaver dams also stabilize water levels, allowing a lodge to be safely built and maintained along the shoreline. Fluctuating water levels that expose the underwater entrance to a lodge compromise a beaver's safety.

Stable water levels also provide deep enough water in winter, guaranteeing freedom of travel under the ice to reach the winter cache of food. Storing food without guaranteed access would not advance one far along the evolutionary scale.

If you'd like to watch this engineering marvel, dam building usually takes place at night in early spring and late summer. Beavers don't always build in the best location. Wise site selection appears to be a function of age—older beavers make better site selections than younger ones.

Dams are built from the bottom up, and it's not unusual for some dams to reach over 10 feet tall and span hundreds of yards across a river.

Beavers also build a series of canals, channels, and plunge holes that are often easily observable along the shore. They like to have a network of

Beaver dam

regular travelways that act as "roads" from the lodge to heavily used areas. Beavers dredge these channels frequently to keep them open. Some canals are actually built on land to extend water travel safety to feeding sites. Bogs are commonly traversed in this way to reach upland trees.

Read more about beavers at the end of the chapter.

≈ ≈ ≈ ≈ ≈

Kingfishers and great blue herons will likely accompany you along much of this stretch of river. Kingfishers consistently wait until you're about 100 feet away, then drop from their perch branch and fly ahead to the next good perch branch overhanging the river, giving their rattling call as they fly. They'll typically lead you about 500 to 1,000 yards downriver, then reverse course and wing back to their original perch, having politely escorted you out of their territory.

The steep banks along the river here provide excellent nesting opportunities for the kingfishers. They usually excavate burrows in near-vertical sandbanks, though the banks needn't be by water. In fact, gravel pits have supported many a kingfisher pair. The bank soil though must be soft enough to dig into, but compact enough to not cave-in, so a little clay in the banks is highly desirable.

We don't often think of birds as nest diggers. We expect such actions from badgers, skunks, chipmunks and the like. But kingfishers have evolved short legs and syndactyl feet, meaning their third and fourth toes are partially fused. This adaptation gives a little more blade width to each shovel of soil they remove during excavation.

The male and female work alternately at excavating the burrow, a good division of labor, though the male usually digs about twice as long as the female. Kingfishers dig with their bills, and eject the soil with their

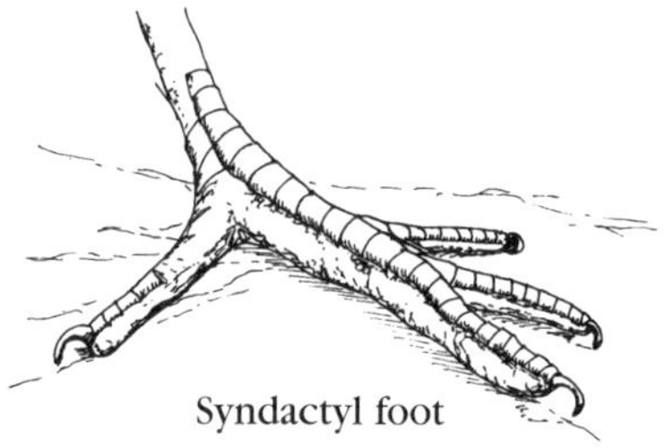
Syndactyl foot

feet, shooting a plume of sand behind them. Burrows typically are hollowed out near the top of the bank, extending 3 to 6 feet inward and slightly upward. The burrow ends at a circular, dome-shaped chamber about 10 inches in diameter and 7 inches high that serves as the nest site.

Kingfishers often dig multiple burrows in a bank, making each hole 3 to 4 inches in diameter. If you see a bank with multiple holes, look for a hole with two grooves in the sand at the entrance to indicate an active burrow. The birds scrape the grooves with their feet as they enter and exit.

The quick, riffly shallows seen frequently in this stretch of river provide excellent fishing for kingfishers because small fish and minnows tend to gather in riffles. Kingfishers need clear, shallow waters in order to see their prey. They typically take fish 3 to 4 inches in length.

≈ ≈ ≈ ≈ ≈

During the next half-hour of paddling, several houses protrude right near the river's edge. Their presence detracts considerably from the aesthetics of the river, though most of us, if we were honest, would love to trade homes with the owners. Therein, of course, lies much of the contradiction between our beliefs and our actions.

The classic property rights issues raise their ugly heads in these instances. Studies of regular people who recreate in the Northwoods show a very high percentage equating

Kingfisher and bank nest

Kingfishers—*Megaceryle alcyon*

I recall few Latin bird names, but the belted kingfisher's scientific name *Megaceryle alcyon* sticks with me, because it tells the story of Alcyone, one of the seven sisters of the Pleiades. The story goes that her husband Ceyx drowned in a shipwreck, and Alcyone, in her grief, threw herself into the sea and also drowned. She and Ceyx were later transformed into kingfishers through the mercy of Thetis.

The story continues that kingfishers were believed to be seabirds that made nests of fish bones on the ocean surface in December. To allow the eggs to incubate, Zeus declared that he would quell the seas for Alcyone during the winter solstice period. And so calm, peaceful, carefree weather came to be known as "halcyon" days. If you are blessed with good weather on your paddles, and believe in Greek mythology, you have the mythical kingfisher to thank.

The male kingfisher assumes an equal function with the female in nest building, incubation, and care of the offspring. In fact, kingfisher males typically display a significant degree of sex-role reversal, where they give more time and energy to these activities than the females. Interestingly, the female is the more conspicuously colored of the two, she showing a rust-red chest-band that the male doesn't possess. Perhaps the male kingfisher should be called a true Renaissance bird given his parenting efforts, though some species of shorebirds outdo the kingfisher, showing almost complete role reversal where the male assumes nearly all parental activities.

Kingfishers always appear top-heavy to me, like they should tip right over on their oversized beak. Their long, heavy bill and large blue-gray head with a conspicuous crest contrasts greatly with their small body, short tail, and tiny feet. They remind me of punk rockers.

When a kingfisher spots a fish, it typically hovers on beating wings and then plunges headlong into the water, emerging with a fish, and then flying back to its perch. There it beats the poor fish into submission, tosses it into the air, and swallows it headfirst, a nifty trick.

Kingfishers get a bad rap from jealous, and usually poorly skilled, anglers. The truth is the kingfisher's fish predation typically benefits the stream by thinning out the populations of small fish. Kingfishers are much too small a bird to take gamefish-sized species.

In excellent habitat like stream riffles, a kingfisher may eat about 10 fish a day, most of which are in the 3 to 4-inch-long range. As opportunistic feeders, they also consume their fair share of tadpoles, frogs, crayfish, mussels, insects, and small snakes.

their image of the Northwoods with natural, unobstructed shorelines on lakes and rivers. I say "regular people" because there's often an effort to discount people who want development restrictions along shorelines as "radical environmentalists." T'ain't so. Enough people have seen what lake development has done in the southern part of the state to know what it can look like if unchecked in the Northwoods. Riparian owners rank most favorably shorelines that have enough plants to screen shoreline structures. The Wisconsin Supreme Court has even ruled (*Muench v. PSC*, 261 Wis. 492, 1952; *Claflin v. DNR*, 58 Wis. 2d 182, 1972) that the public's right to "natural scenic beauty" can be the basis for the state denial of a permit for lakeside construction.

More on this in Chapter 4.

≈ ≈ ≈ ≈ ≈

As you paddle along, you will see at times thick mats of underwater plants that look like impenetrable dense carpeting. The creator of such luxuriant weaving, or of such a tangled nuisance depending on your perception, is common waterweed (*Elodea canadensis*). Pick a stem (this plant can afford to be thinned), and note that the small, dark-green, lance-shaped leaves occur in whorls of three, and how they bush out near the tip of the stem.

Elodea's colonial success can be attributed to its ability to spread vegetatively by stem fragments, to its disease resistance, and to its capacity to overwinter as an evergreen plant and continue photosynthesizing even under the ice. Elodea is remarkable in its ability to photosynthesize in low light conditions under water. It has been recorded living in water deeper than 25 feet where the amount of light is virtually nil.

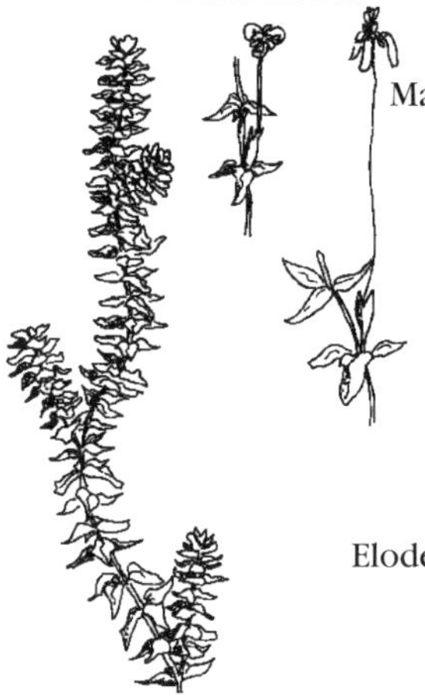

Elodea

Elodea has well-earned its moniker of "waterweed." The plant grows so quickly and roots so readily that it can easily become a nuisance. In an interesting scenario where the shoe is on the other foot, Elodea was imported to Europe in 1850, and is now considered by Europeans to be a very aggressive American exotic. It's the subject of nuisance control programs there. We always get so worked up about nonnative plants like Eurasian water milfoil and purple loosestrife invading our wetlands, but we forget that "our" plants can act just as invasively outside our borders.

In moderation, Elodea provides excellent shelter for fish and many aquatic invertebrates. One study ranked it second among 12 submersed plant species in the diversity of organisms it supports. The leaves and tiny fruits supply a moderate browse for gadwall, mallard, wigeon, wood duck, blue-winged teal, redhead, lesser scaup, goldeneye, coot, and goose, while muskrat and beaver feed on the leaves as an undistinguished side dish.

As has been said many times before, too much of a good thing can be a bad thing, and beds of Elodea that become too extensive can hinder fish movement. Elodea spreads most readily when props on motorboats chew up a bed and the stem fragments float to new sites where they can root or simply grow in the water as floaters without roots. Elodea also spreads rapidly in waters enriched by nutrients, its overwhelming presence acting as an indicator of overfertilized waters.

It often grows rapidly, then dies back. The good news is it decomposes quickly compared to many aquatic plants, and its high nutrient content makes it a promising food source for captive ducks and chickens. Elodea also readily absorbs phosphorous, mercury, and trace metals, though the issue of disposal still remains.

Waterweed, both in name and deed, brings up the whole notion of "weeds." Weeds are plants in the wrong place at the wrong time in the wrong numbers. "Wrong," of course, is a value-laden term, and one that is highly plastic. A plant can be highly desired today, but next week may no longer be desired, now wearing the title of weed. A tomato in the pea patch is in the wrong place. A tomato growing in early May among the cold-weather plants is in the wrong time. Crowded tomatoes are in the wrong number and need to be thinned out.

Emerson wrote, "A weed is a plant whose virtues have not yet been discovered." I like the further definition of a weed given to me by aquatic botanist and researcher Susan Knight, "A weed is a plant without a press agent." Most aquatic plants need press agents. The public is generally unanimous in its quick pronouncement of aquatic plants as "weeds," and "yucky" ones at that. We like smooth sand for swimming and boating. In gorgeous advertisements we sell white sand beaches as the ultimate vacation paradise. At the opposite end of the paradise scale, I can't recall ever seeing an advertisement (except maybe in a fishing magazine) with a picture of a shoreline tangled by aquatic plants and downed trees. Such scenes tend to make one start thinking about snakes and frogs and fish and insects and other slimy, crawly things.

Nature, in other words.

What is it for you—fascination or revulsion?

I don't suggest that you try swimming in such spots. I'm not keen on muck bottoms and tangly plants all around my toes either. I like to swim with sand under my feet. But I don't need to swim in every part of a lake or river. If we can learn to appreciate aquatic plants for what they do, and not incidentally for their beauty, the stigma of "WEED," the big "W" on the chest of these plants, would evaporate.

≈ ≈ ≈ ≈ ≈

Have I mentioned that there are no campsites and no marked pull-outs between County H to Island Lake? So, you're in it for the long haul, though it's only a two to three hour paddle, depending on your exercise mentality. I get too distracted to go fast. I have RDD: River Distraction Disorder. So this stretch takes me at least four hours.

On river left, there is a very nice, but unmarked pull-out on the boundary between section 27 and 28 where the river changes course from heading southwest to traveling almost due north.

The river changes character almost immediately here, now beginning a sweep through extensive black ash and silver maple swamp. I'm often asked when a wetland is a marsh and when it's a swamp. Whether a wetland is considered a flood plain forest or swamp or bog or marsh, or a host of other wetland types, the designation is based on hydrological processes and the dominant plant community.

Hydrologically speaking, this area is a temporarily flooded wetland along a river system, meaning it is typically flooded for several weeks in the spring, and then the water table drops to below the soil surface through the rest of the year. In a "normal" summer, you can walk with generally dry feet here, but in wet years, wear your rubber boots.

Read more about wetland plant communities at the end of the chapter.

A wooded swamp, coniferous bog, or flood plain forest are all defined as wetlands with mature trees present that form closed stands on wet, lowland soils. (Mature trees are defined here as having a diameter of 6 inches or more. Closed stands are defined as more than 17 trees per acre with more than a 50 percent canopy cover.) The dominant trees in this particular flood plain forest are black ash and silver maple, though at times white cedar stands dominate. Using the key in the sidebar, we could also choose to call this area a lowland

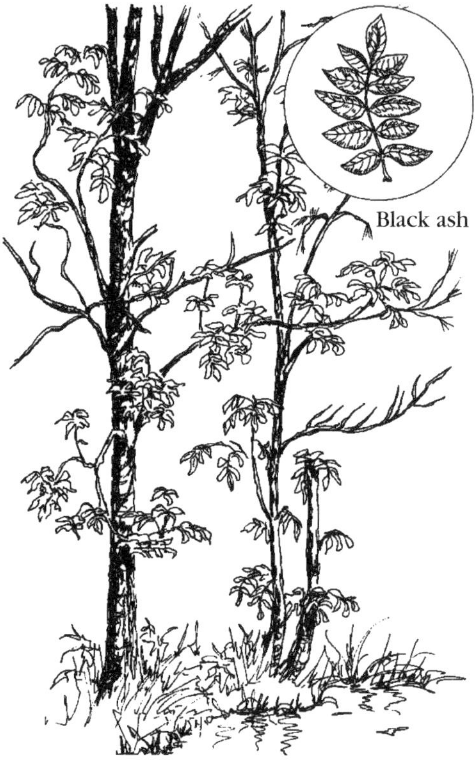

hardwood swamp rather than a flood plain forest.

Whatever technical definition you choose, you can see there are many scraggly black ash and stately silver maple. Numerous sweepers (overhanging branches that can sweep you out of your boat) and downed trees and branches also occur through here, slowing your progress a little or a lot, depending on the year and the water depth.

These downed trees and branches are absolutely a nuisance for river travel, but are also an absolute boon to aquatic life. Since you are a temporary visitor here, avoid the human propensity for "cleaning this area up!" and leave your chain saw at home.

Plants living partially or fully submerged require a series of remarkable adaptations. Let's look at a number of these. One obvious and immediate issue is that it's hard for roots to breathe in saturated soils under water. In flooded and saturated soils, bacteria quickly deplete the oxygen from the soil.

Gills would help. But lacking those, *hydrophytes*—plants that can tolerate long periods of flooding or saturated soil conditions—have developed mechanisms to regulate the amount of water entering their cells and to tolerate oxygen-deprived soils.

For instance, many wetland plants have evolved morphological, or structural, adaptations such as *aerenchymous* tissue in their roots and stems. Giant bur-reed, soft-stemmed bulrush, watershield, spikerush, soft rush, buckbean, cattail, all utilize this type of air-filled plant tissue in which their cells are unusually large and arranged in a manner that creates air spaces in the plant organ. Such tissue is often referred to as spongy. It usually provides increased buoyancy for support in the water, in addition to serving as the conduit for air that passes to and from the roots. It's as if the leaves, stems, and roots are inflated with air.

Air-filled, tubular passages called *lacunae* run through the stems of white water lilies, conducting gases to and from the leaves. The stomata, or breathing pores, of the leaves draw air into the leaf lacunae. The air is then pumped down under pressure to the roots. In the sediments, carbon dioxide and methane are then diffused back into the lacunae and travel upward, where they are released into the air through the stomata. Twenty-two liters of air pass through a single water lily stem in a day, according to one study. Many other emergent plants contain stem lacunae that transport gases to and from the atmosphere.

Pull out the stem of a white water lily and use it as a straw to blow through into the water. You should be able to make bubbles like the ones you made in your milk when you were a kid. Use this plant for breathing under water when escaping from prison or from bad guys.

Rounded, untoothed, floating leaves and stems, like those employed by watershield, white water lily, and yellow water lily, rise and fall with wave action, offering less resistance to the potential tearing force of waves. The upper surface of water lily leaves bears a heavy, waxy, water-repellent cell layer or cuticle, which regulates the amount of water that can penetrate when it rains. And unlike most terrestrial leaves that have their stomata on the underside, these leaves have their stomata on the topside.

The underwater stems, particularly in watershield, often bear a thick gelatinous coating that likely serves to reduce abrasion and deter aquatic herbivores.

Some emergents like arrowhead and water parsnip develop *polymorphic* leaves. These are leaves of differing shapes that appear on the same plant or in the same species when grown under different environmental conditions. An arrowhead may have long, tapered, grass-like leaves under water, and then develop wide arrow-like leaves above water. The wide leaves can photosynthesize more effectively, while the grassy underwater leaves can "go with the flow."

Fully submersed plants seldom utilize the stout, spongy tissue of emergent plants, instead utilizing a thin "skin" that allows the absorption of gases and nutrients directly through their cell walls. Since submersed plants have no need for supporting tissue, they typically collapse into a soggy mass when removed from the water. In the water, the long, grass-like leafblades wave in the current like a field of oats in the wind.

Wetland trees like silver maple, red maple, northern white cedar, and tamarack typically grow shallow root systems that enable them to survive a high water table for long durations. The roots grow horizontally to the ground surface and only as deep as the seasonal high water table; they often die if subjected to long periods of saturated soils that lack oxygen.

The same trees often develop multiple tree trunks, and the swollen or enlarged bases of trees known as buttressed tree trunks, in response to conditions of prolonged inundation.

Other trees and shrubs like green ash, black willow, cottonwood, and willow develop above-ground *adventitious* roots, located well above the normal position occupied by underground roots. This is yet another response to prolonged inundation.

These same trees and shrubs also often develop *hypertrophied* lenticels, oversized pores on the surface of woody plant stems, through which gases are exchanged between the plant and the atmosphere. The enlarged lenticels increase oxygen exchange to the plant roots during periods of inundation.

One way or another, survival in a habitat than can be under water for weeks but dry on the surface during a drought is amazingly complex and worthy of appreciation.

≈ ≈ ≈ ≈ ≈

The river becomes a way of thinking, ingrained, a way of looking at the world. I listen to its commentary on the rocks and willows that block its way, feel cooled by the touch of spray, and smell all the odors that emanate from it. Judgment and recognition of odors depend so much on familiar reference smells that it is difficult to describe a new smell without recourse to them . . . it smells of itself, an aloof and elusive smell, soft, faintly like clean clay or like wet wash hanging out on a windy day. It is neither sweet nor sharp, acrid nor aromatic, nor distinctly anything ever smelled before or elsewhere unless one has had a river in his or her childhood. It is a smooth, tentative smell. It is light and deep, cool, it comes in curling tendrils and sometimes it is difficult to pick out from other smells. But it is there. And, once smelled, it becomes easier to recognize and soon there is almost a sense of river in the landscape even when it cannot be seen.

— Ann Zwinger

Where there's aquatic life, there are birds that live off that aquatic life. One bird species commonly seen along the shorelines here during spring and fall migration is the American robin. The same lawn bird that is one of the emblems of suburban life likes the mucky shorelines of rivers. In spring and fall, robins are common along the riverbanks, working the exposed soil for invertebrates.

A few bird species, like great crested flycatchers, show a clear preference for life in flood plain forests. Other songbirds, like the American redstart, wood thrush, common yellowthroat, red-eyed vireo, swamp sparrow, ovenbird, cedar waxwing, rose-breasted grosbeak, song sparrow, and veery are common inhabitants but nest in other habitats as well. American woodcock and whip-poor-wills nest on the mucky soils in flood plains, while red-headed, red-bellied, and pileated woodpeckers all nest in the trunk cavities of large, dead trees common to flood plain forests. Wood ducks and hooded mergansers utilize the cavities of mature flood plain trees alongside the water. Red-shouldered hawks, a threatened species in Wisconsin, typically nest in flood plain tree canopies. They usually appear in central and southern Wisconsin, but current research indicates they may also nest in northern flood plains. Green-backed herons also nest in flood plain canopies, and may be seen feeding along the water edge.

Read more about the birds of northern hardwood swamps at the end of the chapter.

Need I add that Wisconsin's state bird, *mosquito giantis*, also enjoys the wet, humid conditions in flood plain forests? You've been warned.

Dense northern white cedar stands appear frequently through this stretch. If you're saying to yourself that those are arborvitae, you're right—white cedars also go by the name of arborvitae, which means "tree of life." The name supposedly derived from the men of Jacques Cartier's 1535 expedition who drank white cedar tea to survive scurvy.

That may well be true (though the brew could as well have been black spruce tea), but today the name seems more fitting because of white cedar's ability to live through the toughest of circumstances. If wind knocks a white cedar over, the once horizontal branches often continue to grow upwards, forming the trunks for a row of new trees. Bend a white cedar over with the weight of another fallen tree, and the trunk will grow sideways until it can again curve skyward in search of sunlight. Cedar forms some of the most remarkable free-form sculptures in the natural world. Add the fact that white cedar grows in the most difficult of habitats—typically either on dry limestone rock or in the muck of a swamp—and you have an ecological basis for calling white cedar "the tree of life."

Extensive white cedar swamps were once quite common in northern Wisconsin, but now even small remnant stands are hard to come by. The combined impacts of draining wetlands, timber harvesting, white cedar's exceedingly slow growth rate (to add an inch to its diameter may take 10 to 20 years), and the highly preferred status of white cedar as a winter deer browse has made cedar reproduction an uncommon occurrence.

Cedar swamps often support unusual orchids, like calypso orchid (*Calypso bulbosa*), ram's head lady slipper (*Cypripedium arietinum*), northern bog orchid (*Platanthera hyperborea*), and large round-leaved orchid (*Platanthera orbiculata*), though the search for them can be trying. Wear good supportive boots for the ankle-breaking roots that protrude everywhere, and take a gallon or two of mosquito repellant to survive the legions that will find you.

Note also the height at which the cedar's foliage begins. Usually it corresponds precisely to the height at

Birds of Cedar Swamps

Cedar swamps have honestly earned their titles of ankle breakers and mosquito havens. It's hard to look up at birds when you need to be looking down at the gnarl of exposed roots and intervening muck. It's equally hard to look up with your binoculars when all of God's mosquitoes have amassed to fly in your ears, up your nose, and between your glasses and your eyes.

Nevertheless, birding, like postal delivery, must go on. Whether we're tough enough to be there to observe them is immaterial to the birds.

Hoffman and Mossmam (*The Passenger Pigeon*, Summer 1993) summarized their breeding bird research on seven bog and swamp sites in northern Wisconsin, and also listed by seven wetland habitats the most common breeders in each habitat. In cedar swamp habitats, Nashville warbler, winter wren, black-throated green warbler, and ovenbird dominated the breeding birds.

Characteristic but less abundant were yellow-bellied flycatcher, red-breasted nuthatch, golden-crowned kinglet, veery, hermit thrush, northern parula warbler, yellow-rumped warbler, blackburnian warbler, black-and-white warbler, Canada warbler, white-throated sparrow, and purple finch.

As with all generalized habitat classifications, variations in actual plant species composition and structure tweaked the associated bird community. In areas with pools and exposed roots of tipped-up trees, northern waterthrushes often resided. Lush sphagnum mats attracted yellow-bellied flycatchers, while tamarack, spruce, and fir enticed Nashville warblers. Toss in a substantial balsam fir canopy, and Cape May warblers became present. A mixture of hardwood trees and shrubs benefited Canada warblers and veeries. Pine siskins were the primary seed eaters of the tiny cedar cones, while rare black-backed woodpeckers found insects under the bark of aging cedar trees. Swainson's thrush often used the fibrous cedar bark for lining their nests.

which white-tailed deer can reach to browse it in the winter. A perfect line of cedars, all browsed to the same height, looks much like a group of women all attending a dance and wearing skirts of the same length. In this case, the dance is that of the wintering white-tailed deer, each nibbling five pounds of woody browse daily to keep their internal furnaces burning during the months of continuous cold.

Note also the lack of sunlight penetrating the dense cedar swamp

Northern harrier

canopy. By contrast, the black ash swamp canopy is so thin and scraggly that sunlight easily reaches the forest floor. Birdlife varies between these two swamp communities, so if you have time to stop and bird each habitat type, you will find different membership in each.

≈ ≈ ≈ ≈ ≈

The river, surrounded by extensive swamps, has a particularly wild feeling here that won't be regained until much farther downstream.

Evidence of beaver activity, ranging from peeled sticks to the sharpened stumps of cut shrubs and trees, should now be easily seen. Several large active beaver lodges lie like beehive huts along the river edge. Northern harriers and red-tailed hawks often patrol the wetland edges in search of a meal.

Filamentous algae appear in the water later in summer, sheer like a green nylon stocking. It's usually suspended in a cloud beneath the water surface or appears as a cottony mat on the surface. This "pond silk" will likely be waving in the modest current, and is quite pretty in small amounts. However, most folks have previous experience seeing algae in prodigious blooms, and that's where algae, the tiniest of green plants with thousands of species worldwide, get a bad rap. These rootless, leafless, and stemless plants thrive in warm, fertile waters. Most folks associate algae with pea soup waters in polluted, urban lakes and in rivers that smell bad, taste bad, and are about as attractive to swim in as India ink.

But algae are another example of the repeated adage that too much of a good thing is a bad thing. In a natural waterbody, algae are exceptionally important aquatic organisms, making up the base of the food chain for freshwater and marine organisms. They constitute most of what is called the phytoplankton, the free-floating microscopic plants that inhabit most waters.

Algae provide a constellation of services to aquatic organisms. Insect larvae, crustaceans, tadpoles, minnows, turtles, zooplankton, and waterfowl all eat the cells of algae. Algal mats also provide shelter for an array of insect larvae, zooplankton, adult insects, and fish. Black ducks, wigeons, teal, coots, and pintails all eat the algae and the aquatic organisms within it.

Algae come in lots of sizes and shapes. Blue-green algae are simple one-celled plants. They are considered the most primitive plants on earth because their cells have no well-defined nuclei and no sexual reproduction. In fact, blue-green algae are closely related to bacteria, and are often classified as a bacterium. What they do best is fix nitrogen, making it available in the water. Thus, they are a good indicator of water quality. Water with too much nitrogen and phosphorous, usually the result of human pollution sources, stimulates excessive growth of blue-green algae. When exposed to sunlight, the algal mat typically dies quickly, and bacteria rapidly decompose the mat, causing a decline in oxygen as the bacteria respire.

Green algae are more complex than blue-green algae, with cells that have definite nuclei. Many species appear in a large variety of cell and colony shapes.

Like blue-green algae, green algae become hyper-abundant in nutrient-rich water. Most northern lakes and rivers are nutrient-poor to begin with, so algal blooms remain relatively uncommon in the north, except on hyper-developed lakes that have high nutrient inputs.

The single-celled algae were likely the first sexually reproducing organisms on earth, and are the probable ancestors of all green plants today. Thus, we may all owe our origins to "pond scum," a humbling thought indeed.

≈ ≈ ≈ ≈ ≈

The river flows due north now, and then sweeps west. An eagle's nest was once located on the north side of the channel where the river turns west. Look for a new nest to appear in the territory. Soon the river widens and shallows again, now flowing through an open sedge/leatherleaf plant community.

Baseball-sized stones on the bottom may grind your boat to a halt in this section. The stones form the top layer of what is known as the *hyporheic zone* of a river. Though it may sound like some distasteful medical condition (Oh God, he's going hyporheic!), hyporheic actually is Greek for "*to flow beneath*." A river flows through its bed as well as above it. The hyporheic zone is the area where water moves between the rocks, gravel, sand, and silt that make up its substrate.

The zone may extend several feet beneath and to the sides of the riverbed, depending on the size of the particles involved. For instance, silt particles pack closely together, while stones have lots of spaces between them.

A community of organisms lives in this zone, though these organisms are limited by the declining oxygen levels in the sediments. On the other hand, the hyporheic zone is a great place to hide from predators. Species of flat-

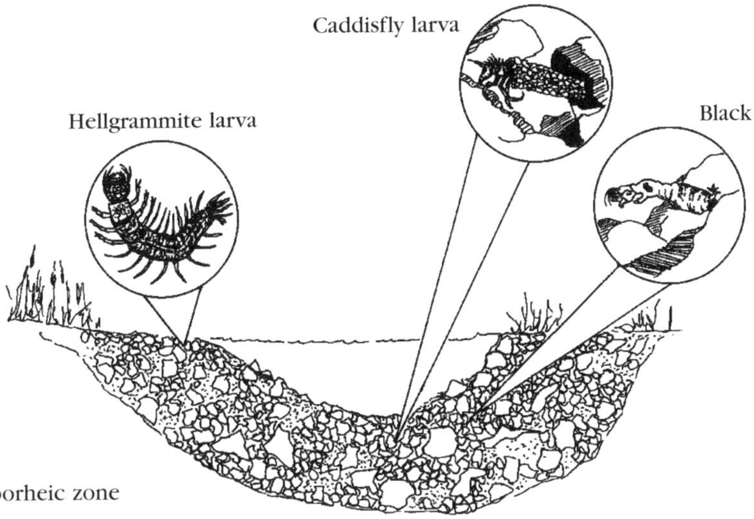

worms, stoneflies, caddisflies, mayflies, scuds, water fleas, copepods, and mites, many of which also live on the bottom of the river, wend their way into the substrate and remain there.

While the hyporheic zone may sound like an inhospitable place, and thus only of minor value, studies have shown that the zone can support huge numbers of aquatic insects. One study found 34,000 organisms per square meter in the 4 inches beneath a stream in Germany. Another study of an Ontario stream found more invertebrates living in the gravel 2 to 6 inches down than on the bottom itself.

In gravel-bed rivers, subsurface waterflow can extend beneath much of the flood plain and may carry as much or more water than the surface waterflow. Normally, the hyporheic zone occupies an area of the same width or just a bit wider than the river channel, but it can be much wider. The flow below the active channels of the Flathead River in Montana averages 3 kilometers wide and 10 meters deep—an unseen river.

In *What the River Reveals*, writer Valerie Rapp calls this unseen river an "interstitial highway." A wide hyporheic zone likely occurs because rivers often shift their course, leaving behind old channels with porous sediments that still carry hyporheic water. Sometimes amidst an open flood plain, a meandering band of trees illustrates clearly the former flow of the river.

If you think about it, a river really doesn't have an absolute bottom, unless it's running through a concrete gutter. Somewhere within the myriad spaces between the stones and sand, most rivers merge with the groundwater that supplies their continual flow. The transition from well-lit, fast-moving river water to dark, barely moving groundwater occurs in the hyporheic zone. This is a true twilight zone, sans Rod Serling and the eerie music.

Hyporheic water typically flows at little more than an inch per hour, contains a good supply of dissolved minerals, is poorly oxygenated, and is cool in the summer and warmer in the winter. The roots of many aquatic plants occupy this zone, while many invertebrates move between this zone and the river itself, often finding refuge in the hyporheic zone in the dry season or during the bottom-scouring that takes place during high water. Like the ground under our feet in a forest, the ground under our feet in a river is something more than a place to trample. It's alive, active, diverse, shifting, and fundamental.

Read more about caddisflies at the end of the chapter.

≈ ≈ ≈ ≈ ≈

A campsite will come into view on the left side of the river, signaling the proximity of the Island Lake boat landing and pull-out. Beds of wild rice suddenly appear, spreading across nearly the entire river channel. Rice has become controversial in recent years. Some see the rice as good for wildlife and human harvest, while others see it as "choking out" the river channel.

From a motorboat viewpoint, the rice certainly makes navigation difficult. From a viewpoint of a canoe, which slips through the stalks rather easily, the rice creates character and refuge, in particular for ducks.

On a warmwater, marshy, northern river like the Manitowish, four species of ducks breed most commonly—mallards, blue-winged teal, wood ducks, and hooded mergansers. On occasion in the summer, I'll see a black duck, green-winged teal, ring-necked duck, red-breasted merganser, or common merganser on the Manitowish, but mallards dominate the breeding population, followed by blue-winged teal, and then low numbers for the rest of the ducks.

Waterfowl form pairs on wintering grounds, and migrate north as pairs. This way the drakes don't have to spend time attracting females, and all their efforts can be directed toward nesting.

The hen takes care of all the incubation. As a general rule, the males leave once the incubation begins, form new pair bonds the next winter, and nest in a different spot the following spring. In contrast, the females often practice nest fidelity. The males apparently have never heard of "family values."

Mallards rule the duck roost because they're the greatest generalists, extremely adaptable in their habitat use, feeding behaviors, and association with humans. The mallard is present in good numbers in all seasons throughout Wisconsin, the only duck that can claim such a distribution. In fact, mallards range throughout all of North America from the Arctic into central Mexico, and are the most abundant duck in the entire Northern Hemisphere worldwide. They only breed, however, in the northern third of the U.S. up to the arctic treeline.

Mallards nest in all wetlands, but optimal habitat includes a permanent marsh surrounded by small, shallow ponds. Marsh habitat with tall emergent vegetation provides excellent cover for raising broods and for protection during the adult's summer molt.

Mallards have adapted to agricultural land use far better than any other duck. They commonly nest in hayfields, helping themselves to adjacent fields of corn, wheat, and barley.

On the river though, they'll forage on the seeds of wild rice, pondweeds, wild celery, and smartweeds, as well as eating aquatic insect larvae and mussels.

It's a little-known historical fact that mallards were mainly prairie ducks until the 20th century, virtual unknowns east of the Great Lakes. Thoreau's journals contain no references to mallards, apparently because he never saw one. The clearing of eastern woodlands encouraged the spread of mallards into the more open landscapes, and now mallards are the first species that comes to just about everyone's mind when ducks are mentioned.

Blue-winged teal form pair bonds late on their wintering grounds, arrive in the Northwoods later in April than most other ducks, and nest later, behaviors attributable to their long flight from their wintering grounds in Central and South America and along the Gulf Coast. No other North American duck winters in greater numbers so far south.

The drake's white crescent moon marking in front of the eye easily identifies him, while the mottled-brown female offers little in the way of clear markings to differentiate her from other female teals.

In flight, both genders show a gray-blue patch on the forewing. The blue-wings distinctive twisty-turny, low flight over the water makes them appear to be flying at high speeds.

Optimal nesting habitat for blue-wings occurs in a ratio of half open water to half emergent marsh vegetation. The nests are located usually on dry, grassy ground (old muskrat houses work well). As with most ducks, extensive marsh vegetation provides necessary cover during the summer molt when the birds are flightless.

A desirable diet includes filamentous algae, pondweeds, duckweeds, and the seeds of water lilies, bulrushes, smartweeds, and various sedges, along with generous portions of aquatic insects and snails.

Wood ducks prefer to inhabit forested wetlands, so in spring, they're most commonly seen along the extensive silver maple and black ash flood plains of the Manitowish. As cavity nesters, wood ducks require older and larger trees that have been holed by woodpeckers, favoring cavities about 25 feet up.

In early summer, when wood ducks are raising and feeding their broods, shrubby thickets along streams with emergent and overhanging vegetation provides ideal safe cover. Look for woodies in the autumn wild rice beds along the Manitowish.

The woodies' resplendent breeding plumage, which is arguably the showiest of all ducks and was likely made during God's (or evolution's) most creative moment, makes identification a snap. The female expresses herself more subtly in a dull gray-brown plumage with a highly distinctive white eye-patch. Both genders have a swept-back crest, looking a bit like those flashy bike-rider's helmets used in the Olympics.

Woodies have strong, sharp toenails for perching in trees, and the largest eyes of any waterfowl, an indicator of exceptional eyesight. Their broad wings and long tail undoubtedly help them navigate flights through trees in search of food and a nesting cavity.

A woodie's omnivorous pursuit of food is remarkably diverse, including

foraging in upland woods for oak acorns, feeding in harvested fields for waste corn, dabbling in wetlands for duckweed, pondweed, the seeds of water lily and watershield, and capturing aquatic insects, tadpoles, small frogs, and Lord knows what else.

Widespread market hunting led to near extinction around the turn of the century, but a complete ban on wood duck hunting seasons from 1918 through 1941 helped bring the woodie back. And now with the placement of wood duck boxes long many river and lake edges, woodies rank second in number only to the mallard in most eastern states.

The hooded merganser may be the duck I enjoy seeing most on the river, because the male's plumage is so unique. He can raise and lower a fan-shaped white crest behind his otherwise all-black head, and his white breast with two black bars, his cinnamon flanks, and his spike-like bill make him one-of-a-kind. The female's punk rocker look of rusty feathers bushed out in a loose crest from the back of her head makes her distinctive as well. Sometimes a hoodie will just sink down in the water until its neck and head are about all that can be seen.

Hoodies like fast-running water, and are commonly seen on rivers during spring migration. They nest in tree cavities, sometimes as high up as 40 feet, and sometimes as low the open hollow of a stump.

Hoodies are expert divers and fast underwater swimmers, thus feeding extensively on small fish, crayfish, aquatic insect larvae, frogs, and snails, as well as a few roots and seeds of aquatic plants.

Broods of young hoodies in tow

behind females are a relatively common sight along the Manitowish.

≈ ≈ ≈ ≈ ≈

To live by a river is to be kept in the heart of things. — John Haines

The Island Lake boat landing soon appears on the left. Rice may surround the landing on either side.

If you're looking for a place to camp for a night, a campsite is just a few hundred yards farther downstream on the right.

Remember the rice and the wildness of the river throughout this last section, because things are about to change.

The Plow Follows the Ax—Ya, Sure

At the time of Wisconsin's statehood in 1848, nearly all of northern Wisconsin was covered by forest. Seventy years later, virtually the entire forest had been scalped, either by ax or fire, or the combination of the two. Less than 1 percent of the pre-European forest remains today.

In 1855, Increase Lapham wrote, "It is much to be regretted that the very superabundance of trees in our state should destroy, in some degree, our veneration for them. They are looked upon as cucumbers of the land; and the question is not how they shall be preserved, but how they shall be destroyed."

The cucumber analogy was appropriate, for that's the image that was sold to the people pouring into the region. The pleas for forest conservation, what few there were, were perceived as voices of fanatics and intellectuals.

Dean William Henry of the University of Wisconsin College of Agriculture wrote *Northern Wisconsin: A Handbook for the Homeseeker* in 1896. This handbook was produced due to a mandate from the Wisconsin legislature, which had established a new Board of Immigration for the purpose of promoting the sale of northern Wisconsin land in Europe. Since most immigrants couldn't read English, the book was filled with pictures of prosperous farms. Some 50,000 copies were distributed, printed in English, German, and Norwegian.

The lumber companies, railroads, and speculators all added their promotions, because more people meant more money.

Wisconsin Central Railway had 1 million acres to sell, so they distributed brochures such as "A Farm in Wisconsin Will Make Money for You From the Start: Crops Never Fail," accompanied by pictures of thriving farms.

Land companies acquired cutover lands with hopes of selling them at a big profit to settlers. Blue Grass Land Company of Minneapolis advertised land for a settlement near Eagle River, describing the area as "suitable for all kinds of farming . . . the land is rich, clay-bottomed, making it the most productive hay land in America, just as good as those famed 'blue grass' lands of Kentucky."

Blue Grass Land Company's town, Farmington, eventually changed its name to St. Germain.

By 1910, the profitable pineries had been exhausted, and the lumber companies had moved west. The stumps left behind were seen as the precursors to cornfields, fitting the perception that "the plow follows the ax." Still, even with all the sales pitching, only 13 percent of Oneida County and only 3 percent of Vilas County were under cultivation by

Continued

The Plow Follows the Ax—Ya, Sure (Continued)

this time. One reason was the stumplands proved to be a curse to remove.

Dean Henry was succeeded by H.L. Russell as dean of the College of Agriculture, and in 1915 the college opened a new land-clearing branch in the Department of Agricultural Engineering with a major focus on techniques for stump clearing. The college even took its operation on the track, outfitting a train to ride through the cutover. Instructors gave demonstrations on the state-of-the-art stump pulling and the use of dynamite. The train was aptly called the "Land Clearing Special."

By 1920, the train and booklet proselytizing proved ineffective, and farm clearing was nearly at a standstill. Despite everyone's efforts, only 6 percent of northern counties were cleared and planted in cultivated crops. The land companies went bankrupt, farms were abandoned, and local governments, which depended on property taxes, were in dire straits. When landowners couldn't pay taxes, the land reverted to the county. In 1927, one-quarter of the land in Wisconsin's 17 northern-most counties had been offered for sale as tax delinquent, and over four-fifths of it went unsold.

The only real crop that thrived in these sandy soils was the potato, as well as cranberries in the bogs. In the true pinery, the soils were fundamentally too acidic for growing anything profitably but blueberries and new pine. Add to the equation a growing season often less than 90 days in Vilas County (typically 65 days at my home in Manitowish), and farming was clearly "The Big Lie."

Beaver Dams—Their Ecological Impact

"Where beaver remain largely unexploited their activities may influence 20-40 percent of the total length of 2nd to 5th order streams with the alterations remaining as part of the landscape for centuries." So reads a Canadian study on beaver activity in small streams. It's quite a statement, because it means somewhere around one-third of the total length of all undisturbed small to moderate-sized northern rivers are typically altered in some manner by beaver dams. "Undisturbed" is, of course, a key word, since few of our rivers today are undeveloped or unaltered.

Continued

Beaver Dams—Their Ecological Impact (Continued)

The researchers studied two pristine rivers in Quebec that had no roads leading into them, no timber harvest, and very limited trapping.

They found that beavers altered the main channels of second to fourth order streams, rarely building dams in tiny first order streams or building in fifth order and above streams.

Beavers built an extraordinary number of dams; active dam density ranged from 8.6 dams/km to 16 dams/km. The mean was 10.6 dams/km (or about 17 dams per mile), a paddler's nightmare.

According to the study, beaver activity results in numerous changes in a river system. Beaver:

- modify the channel shape and water flow.
- increase the retention behind dams of sediments and organic matter that would otherwise flow farther downstream.
- create and maintain wetlands.
- modify the process of nutrient cycling by keeping soils wet, by altering the system of water flow, and by creating anaerobic zones.
- modify the riparian zone, including the species composition, their growth form, their chemistry, and the quantity of leaves and woody debris that are dropped in the river.
- influence the character of the water and materials transported downstream.

Beaver left lots of wood debris in the rivers—mostly speckled alder, balsam fir, paper birch, aspen, willow, and spruce.

Beaver also heightened the role of erosion and wind in the transfer of wood to streams. More erosion occurred in areas that became covered with water that were not usually subject to water flow. When the beaver dams raised water levels, the high water killed trees, resulting in many more trees falling into the water. Killing trees had the positive effect of increasing primary production (plant growth) by opening up the leaf canopy along the streams.

Beaver activities also resulted in extensive accumulation of anaerobic organic matter suitable for methane-producing organisms. Methane rates were 15 times greater in beaver ponds than in riffle sites.

Most obviously, beaver alter headwaters from typically narrow, heavily shaded channels, to slower, sun-dappled rivers more characteristic of middle order streams.

Given all the changes in the last century, it's remarkable that beaver meadowlands can still be seen today in many areas even after the beaver's extirpation, a testimony to their profound influence on the landscape.

Wetland Plant Communities

Wetlands comprise extremely varied landscapes, from mudflats to seasonally flooded farm fields, from cattail marshes to sedge meadows, from prairie-fen complexes to open bogs, from conifer bog forests to alder thickets, from black ash swamps to the enormous flood plain forests along the lower Wisconsin and Mississippi Rivers. With all those categories, it's easy to be quite confused as to the classification types of wetlands.

Wetlands are defined as areas under water or saturated by water that under normal circumstances support vegetation adapted to living in saturated soils or under water. Wetlands can be ephemeral, however, drying out in summer, but still supporting wetland plants. It's always a good idea when purchasing land, or when deciding on its zoning use, to visit the site in spring. Many a dishonest soul has led the unwary astray in August.

I learned wetland classifications originally by using the *Vegetation of Wisconsin* by John Curtis, the "bible" for all botanically inclined Wisconsinites. But nowadays, I like the "Key to the Wetland Plant Communities" from the book *Wetland Pants and Plant Communities of Minnesota and Wisconsin*. This key should help you appropriately identify what type of wetland you are in.

To make matters more difficult, however, numerous wetland classification systems are in usage. To that end, I have included a "Comparison of Wetland Classification Systems," also from *Wetland Plants and Plant Communities of Minnesota and Wisconsin*. Note the array of terminology to torture your tongue.

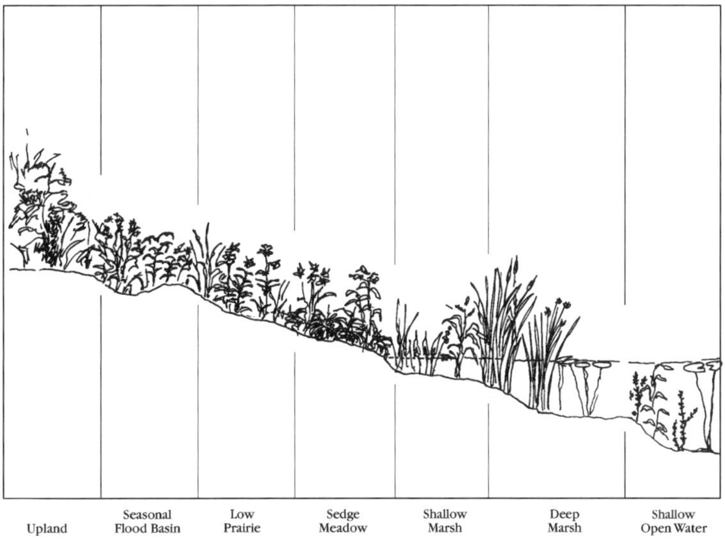

Benthic Insect Classification by Physical Habitat and Movement, or Substrate Trekking

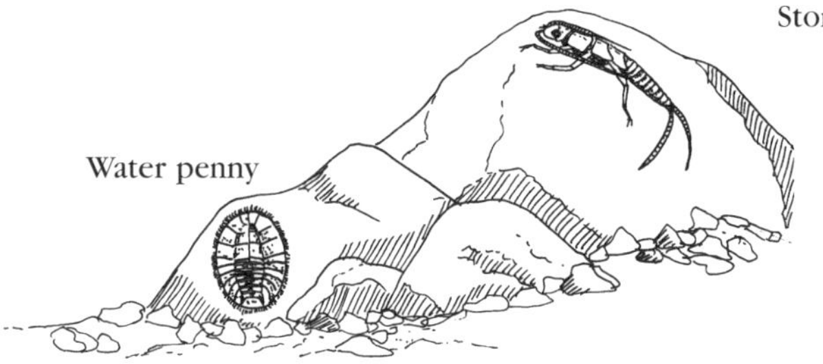

Clingers, sprawlers, climbers, and burrowers. It sounds like an odd array of behavioral categories, but if you could put yourself in the wet shoes of an aquatic insect, these behaviors make great practical sense. While the Manitowish possesses a slow current, it still packs enough velocity to carry something a long distance that only weighs one-twentieth of an ounce. So, how do evolved behaviors help invertebrates to stay in one place? Here are the ways, and who does them.

Clingers: Some clingers grasp and hold onto rocks or plants with claws or hooks at the end of their abdomens (the larvae of some net-spinning caddisflies, midges, and dobsonflies), while some cling using claws at the end of their legs (some larval mayflies and some adult riffle beetles). Other more specialized clingers have an abdominal disc that allows them to grip rocky substrates in rapids (some mayfly larvae). Some clingers make and live in cases that they cement to the substrate (some caddisfly larvae).

Many clingers that live in fast water have evolved a low profile or streamlined body, and orient their bodies to reduce friction from the current (some mayfly larvae).

Sprawlers:These benthic insects could just as easily be called "Crawlers" since they crawl about on various substrates like rocks, wood debris, and leaf packs. Many sprawlers live on the underside of rocks (some mayflies, stoneflies, and caddisflies). Picking up river rocks and examining their underside is an easy way to observe numerous aquatic invertebrates. Some sprawlers are active and predacious, like horsefly larvae and some dragonfly and damselfly larvae.A number of sprawlers live in the sand bottom (a few case-making caddisflies, midges, mayflies, and dragonflies).

Continued

Benthic Insect Classification by Physical Habitat and Movement, or Substrate Trekking (Continued)

Climbers: Climbers typically reside on aquatic plant stems, filamentous algae, and plants/roots that reach into the water along the bank. They like slow water, so are found along the edges and backwater sloughs among the myriad aquatic plants. Many dragonflies and damselflies fit into this category, and can be found in enormous numbers.

Burrowers: Burrowers live in "interstitial" habitat, like the spaces between coarse sediments and rocks, or burrow into soft substrates like silt and clay. Examples include the larvae of burrowing mayflies, numerous midges, and some dragonflies. Burrowing mayflies often possess digging adaptations, like broadened forelegs, for efficient excavation.

A number of aquatic insects (some craneflies, midges, and shoreflies, for instance) bore into plant stems and live inside there, so they may be classified as burrowers, too.

Another three groups of aquatic insects don't utilize the substrate, plant stems, or leaf detritus within the water, but instead simply live freely in the water. "Floaters" may live at or near the water surface, or live at significant depths. Mosquitoes may be the best known floaters. The pupae are known as "tumblers" because they live near the surface and swim or dive with a tumbling motion. Another group, the "Swimmers," periodically surface for air, have streamlined bodies often with oar-like legs, and are adept swimmers. Most people are familiar with adult whirligig beetles, which swim in groups on the surface.

And finally, the "Drifters" are the James Deans of the aquatic insect world, engaging in "behavioral drift," floating downstream a short distance in a periodic manner. Why benthic insects engage in drift is unclear, but it may have to do with predation pressures, reducing overloaded carrying capacity stress, or some colonizing mechanism.

Personally, I engage in behavioral drift all the time, a time-honored act that's gotten a bad rap by some who see it as "getting off-task" or "impulsive attention deficit." I call it exploring.

For clarity's sake, we shouldn't forget the two categories of aquatic insects that live on top of the water and who walk, skate, or jump around without breaking the surface film of water. The parts of their legs that contact the water have a waxy layer or hairs that are "waterproof" and bend rather than break the water surface. "Skaters" include the water striders, while "Jumpers" include springtails that utilize a spring-like structure on their abdomen to leap around on the water film. We just called them all of these "Jesus bugs" when we were kids since they could walk on water.

Birds of Northern Hardwood Swamps

Avian Richness Along the Bad River
Bottomland Hardwoods

Species	**#**	**#heard at each point count**
Ovenbird	33	2.06
American redstart	26	1.63
Red-eyed vireo	20	1.25
Least flycatcher	15	0.94
Veery	12	0.75
Rose-breasted grosbeak	10	0.63
Song sparrow	8	0.50
Mourning warbler	7	0.44
White-breasted nuthatch	6	0.38
Cedar waxwing	5	0.31
Common yellowthroat	5	0.31

Breeding birds of northern hardwood swamps differ considerably from those of all other northern bogs and swamps, because the habitat is mainly comprised of hardwood canopy and wet, shrubby areas. A paper written by Hoffman and Mossman (*The Passenger Pigeon*, Summer 1993) described the most common bird species in northern swamps as red-eyed vireo and veery, and in order of decreasing abundance, great-crested flycatcher, ovenbird, common yellowthroat, American redstart, song sparrow, black-and-white warbler, and northern waterthrush.

A later study, also published in *The Passenger Pigeon* (Joan Elias, Spring 1997), summarized the most common bird species found in five main habitat groupings along the Bad River in Ashland County, northern Wisconsin. Two of the five habitat groupings described types of northern hardwood swamps—black ash and bottom land hardwoods. Elias defined black ash swamps as dominated by black ash, with red maple, basswood, balsam fir, white spruce, and white cedar. Shrub species included downy arrow-wood, dogwood species, and speckled alder.

Dominant birds in black ash swamps in decreasing order included red-eyed vireo, swamp sparrow, common yellowthroat, great-crested fly

Continued

Birds of Northern Hardwood Swamps (Continued)

catcher, ovenbird, veery, American redstart, song sparrow, cedar waxwing and rose-breasted grosbeak.

Elias defined bottom land hardwoods as a relatively dense canopy of sugar maple, silver maple, basswood, box elder, and yellow birch, with fewer numbers of white spruce, hemlock, and white cedar. The shrub layer was predominately alternate-leafed dogwood, nannyberry, downy arrow-wood, and American elm. Birds observed in decreasing order in this habitat were: ovenbird, American redstart, red-eyed vireo, least flycatcher, veery, rose-breasted grosbeak, song sparrow, and mourning warbler.

Interestingly, the 10 most common species throughout all five habitats within the corridor (one mile on either side of the river) observed in decreasing order of abundance were: ovenbird, red-eyed vireo, Nashville warbler, white-throated sparrow, veery, chestnut-sided warbler, common yellowthroat, black-throated green warbler, American redstart, and great-crested flycatcher. These 10 species accounted for over 50 percent of the total number of individuals.

The Bad River varies from the Manitowish River in that its shorelines typically are comprised of a richer soil, thus producing fewer pines and more hardwoods. The Bad also has a significantly greater gradient, tumbling toward Lake Superior in a much more rapid fashion than the Manitowish. Nevertheless, Elias' habitat typing and associated bird communities compares favorably to much of that found in the Northern Highlands State Forest.

Avian Richness Along the Bad River
Black Ash Swamp

Species	#	**#/PT**
Red-eyed vireo	33	1.65
Swamp sparrow	30	1.50
Common yellowthroat	21	1.05
Great-crested flycatcher	20	1.00
Ovenbird	19	0.95
Veery	13	0.65
American redstart	12	0.60
Song sparrow	12	0.60
Cedar waxwing	10	0.50
Brown creeper	9	0.45
Northern waterthrush	9	0.45

Caddisflies—the RVers of the Aquatic Insect World

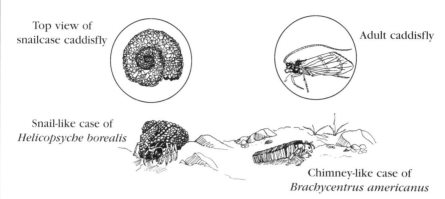

Top view of snailcase caddisfly

Adult caddisfly

Snail-like case of *Helicopsyche borealis*

Chimney-like case of *Brachycentrus americanus*

Henry David Thoreau watched caddisflies and wrote:

"Stretched over the brooks, in the midst of the frost-bound meadows, we may observe the submarine cottages of the caddis-worms . . . their small cylindrical cases built around themselves, composed of flags, sticks, grass, and withered leaves, shells, and pebbles, in form and color like the wrecks which strew the bottom."

His description of their homes as "submarine cottages" fits, because the homes of the case-building caddisfly species are mobile trailers, hauled around by the caddisfly larvae as they wander stream and lake bottoms in search of food. Woody Guthrie would have liked these hobos, wandering the aquatic countryside, though caddisflies may be more akin to modern-day travelers driving around inside Winnebagos or pulling silver Air-Streams.

Case-building caddisflies construct their homes by weaving an intricate tube of silk threads that is closed at one end. They then glue pieces of small sticks, leaf grains, tiny pebbles, or sand grains to the silk tube, often in a spiral pattern. The silk is not spun into a thread, but is poured forth in a glue-like sheet upon the materials to be cemented together. Each species uses a specific set of materials and builds its own style of home. Cases are well-adapted for specific habitats. For example, in swift currents, streamlined cases may be equipped with ballast stones or with trailing twigs that act as a rudder.

Cases come in various architectural forms, from perfect tubes to long four-sided cases (the log-cabin caddisfly) to snail-like forms. The camouflage of each case is exquisitely performed. For instance, wooden cases appear like little more than a tiny piece of wood, easily overlooked unless you spend a few minutes observing them and note that the wood actually walks around on the stream bottom.

Continued

Caddisflies—the RVers of the Aquatic Insect World (Continued)

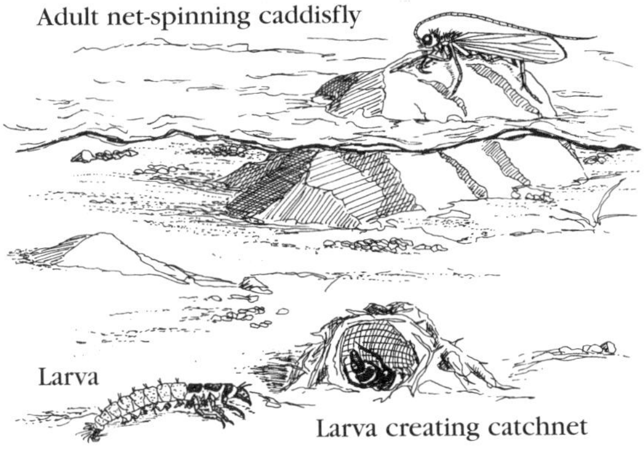

Inside the case lives a caterpillar-like or worm-like caddisfly larva, who walks about in the case with only the front end of its body and legs projecting from the case. Two tiny prolegs, or hooks, help to anchor it in place inside the case. To ensure sufficient oxygen reaches them inside the case, caddisflies typically increase the amount of dissolved oxygen passing over their gills by undulating their abdomen within the case and thus increasing the stream of water that flows over them.

Not all caddisflies construct cases. Every bit as inventive and unique as the case-builders are the netweaving, or netspinning, caddisflies. The netspinners capture both plant and animal materials that are floating by using fixed, silken nets. Most construct a catchnet in front of or adjacent to fixed objects like rocks or woody debris. The nets catch tiny particles of food suspended in the water. Differences in the catchnet area, strand size, and mesh size are typically related to necessary functional adaptations, like responses to high or low stream current speeds, and the amount and size of suspended particles desired to be caught. *Neureclipsis* spins a tubenet that looks like a tiny French horn and catches small particles, while *Arctopsyche* builds a catchnet that appears to be supported by sticks and can catch larger stuff.

Netspinners are the most common caddisflies in many parts of North America. Some 150 species ply their spider-like trade under water. Caddisfly larvae are mostly herbivorous, eating moss, algae, and leaves, but many species are omnivorous, and some are predatorial.

The larvae eventually pupate, attaching their homes permanently to a rock and closing themselves in. They typically emerge above water several weeks later as a moth-like adult with long antennae and large wings held over their backs like a tent. The adults mate in a few days, and shortly thereafter, the female lays eggs on the stream bottom or on the water surface. In Midwestern rivers, caddisflies yield tremendous numbers of adults that are often attracted to lights and become a nuisance.

Key to Wetland Plant Communities

from *Wetland Plants and Plant Communities of Minnesota and Wisconsin*
Used with permission.

1A. Mature tree (dbh of 6 inches or more) are present and form closed stands (more than 17 trees per acre; more than a 50 percent canopy cover) on wet, lowland soils (usually floodplains and ancient lake basins).

2A. Hardwoods trees are dominant; usually alluvial, peaty/mucky, or poorly drained-mineral soils.

3A. Silver maple, American elm, river birch, green ash, black willow and/or eastern cottonwood are dominant; growing on alluvial soils associated with riverine systems... FLOODPLAIN FOREST

3B. Black ash, yellow birch, silver maple and/or red maple are dominant; northern white cedar may be subdominant; growing on poorly-drained mineral or peat/muck soils, often associated with ancient lake basins... HARDWOOD SWAMP

2B. Coniferous trees are dominant; soils usually peaty.

4A. Tamarack and/or black spruce are dominant; growing on a continuous sphagnum moss mat and acid, peat soils ... CONIFEROUS BOG

4B. Northern white cedar and/or tamarack are dominant; continuous sphagnum moss mat absent; usually growing on neutral to alkaline peat/muck soils ... CONIFEROUS SWAMP

1B. Mature trees are absent or, if present, form open, sparse stands; other woody plants, if present, are shrubs or saplings and pole-sized (dbh less than 6 inches0 less than 20 feet high and growing on wet, lowland, or poorly-drained soils, or in groundwater seepage areas.

5A. Community dominated by woody shrubs.

6A. Low, woody shrubs usually less than 3 feet high; sphagnum moss mat layer may or may not be present.

7A Shrubs are ericaceous and evergreen growing on a sphagnum moss mat layer; peat soils are acidic.... OPEN BOG

7B. Shrubs are deciduous, mostly shrubby cinquefoil, often growing on sloping sites with a spring-fed supply of internally flowing, calcareous waters; other calciphiles are also dominant; sphagnum moss mat layer absent; muck/poorly-drained mineral soils are alkaline.... CALCAREOUS FEN

6B. Tall, woody deciduous shrubs usually greater than 3 feet high; sphagnum moss mat layer absent.... SHRUB SWAMPS

8A. Speckled alder is dominant; usually on acidic soils in and north of the vegetation tension zone.... ALDER THICKET

8B. Willows, red-osier dogwood, silky dogwood, meadowsweet and/or steeplebush are dominant on neutral to alkaline poorly-drained muck/mineral soils; found north and south of the vegetation tension zone.... SHRUB-CARR

5B. Community dominated by herbaceous plants.

9A. Essentially closed communities, usually with more than 50 percent cover.

10A. Sphagnum moss mat on acid peat soils; leatherleaf, pitcher plants, certain sedges, and other herbaceous species tolerant of low nutrient conditions may be present. .. OPEN BOG

Continued

Key to Wetland Plant Communities (Continued)

10B. Sphagnum moss mat absent; dominant vegetation consists of sedges (Cyperaceae), grasses (Gramineae), cattails, giant bur-reed, arrowheads, forbs and/or calciphiles. Soils are usually neutral to alkaline poorly-drained minerals soils and mucks.

11A. Over 50 percent of the4 cover dominance contributed by the sedge family, cattails, giant bur-reed, arrowheads, wild rice, and/or giant reed grass (*Phragmites*).

12A. Herbaceous emergent plants growing on saturated soils to areas covered by standing water up to 6 inches in depth throughout most of the growing season.

13A. Major cover dominance by the sedges (primarily genus *Carex*).... SEDGE MEADOW

13B. Major cover dominance by cattails, bulrushes, water plantain, *Phragmites*, arrowheads, and/or lake sedges.... SHALLOW MARSH

12B. Herbaceous submergent, floating and emergent plants growing in areas covered by standing water greater than 6 inches in depth throughout most of the growing season....DEEP MARSH

11B. Over 50 percent of the cover dominance contributed by grasses (except wild rice and *Phragmites*), forbs and/or calciphiles.

14A. Spring-fed supply of internally flowing, calcareous waters, often sloping sites; calciphiles such as sterile sedge, wild timothy, Grass-of-Parnassus and lesser fringed gentian are dominant.... CALCAREOUS FEN

14B. Water source(s) variable; calciphiles not dominant.

15A. Soils saturated to inundated during the growing season; prairie grasses such as big bluestem, prairie cordgrass and/or Canada bluejoint grass are usually dominant, and various species of lowland prairie forbs are present.... WET TO WET-MESIC PRAIRIE

15B. Site rarely inundated, but soils are saturated for all or part of the growing season; dominated by forbs such as giant goldenrod and/or grasses such as redtop and reed canary grass..... FRESH (WET) MEADOW

9B. Essentially open communities, either flats or basins usually with less than 50 percent vegetative cover during the early portion of the growing season, or shallow open water with submergent, floating and/or floating-leaved aquatic vegetation.

16A. Areas of shallow, open water (to 6.6 feet in depth) dominated by submergent, floating and/or floating-leaved aquatic vegetation.... SHALLOW, OPEN WATER COMMUNITIES

16B. Shallow depressions or flats; standing water may be present for a few weeks each year, but are dry for much of the growing season; often cultivated or dominated by annuals such as smartweeds and wild millet....SEASONALLY FLOODED BASIN

Comparison of Wetland Classification Systems

Wetland Plant Community Types (Eggers & Reed)	Vegetation of Wisconsin (Curtis 1971)	Wisconsin Wetland Inventory	Classification of Wetlands and Deep Water Habitats of the United States (Cowardin et al. 1979)	Fish and Wildlife Svce. Circular 39 (Shaw and Fredine 1971)
Shallow, Open Water	Submergent aquatic community	Aquatic bed, submergent and floating	Palustrine or lacustrine, littoral; aquatic bed; submergent, floating, and floating-leaved	Type 5: Inland open fresh water
Deep Marsh	Emergent and submergent aquatic community	Aquatic bed, submergent, and floating; and persistent and nonpersistent, emergent/wet meadow	Palustrine or lacustrine, littoral; aquatic bed; submergent, floating, and floating-leaved; and emergent; persistent and nonpersistent	Type 4: Inland deep fresh marsh
Shallow Marsh	Emergent aquatic community	Persistent and nonpersistent, emergent/wet meadow	Palustrine; emergent; persistent and nonpersistent	Type 3: Inland shallow fresh marsh
Sedge Meadow	Northern and southern sedge	Narrow-leaved persistent, emergent/wet meadow	Palustrine; emergent; narrow-leaved persistent	Type 2: Inland fresh meadow
Fresh (Wet)	—	Broad- and narrow-leaved persistent, emergent/wet meadow	Palustrine; emergent; broad- and narrow-leaved persistent	Type 1: Seasonally flooded basin or flat; Type 2: Inland fresh meadow
Wet-to-Wet Mesic Prairie	Low (wet to wet-mesic) prairie	Broad- and narrow-leaved persistent, emergent/wet meadow	Palustrine; emergent; broad- and narrow-leaved persistent	Type 1: Seasonally flooded basin or flat; Type 2: Inland fresh meadow
Calcareous Fen	Fen	Narrow-leaved persistent, emergent/wet meadow; and broad-leaved deciduous, scrub/shrub	Palustrine; emergent; narrow-leaved persistent; and scrub/shrub, broad-leaved deciduous	Type 2: Inland fresh meadow
Open Bog	Open bog	Moss; and broad-leaved evergreen, scrub/shrub	Palustrine; moss/lichen; and scrub/shrub; broad-leaved evergreen	Type 8: Bog
Coniferous Bog	Northern wet forest	Broad-leaved evergreen, scrub/shrub; and needle-leaved evergreen and deciduous, forested	Palustrine; forested; needle-leaved evergreen and deciduous; and scrub/shrub; broad-leaved evergreen	Type 8: Bog
Shrub-Carr	Shrub-carr	Broad-leaved deciduous, scrub/shrub	Palustrine; scrub/shrub; broad-leaved deciduous	Type 6: Shrub swamp
Alder Thicket	Alder thicket	Broad-leaved deciduous, scrub/shrub	Palustrine; scrub/shrub; broad-leaved deciduous	Type 6: Shrub swamp
Hardwood Swamp	Northern wet-mesic forest; southern wet to wet-mesic forest	Broad-leaved deciduous, forested	Palustrine; forested; broad-leaved deciduous	Type 7: Wooded swamp
Coniferous Swamp	Northern wet-mesic forest	Needle-leaved deciduous and evergreen, forested	Palustrine; forested; needle-leaved deciduous and evergreen	Type 7: Wooded swamp
Floodplain Forest	Northern and southern wet-mesic forest	Broad-leaved deciduous, forested	Palustrine; forested; broad-leaved deciduous	Type 1: Seasonally flooded basin or flat
Seasonally Flooded Basin	--	Flat/unvegetated wet soil; and persistent and non-persistent, emergent/wet meadow	Palustrine; flat; emergent; persistent and non-persistent	Type 1: Seasonally flooded basin or flat

Chapter 4
Our Course:
The Manitowish Chain of Lakes from Island Lake to the Rest Lake Dam

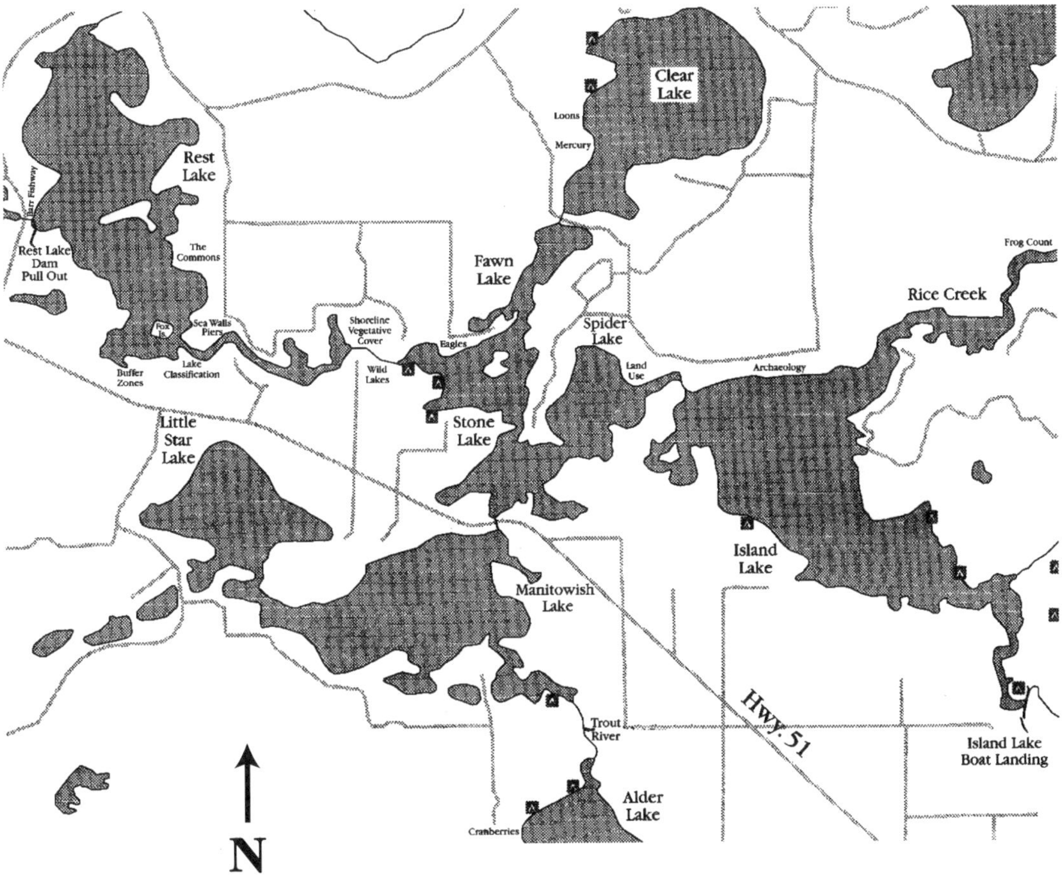

Mechanized man, having rebuilt the landscape, is now rebuilding the waters. The sober citizen who would never submit his watch or his motor to amateur tamperings freely submits his lakes to drainings, fillings, dredgings, pollutions, stabilizations, mosquito control, algae control, swimmer's itch control, and the planting of any fish able to swim. So also with rivers. We constrict them with levees and dams, and then flush them with dredgings, channelizations, and the floods and silt of bad farming . . .

Thus men too wise to tolerate hasty tinkerings with our political constitution accept without a qualm the most radical amendments to our biotic constitution.

—Aldo Leopold

Chapter 4

Of Archaeology, Mercury, Cranberries, Buffers, and Dams

If we lose the view that water is a miracle, we lose also some of our ability to look upon the world around us with passion and gladness.
— Peter Steinhart

Allow three hours or so for this stretch. Paddling time depends on the wind, since this route is nearly all lake crossings.

Many options exist for paddling routes through here:

•Paddling Island to Spider to Stone to Rest lakes—the direct path through the main channel.

•Heading north out of Stone Lake and into Fawn and Clear lakes.

•Taking a side trip up Rice Creek out of Island Lake.

•Paddling south out of Spider into Manitowish Lake and into the Trout River.

•Taking another side trip north out of Manitowish Lake into Little Star Lake.

One way or another, this is the Manitowish Chain of Lakes, or what may better be called the Manitowish Gold Coast. Here stands some of the most expensive property in Vilas County—often $1,000 a foot for shoreline at the end of the century—a price that, remarkably, may appear low in a few decades. Development along some shorelines in this section rivals the excesses of southern Wisconsin. A few parcels of state-owned lands provide some relief from development, and some owners have preserved shoreline frontage and built well off the water. But concrete seawalls and lawns jutting down to the water are often the shoreline norm.

In summer, waterskiing, jet-boating and high-speed motorboating transform the chain into a recreational dynamo, reducing the Manitowish into at best a scenic waterbody, and at worst a racetrack.

Still, the lakes possess a fascinating cultural history, and in the off-seasons, offer serenity and beauty.

In 1840, Thomas Jefferson Cram, an engineer, was given the job of charting the Montreal River in order to delineate the border between the state of Michigan and Wisconsin Territory. In the process he traveled the Manitowish River, and wrote:

"There is no direction . . . which will not lead into a series of small but exceedingly beautiful lakes in this part of the country. These little lakes so beautifully diversified in size, shape and scenery are but the limpid springs which form the summit reservoirs that nature seems to have furnished with admirable foresight, for a never failing supply to the Chippewa [which includes the Manitowish], the Ontonagon, and several smaller streams such as the Montreal, the Carp, Iron, etc."

He spoke, of course, of lakes like Island Lake, and all the others in the Manitowish Chain. Each lake speaks to the traveler in a different voice, in a lilt fashioned from its unique interplay of land and water. Island Lake's vocabulary could be written in Welsh for all its nonlinear scribbling of shoreline bays and peninsulas, as well as its islands. This lake has personality, though its current-day expression is dampened considerably by its development and recreational buzz.

Nevertheless, eagle, osprey, great blue heron, and loon should be common sightings. An eagle's nest is located on a tiny island within a half mile of putting-in on Island Lake.

For the angler, Island's 1,023 acres and 35-foot maximum depth offer diverse fishing opportunities, including good numbers of muskie, northern pike and walleye, as well as a reasonable population of large and smallmouth bass.

Upon launching from the boat landing, if the water is low, the first notable feature of Island Lake will be the stumpfield immediately on the left. If the water is high, these stumps may be just under water, and constitute a definite motor hazard. The topographical maps show the elevations of Island, Spider, Stone and Rest lakes all as 1,601 feet, but elevations tell very little of the water level story here.

The stumps are almost certainly the result of high waters backed up from the Rest Lake Dam at the far end of the Chain. The dam can raise or lower the water levels at least 5 feet on the Chain, and typically will do so in any given year, filling up the lakes in the spring and drawing them down in the fall.

But again, that's only part of the story. In 1993, a geomorphological study was conducted by the University of Wisconsin-Milwaukee to locate the pre-logging shoreline and determine how much water the Rest Lake Dam holds back. The researchers found the high water level on Rest Lake to be 9.2 feet above the current elevation of the river. Winter water levels are drawn down to 3 to 4 feet above the pre-dam lake level.

This head of water significantly inflates the acreage of the lakes on the Chain. Historically, it drowned shoreland trees, creating stump pastures like this. As with most human interventions in nature, there's good news and bad news as a result. The good news is that the deeper water has made boat travel easier between the lakes. And the stumps, as discussed earlier, are good medicine if you're an angler. Structure, structure, structure—that's the angler's mantra.

The bad news is that the raised water levels have inundated numerous archaeological sites. A two-year archaeological shoreline survey of the Manitowish Chain conducted from 1992 to 1993 found nine submersed sites. Both prehistoric and historic artifacts were found on the lake bottoms, suggesting that a wide array of archaeological sites were affected.

Where these lake shorelines may once have been is easy to see in the fall, when water levels are dropped.

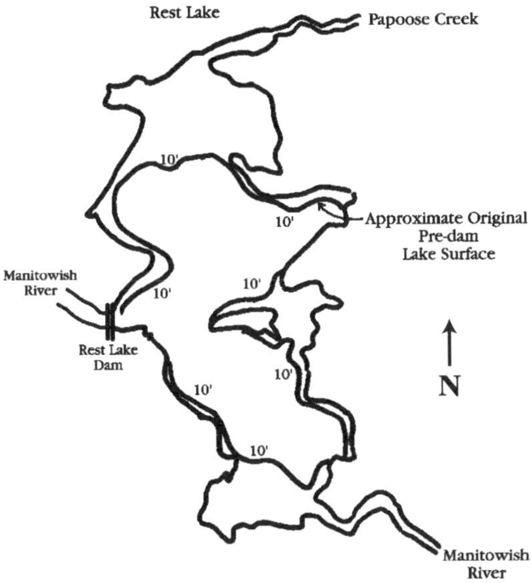

Rest Lake pre-dam levels

Still, as the study notes, the water levels were probably another 3 feet or more lower than the current low-water level in the fall. Long-time residents say that before the dam was built, there was a rapids between Rest Lake and its outlet into the river. The height of that drop may or may not have been significant, and would obviously affect the calculations of the estimated increase in water levels.

One way or another, you'll be paddling on much higher waters than those traveled by the prehistoric American Indians, the historic Ojibwa, and the fur traders.

Paddling the eastern shore of Island Lake offers the least shoreline development. Rice Creek, which broadly enters the lake several miles farther up the shoreline, is worth a side trip. In September, the creek supports an enormous growth of wild rice. An eagle's nest stands sentinel on the south shoreline just a short ways up the creek. This nest has been active and consistently productive since the mid-1980s, despite all the summer lake activity.

Rice Creek rises in Round Lake, flows southwest into Big Lake, and then meanders farther southwest into Island Lake. Water flow is usually more than sufficient for canoe travel. The creek becomes shallow and narrow only for a short stretch between County K and Big Lake. Another eagle's nest is active on the southern shoreline just into section 9.

Rice Creek flows almost entirely through public lands, and remains in pristine condition. The Wisconsin Bureau of Endangered Resources has called for strong protective designation of a 2-mile stretch of the creek above Big Lake. The creek flows through a large, diverse conifer swamp of white cedar, balsam fir, black spruce, and tamarack, along with several stands of old growth hemlock-hardwoods that support supercanopy white pines. Lush beds of aquatic plants clothe the creek shallows and surrounding wetlands,

Biotic Inventory and Analysis of NH/AL

From 1992 to 1996, the Natural Heritage Inventory section of the Wisconsin Department of Natural Resources' Bureau of Endangered Resources inventoried and analyzed the biotic resources of the Northern Highlands/American Legion State Forest. The purpose of such an in-depth analysis was to provide an accurate baseline of data from which informed decisions could be made regarding the development of a new master plan for the NH/AL.

The finished document identified and evaluated the forest's natural plant and animal communities, focusing on rare and/or otherwise significant communities. Their goal was to identify appropriate sites for restoration of lost or declining communities, and to emphasize protection and management opportunities on the forest property.

Sixty-five sites were termed "Primary Sites," and generally include the best examples of both rare and representative natural features that the NH/AL has to offer. Thirty-six of the 65 were ranked "high" in all of the BER's evaluation categories. The categories include criteria such as the current quality and condition of the site, size of the community, and the representation of rare and characteristic species within the community. The BER recommends that each of these 36 sites be given "strong protective designation."

Four major "macrosites" were also identified. These sites contain two or more of the primary sites in close proximity, where the collective attributes of the larger site have a much greater ecological value than the individual sites alone. The Lower Manitowish River (from Circle Lily Creek downstream, as well as associated uplands and wetlands) was listed as one of these macrosites, and contains five of the primary sites, all of which were evaluated as "high" in value.

The book is loaded with maps and charts and tables. Each of the 65 primary sites is given a site description, a paragraph on the site's significance, and a discussion of the management considerations, all of which are followed by a color map of the site. The book also includes numerous maps detailing features such as pre-settlement vegetation, ecoregions, and current vegetation.

"Knowledge is power," said Francis Bacon, he who invented dynamite, apparently in case knowledge wasn't sufficient. For the individual who wishes to better understand the working ecology of the Northern Highlands, this guide provides essential background. Public forests belong to the public, and forest management should be guided by an informed public.

Order a copy (PUBL-ER-093 99—cost is $10) from the Wisconsin Natural Heritage Inventory Program, Department of Natural Resources, P.O. Box 7291, Madison, WI 53707.

including a high concentration of rare species such as showy lady's slipper and swamp pink.

A rich bird and amphibian community also thrives here. For many years, as a participant in the WDNR's annual statewide frog count, I've counted calling frogs where County K crosses Rice Creek. Bullfrogs, leopard frogs, Eastern gray treefrogs, wood frogs, spring peepers, green frogs, and American toads all chorus with gusto here.

The state DNR implemented this volunteer population survey in 1984 because frogs are thought to be environmental indicators. With their permeable skins, frogs are quite sensitive to environmental changes. Thus, one way to determine just how good or bad the health of our wetlands, lakes, and rivers has become is to watch the population trends in frogs and toads. A decline in numbers would suggest a decline in health of our aquatic ecosystems, while a stable or increasing number would suggest generally good health.

Over the 16 years that volunteers have sampled nearly 130 routes statewide, Wisconsin's frog and toad populations show a positive trend for wood frogs, American toads, Eastern gray and Copes gray treefrogs, and green frogs. Stable populations appear to be the case for chorus frogs, mink frogs, and the scattered populations of bullfrogs. Spring peepers have shown a fairly constant small reduction, while leopard and pickerel frogs have suffered significant declines.

In north-central Wisconsin, statistically significant positive trends are being seen for wood frogs, American toads, and green frogs, while statistically negative trends appear for spring peepers and leopard frogs.

Four factors can affect the accuracy of such sampling. One, volunteers exhibit different levels of skill in differentiating frog and toad calls. The increase in wood frogs, for instance, may reflect the volunteers' increasing ability to hear this easy-to-miss species.

Two, timing can be everything. Wood frogs only breed for about two weeks. If the survey takes place around, but not within, that time frame, wood frogs may not appear present when, in fact, the counter simply missed them. Counters do three different surveys along the same route each year to help prevent this timing error from happening, but it still can, given that the survey occurs over a three-month span.

Three, seasonal climatic variations affect singing frogs and toads. A temporary drought dries up breeding sites, greatly reducing or eliminating the breeding songs of the male frogs and toads in affected areas. Conversely, a warm, wet spell can work frogs into a breeding frenzy.

Four, daily weather variations—a very cold night, for instance—can greatly inhibit calling frogs. Counters are supposed to sample when water and air temperatures are optimal, but volunteers can't always get out to count when the times are best.

Still, a long-term survey will usually overcome these short-term kinds of misinformation, and the data eventually even out. So, the longer this survey can be continued, the more faith one can put into its accuracy and its conclusions.

What to make of these trends in terms of overall ecosystem health is

Frog and Toad Survey 1984-99

+ = Positive trend
- = Negative trend

	Wisconsin Ecoregions					
	North-west	North-central	Drift-less	Central Sands	East	South-east
Wood Frog	-	+	+	+	+	-
Chorus Frog	-	+	+	+	-	+
Spring Peeper	-	-	+	+	+	-
Leopard Frog	-	-	+	+	-	-
Pickerel Frog		-	-		-	+
American Toad	+	+	-	-	+	+
Eastern Gray Treefrog	-	-	+	-	+	+
Copes Gray Treefrog	-	-	+	-	-	+
Mink Frog	+	+				
Green Frog	+	+	+	-	+	+
Bullfrog		-	-		-	+
Total +/-	3/6	5/6	7/3	4/4	5/5	7/3
No. Routes	3	38	29	3	14	44

another much larger question, and one that researchers are still working to answer.

≈ ≈ ≈ ≈ ≈

Follow the north shore of Island Lake to the west. A narrow channel opens into Spider Lake. Boulders line parts of the channel, and large stones dot the streambed. Old-timers say a rapids was once here before the dam heightened the water levels and drowned it.

Private land ownership completely surrounds Spider Lake. But while the lake's natural state has been significantly compromised, its historical context remains rich. In early August of 1845, surveyor A.B. Gray paddled though the Manitowish Chain in an attempt to map the mineral region of Lake Superior. He wrote:

"In the evening we entered 'Cross' Lake from the river—so called by the Indians from its resembling a cross in shape—and encamped upon a high point of land jutting out and forming one of the arms of the cross. Upon this point are two large wigwams and several acres of ground cleared and cultivated, being the summer residence of 'White Thunder,' a tall and athletic looking Indian. . .

"Our course up the river for about 10 or 15 miles, to this lake, was easterly, although the stream curved around in every direction, occasionally opening into small and picturesque lakes, surrounded by high land, with excellent pineries, and narrowing again to a width barely sufficient for the passage of a canoe."

It is unclear if the reference to "Cross" Lake represents Spider or Stone Lake today. The increased water levels created by the Rest Lake Dam change the appearance of the lakes so much that one can only conjecture where White Thunder made his home. Whatever the location, Gray's letter is one piece of evidence attesting to the importance of the Manitowish Waters area as a water trail utilized extensively by American Indians and Europeans.

Gray was certainly not the only one to describe or refer to this route. J.G. Norwood refers to it in 1847, as does Thomas Jefferson Cram in 1841, Charles Penny in 1840, James Doty in 1820, Francois Malhiot in 1804, and Perrault in 1792. From Spider Lake, the water trail could be taken a variety of directions depending on your desired destination:

- Downstream west, and then north through the Turtle River and over the famous Flambeau Trail to Lake Superior.
- Downstream west and south to the confluence with the Bear River, and then upstream on the Bear to Lac du Flambeau.
- Downsteam west and south into today's Chippewa River.
- Downstream a very short distance into Manitowish Lake, and then upstream through the Trout River to a series of portages that led to the Wisconsin River and the village at Lac Vieux Desert.

The most complex route was via the Trout River to the Wisconsin River. An Indian village existed where the Trout River emerged from Trout Lake.

In 1847, geologist J.G. Norwood described the village as only occupied during the summer and fall months. "They have gardens for corn and potatoes at this place, though their principal dependence for food is upon the lake, which yield them a plentiful supply of fine fish. We received from an Indian here a lot of very fine potatoes, a most acceptable present, as more than two-thirds of the provisions we had brought from La Point [Madeline Island] were consumed, and we had not yet performed more than one-third of our journey."

Norwood describes in detail the portage route to the Wisconsin River. I have placed in brackets [] today's names of the lakes he refers to.

"Trout Lake is seven or eight miles long by four miles wide, and contains a number of small islands. It is surrounded

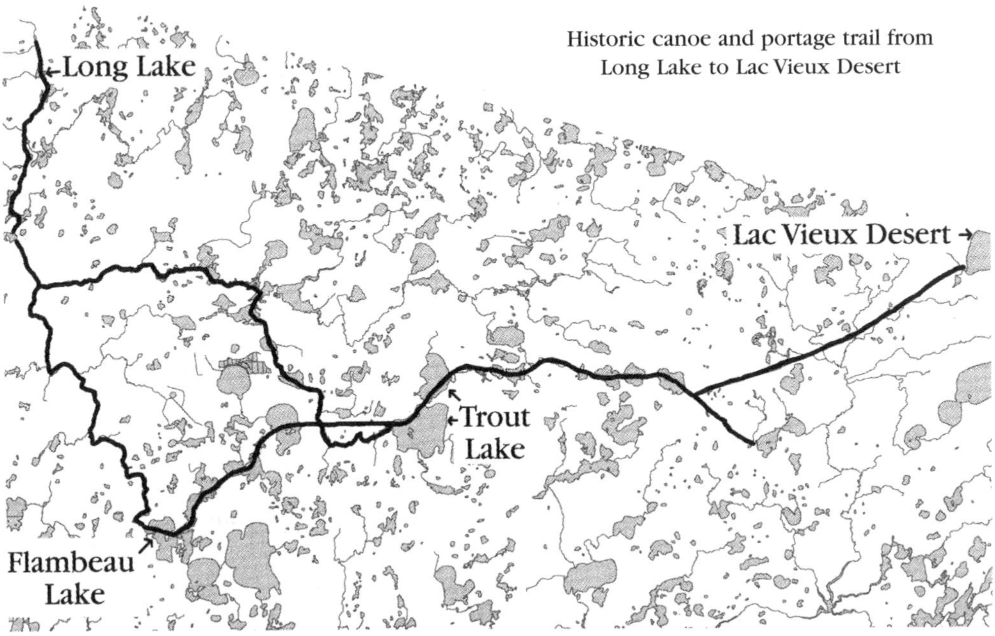

Historic canoe and portage trail from Long Lake to Lac Vieux Desert

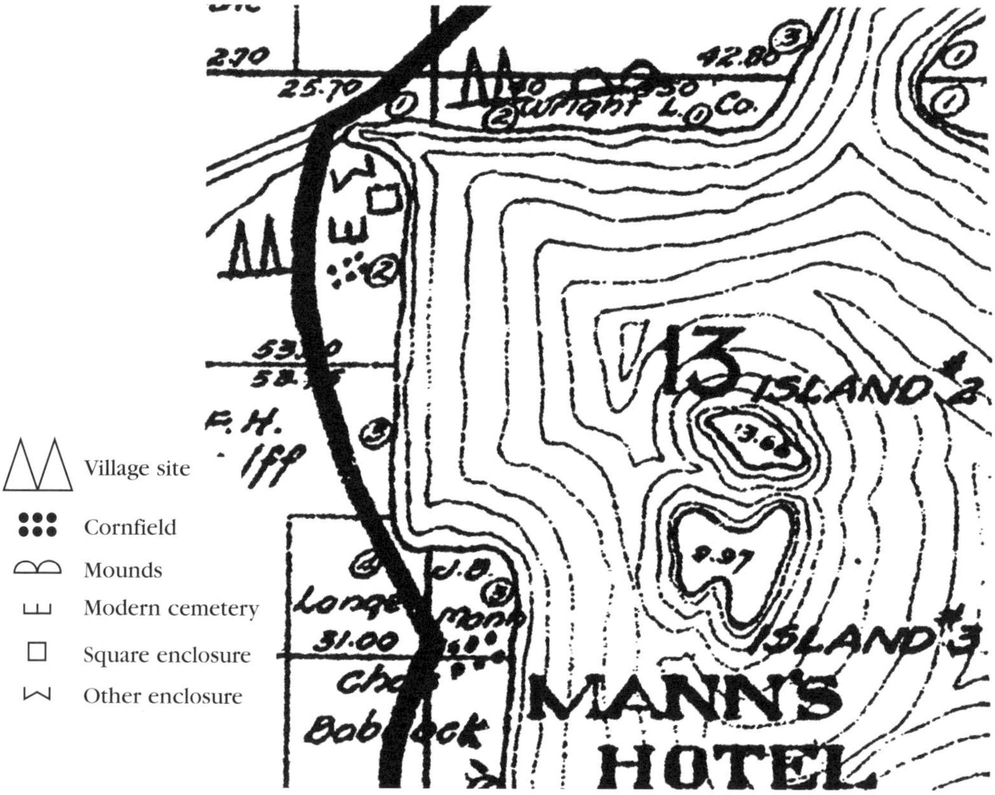

Trout Lake Historical Map—Charles E. Brown, 1924 Platbook of Wisconsin, Wisconsin Archaeological Society. Courtesy of State Historical Society of Wisconsin

by drift hills, from twenty five-to forty-five feet high, supporting a sparse growth of small pines and birch. Our course across it was northeast, to a trail leading to Lower Rock Lake [Pallette Lake—likely the trail followed along the north side of Stevenson Creek]. We encamped on the trail, a short distance from the lake. At six o'clock, P.M., the thermometer stood at 31°F and our tent and baggage, which had got wet in crossing the lake, were frozen.

"September 30. Ice formed one-fourth of an inch thick last night. The portage between Trout and Lower Rock Lakes is about two miles and a quarter in length, and runs along the base of drift hills. These lakes are connected by a small stream [Stevenson Creek], not navigable for canoes. The Lower Lake is about half a mile in diameter. A portage of three hundred yards leads to Upper Rock Lake [Escanaba Lake], which is one mile in its largest diameter. These lakes are also connected by a small stream. They derive their name from the immense number of boulders which line their shores, and show themselves above the water in the shallow parts. . .

"We had great difficulty in finding the portage from this lake. It begins on the northeast shore, and is about two and a half miles long. Its course is nearly due east, passing a good part of the distance along the margins of cranberry marshes. Three small ponds were passed in the first two miles. They are connected by a small stream [Birch Creek] flowing into Upper Rock Lake, which is navigable for canoes up to the second pond. From this point a portage of everything has to be made to

Lower White Elk Lake [White Birch Lake]... The soil is thin, but supports a growth of small pine, poplar, birch, spruce, hemlock, fir, a few oaks, and some basswood ...

"Lower White Elk Lake, where we camped, is about three-quarters of a mile long and a quarter of a mile wide. Here were found a number of deserted wigwams and the remains of a garden. The lake affords great numbers of fish, and the quantity of their remains scattered around shows they are the principal article of food among the Indians who occasionally inhabit it.

"October 1 ... We crossed First White Elk Lake, and by a stream twenty feed wide and a quarter of a mile long, passed into Second White Elk Lake [Ballard Lake], which is about two miles long and one mile wide. From this we passed into Third White Elk Lake [Irving Lake], by a stream ten yards wide and three hundred yards long. This lake is nearly circular, and about one mile in diameter. It is very shallow, not having a depth of more than three feet at any point, and has a mud bottom. We noticed here a phenomenon, not hitherto observed in any of the great number of small lakes we have seen in the territory. The whole surface of the lake was covered with bubbles of light carburetted hydrogen gas, which were constantly ascending from the bottom.

"From this lake, a portage of a quarter of a mile brought us to the Fourth White Elk Lake [Laura Lake]. The portage leads due east, over drift, covered with a better soil than any met with for several days past. It supports a tolerably good growth of sugar maple, birch, oak, poplar, and a few pines. This lake is a beautiful sheet of water, about one mile long and three-fourths of a mile wide. The bottom is covered with pebbles and the shore with boulders, some of which are very large; one of them being over fifty feet in circumference....

"The portage to the headwaters of the Wisconsin River starts due east from this lake. In about half a mile the trail divides, the left-hand branch leading directly to Vieux Desert Lake, the other to a small lake [Upper Buckatabon Lakes] which discharges its waters into the Wisconsin, about ten miles in a direct line south of Vieux Desert ..."

I take you through all this detail, because it's important to understand just how interconnected these trails were, and thus how interconnected people were at the time. One writer refers to ancient water trails as "elongate archaeological sites." Your travel here is no less than an exploration of a spiderweb of archaeological sites, all linked together by lengthy canoeprints.

Norwood's journal is also invaluable for its description of the plant communities he encounters along the portages. His failure to find archetypal old-growth stands in many areas says, as do the surveyor records a decade and more later, that the Northwoods was a mosaic of different-aged plant communities. Natural and deliberate fires, insects and disease, windstorms, snow-throw, floods, all conspired to knock back plant succession. The myth of squirrels traveling from branch to branch and never having to touch the ground from the Atlantic Ocean to western Minnesota has always failed to convey the realities of a natural landscape disturbed by natural and human forces.

Read more about pre-settlement vegetation at the end of the chapter.

Of lesser significance, but not of less interest, is Norwood's naming of

J. G. Norwood "Provisional Geological Map of Part of the Chippeway Land District of wisconsin with part of Iowa & of Minnesota Terratory." 1847 (courtesy State Historical Society of Wisconsin)

four consecutive "White Elk Lakes." Throughout the 20th century, the Boulder Junction area has been well-known for its population of white-colored (not albino) white-tailed deer. To this day, it's not uncommon to see white deer in this area. Could these white deer have mistakenly been identified as elk, or colloquially called elk? Given the naming of these lakes, could this white deer population have originated in the 1800s or before?

To further complicate matters, Hartley Jackson writes in his book *Mammals of Wisconsin* that, "Europeans who first came to American knew about the European moose (called 'elk' in Europe) and were also familiar with the European red deer, a mammal somewhat similar to the American elk."

So, is it also possible that white moose were once seen here, and called by the first Europeans "elk"?

Elk were extirpated in Wisconsin sometime before 1875, but are thought to have probably lived throughout the state, though more abundantly in the south and western two-thirds of the state. So, it's also possible that these lakes could actually have been named after white elk.

In his 1845 journal, A.B. Gray is less exacting than Norwood in his description of passing through this route, but he has this to say upon leaving Trout Lake:

"The next evening we reached 'White Deer' lake with our canoes, after making several portages and passing up a small and crooked branch, with difficult swamps, through which we pushed ourselves. This small sheet of water—so called, according to tradition, from the circumstance of a white deer having been seen upon its bank—was supposed by us to empty into a branch of the Wisconsin River."

Thus, the historic name of at least one of these lakes is either White Deer or White Elk. And the sightings of white elk, white moose, or white deer are all possible explanations for the original names of these lakes. However, I find the remarkable coincidence of both historic and current white deer populations in this area to offer the most likely of all of the explanations for the naming of these lakes.

≈ ≈ ≈ ≈ ≈

How many people were actually using the Manitowish River for travel and trade is impossible to say. According to historian Jim Bokern, by the early 1800s, six bands of Ojibwa were clearly established in the Lac du Flambeau District—the Lac du Flambeau, Pelican Lake, Lac Vieux Desert, Turtle Portage, Trout Lake, and Wisconsin River bands.

Lac du Flambeau operated as the central village for this district. Two fur-trading posts were built by the French on Flambeau Lake, the XY Post and the North West Post, and the other five bands would travel there to trade. Ojibwa families also lived outside of the six villages on many of the lakes.

Read more about the Bands of the Lac du Flambeau district at the end of the chapter.

By the official census in 1843, 1,016 Chippewa lived and traveled in this general area, though the number was probably very inaccurate given the fluid nature of the bands.

Band	**Total Number**	**Families**
Lac du Flambeau	274	58
Pelican Lake	134	38
Lac Vieux Desert	213	53
Trout Lake	82	16
Turtle Portage	68	16
Wisconsin River	245	82

The number of French traders who may have traveled through here in any given year is more difficult to say. For a general perspective on how active the entire Upper Midwest was in the fur trade, 2,341 voyageurs were recorded in licenses obtained at Montreal and Detroit in 1777. Translating that figure into the number that traded in any given year in this specific area is virtually impossible.

However, we do have the 1804-05 journals of Francois Victor Malhiot, the clerk in charge of the North West Fur Company's trading post at Lac du Flambeau. Malhiot kept careful

Archaeological Protection Act of 1979

Many people are loathe to pipe up if they find a historical artifact, particularly on their own land, for fear that BIG BROTHER GOVERNMENT will swoop down, take the artifact, kick them off their land, and declare the land a historical monument. Well, while paranoia is rooted in innate survival needs, it's misplaced in this particular here and now. The 1979 Archaeological Protection Act says two major things:

1- If you find an artifact on public property, it's illegal to take it. You have to leave it there, just like you have to leave any trees, flowers, birds, or anything else that you find on public land. Since the land and its contents and inhabitants belong to everyone, this shouldn't be a hardship.

On state-owned lands like the NH/AL, it is illegal to remove artifacts or otherwise disturb archaeological sites without a permit under the Field Archaeology Act, Section 44.47 of the Wisconsin Statutes. The law applies equally to upland and submerged sites under lakes and rivers. Permits are administered by the Office of the State Archaeologist, but typically are given only to professional archaeologists.

2- If you find an artifact on your private property, you can keep it, but you are asked to report your find. Better yet, allow it to be photographed on site before you remove it in order to add a piece to the historical puzzle of your area. Best of all, before doing anything, call the State Historical Society, the local DNR, or the Federal Forest and get their suggestions on how best to honor the significance of your site.

On your private land, the state offers a property tax incentive/exemption to landowners who agree to protect archaeological sites. To obtain the tax exemption, the landowner must agree to place a permanent covenant on the site.

Landowners can become true stewards of Wisconsin's past, and save some tax money in the process. Seems like a win/win deal to me.

I highly recommend the Web site for the State Historical Society of Wisconsin (<www.shsw.wisc.edu>) for information on such topics as archaeology programs, research, preservation, laws, and property tax exemption.

For those of you who want to pursue your local history, the SHSW has established the Office of Local History, which serves more than 300 county, local, and specialized historical organizations in the state. The Office works with the Wisconsin Council for Local History, which represents more than 50,000 people throughout the state who belong to affiliated historical organizations.

records of the goods arriving, their trade to the Ojibwa, and the goods shipped back to Montreal. He documented the comings and goings of his men in his account book, showing what goods were taken to a band, and what furs were brought back in exchange. For instance, a considerable amount of goods was sent to the "Ouisconsaint" on October 4, 1804, and February 22, 1805, in exchange for furs brought back on May 21,1805.

The fur trade represents only a brief moment in the history of the Manitowish Chain. The previously mentioned archaeological study of the Manitowish Chain surveyed over 29,000 feet of shoreline and found one site spanning 8,000 years of prehistory, the period from which no written records exist.

Read an accounting of the actual goods traded by Malhiot at the end of the chapter.

Twenty-six sites were newly discovered and seven previously reported sites were revisited during the field survey. Sixteen of these 33 sites are on Island Lake alone. From prehistory times, one Paleo Period (roughly 8,000 to 10,000 years before present) site was found on Spider Lake. Four sites show occupation during the Archaic Period (roughly 2,000 to 8,000 years before present). Twelve sites have been dated to the Woodland Period, running generally from 500 to 2,000 years ago. Another 16 sites can only be said to be "unknown prehistoric" sites.

From historic times, represented by written records from both Indian and European-American cultures (roughly 100 to 400 years ago), seven sites were found.

If you add these sites together, they total 39, not 33. Five sites showed occupation spanning several of the periods, and one spanned all three prehistoric periods.

Nearly all sites are located today on private land, and property owners have been made aware of their legacy. Sites include prehistoric mounds and burials, prehistoric campsites, historic Indian burials, historic sugarbushes, historic logging and boom camps, a historic canal, a historic pitch-making site, and a historic resort. Discretion and respect for private property owners prevent a detailed listing of site locations in this book.

The high number of sites in the Manitowish Waters area was likely due to the excellent fishing opportunities, the number of rivers that cross watersheds, and the rich opportunity for extensive trading and travel. But who really knows for sure? What matters most may not be the whys and whos of all these sites, but the respect, if not honor, we offer the area today.

We'd do well to consider Aldo Leopold's famous line in the "July" chapter of *Sand County Almanac*, "If I were to tell a preacher of the adjoining church that the road crew has been burning history books in his cemetery, under the guise of mowing weeds, he would be amazed and uncomprehending. How could a weed be a book?"

In the Manitowish Chain we've done the same as that road crew, either drowning our history books under waters from dams, defacing their pages by racing across them with jet-skis, or burying the books under the finery of modern homes and Kentucky bluegrass. Would we not do better to save at least a few of

Prehistoric Archaeological Sites on the Manitowish Chain of Lakes

The one Paleo site on the Chain represents one of the richest sites in northern Wisconsin, and probably one of the largest stone tool assemblages anywhere! This site actually spans the Archaic and Woodland time periods too, with artifacts showing occupation for nearly 8,000 years up until historic (written) times.

Northern Wisconsin looked very different 8,000 years ago. The glacier had just receded, and the land lay barren, with high winds carrying fine sands left behind in the drift. Small bands of Paleo Indians are thought to have followed the retreating glacier, hunting megafauna like mastodons, giant beaver, and giant bison.

The later Archaic Indians wandered far and wide too, but are believed to have been more regional than the Paleo people. Early in this period the climate was quite cold, but it warmed significantly in the middle Archaic period. By the late Archaic, copper-culture people were fabricating fish hooks, knives, spear points, and ornaments from cold-rolled and pounded copper.

Of the four Archaic sites on the Chain, three are located on Island Lake, and one on Spider.

The prehistoric Woodland Indians resembled the historic Indian tribes that were written about by early explorers and traders. But no prehistoric tribe has been related directly to a historic tribe in Wisconsin.

Woodland people were less nomadic, but still followed seasonal traditions like maple sugaring and ricing that required mobility to the best sites. They used cultivating tools for minor agriculture, were more developed artistically, developed fishing villages and trade networks, built burial and effigy mounds, and made distinctive cord-wrapped pottery.

Of the 12 Woodland sites on the Chain, six are located on Island Lake, two on Spider Lake, two on Fawn and Clear Lakes, one on Stone Lake, and one on Rest Lake.

Seven sites on Rest Lake, seven on Island Lake, one on Little Star Lake, and one on Stone Lake are of unknown prehistoric origin, most showing lithic debris from toolmaking that is currently impossible to date.

these sites so we, the descendants of all those who came before us, could sit on the very same spots and dream of what life here may have looked like 8,000 years ago?

The river is its own memory and needs neither label nor monument, unless it be the sun, beyond yawning.
— Justin Isherwood

≈ ≈ ≈ ≈ ≈

Natural lakes vary just like natural forests—one size doesn't fit all. I've never forgotten the time I asked one of the senior research scientists for the University of Wisconsin Trout Lake Limnology Station just what they were trying to learn with their "Long-Term Ecological Research" studies on these lakes. His response, after a short pause, was simply, "We're trying to figure out what's normal with lakes. We're still not sure."

The University of Wisconsin Trout Lake Station has been in the research business since the 1920s, arguably doing the longest and best research on fresh water in North America. Some would go so far as to say it's the best in the world. If they're not sure what's normal for a lake, that says a lot.

Two lakes right next to each other, born from the same glacier, the same soils, bathed by the same climatic conditions, managed by humans in recent history in the same manner, can be remarkably different. What's "normal" for one clearly is not "normal" for the other. The question is why. What makes lakes naturally different, and how do we learn to understand the variables and the interplay that create the differences?

The question is much bigger and more complex than this book on rivers can handle. Consider this partial list of variables and a general example of how each variable plays a role in the "normal" life of a lake:

- The position of the lake in the landscape. A lake positioned high in the landscape of a watershed receives far fewer nutrients from groundwater and runoff than a lake lower in the landscape.
- The quality of the watershed. A lake being fed runoff water that crosses open, fertilized agricultural lands or lawn will receive far more nutrients than a lake fed from runoff crossing intact woodlands.
- The depth of the lake basin. Shallow lakes heat up faster than deeper lakes, and since warm water grows plants and other organisms bigger and faster than cold water, colder lakes tend to have clearer water.
- The volume of the lake basin. Big volumes of water can dilute the same amount of nutrients better than small volumes.
- The shape of the basin. An irregular, convoluted shoreline provides more runoff area and groundwater input than a simple, regular shoreline, increasing potential nutrient flow into a lake.
- The lake's biological community. The more predator fish, the clearer the water. Follow the food chain in this example: High numbers of predator fish (walleye, muskie, northern pike, bass, and others) eat the prey fish (bluegills, various minnows), reducing their numbers. Fewer prey fish are then available to eat zooplankton (tiny

Long-Term Ecological Research—North Temperate Lakes

The National Science Foundation began the Long-Term Ecological Research (LTER) program in the late 1970s recognizing that most ecological research was being done along timelines of only a few years. This is considered far too short a time frame to generate real answers about the way ecological systems work. Time scales of decades or centuries are necessary to understand not only cyclic interactions, but long-term changes in ecosystems.

LTER sites were established in areas as diverse as Antarctica, the Arctic tundra, and Konza Prairie in Kansas. The University of Wisconsin Limnology Laboratory became one of the first LTER sites to be funded by the NSF. The project, titled "North Temperate Lakes LTER," sought to gain a predictive understanding of the ecology of northern lakes. In northern Wisconsin, scientists have studied seven primary lakes for physical, biological and chemical characteristics, sampling each lake every two weeks during open water periods, and three times in winter.

Research has been undertaken through two field stations—the Limnology Laboratory on Lake Mendota in Madison, Wisconsin, and the Trout Lake Station in the Northern Highlands State Forest, just south of Boulder Junction, Wisconsin. Interdisciplinary studies have included the efforts of social, physical, and biological scientists who together are attempting to develop a total picture of long-term processes in and on lake ecosystems.

Some of their major findings and initiatives to date of the North Temperate Lakes LTER include:

- Changes in lake-ice phenology are being used as an indicator of global climate change and variability. Studies have shown that the duration of ice cover has decreased in the last 150 years in lakes throughout the Northern Hemisphere.
- The hydrologic position of a lake in the landscape is a concept, similar to the stream continuum concept, which explains many systematic differences among adjacent lakes in northern Wisconsin. Many chemical, physical, and biological (including human development) features and processes vary systematically with the position of the lake in the landscape. Comparison with other lake districts in the Northern Hemisphere suggests this is a recurring pattern of many, but not all, lake districts.
- Water quality is something people in northern Wisconsin understand, care about and are willing to pay for. They understand this concept much better than they do abstractions like biodiversity.

Continued

Long-Term Ecological Research—North Temperate Lakes (Continued)

•Long-term data reveal time lags in effects of invaders on lake communities. In Sparkling Lake in Vilas County, cisco went extinct 16 years after a smelt invasion; in Trout Lake in Vilas County, the loss in aquatic plant diversity was delayed as long as 20 years after the rusty crayfish invasion. The loss of plant diversity appears reversible because plant regrowth occurred in crayfish exclosures.

•Biological recovery from lake acidification lags behind chemical recovery. After seven years of recovery from acidification, the zooplankton community structure in the treatment basin of Little Rock Lake in Vilas County is only now starting to show similarity to that in the reference basin, even though chemical recovery had occurred several years ago.

A book summarizing these studies and many others is currently being prepared with Oxford University Press, and is titled *Lakes in the Landscape: Long-Term Regional Ecology of North Temperate Lakes*, edited by Magnuson and Kratz.

These research efforts will help provide us with true perspectives of lake ecology over the long-term, and help us dispense with the knee-jerk responses we so typically, and inaccurately, make to short-term changes in our environment.

See the North Temperate Lakes Web site for in-depth information on this program: < www.limnosun.limnology.wisc.edu>.

animals like rotifers), increasing the zooplankton community. Zooplankton eat phytoplankton (algae and other microscopic plants), so the more zooplankton, the less phytoplankton. The fewer the microscopic plants, the clearer the water.

•The rock layer in the watershed. Groundwater moving through rock layers high in phosphorous will carry more nutrients, which stimulate algal blooms. Some lakes are naturally fertile due to the surrounding rock, and little will change that.

•The texture and porosity of the surrounding soils. Some soils, like clay, bind together and compact easily, thus washing away more readily in a good rain, and adding nutrients to a lake.

•Daily, seasonal, and yearly variations in weather. A drought year versus a flood year makes a big difference in plant life, which makes a big difference in animal life.

•A veritable host of human land-use issues. Take storm sewers for instance. The necessary choice is to filter out, or better yet, treat all the stuff (dirt, salt, leaves, etc.) that finds its way into a storm sewer.

•The type of vegetative cover in the watershed and around the lake. Leafy trees intercept raindrops, splattering them into smaller droplets, thus reducing the speed at which they hit the ground and the resultant speed of erosion.

•Introduction of exotic species. Consider the effects of the rusty crayfish, rainbow smelt, or the numberless zebra mussels.

The list goes on. Clearly, every lake has its own personality, one that is every bit as changeable, volatile, and often mysterious as a human personality. Individual factors occurring within a lake often combine or react symbiotically, with surprising results. The lake's plant and animal life will vary over the years, but in what predictable manner and method—that bothersome "what is normal" question—is nearly impossible to predict.

Trying to define what is "good" and what is "bad" also creates havoc, in this case political havoc rather than ecological havoc. Good water to a trout angler is bad to a waterfowl hunter. Good water to a swimmer is bad to the muskie angler. Good to the water-skier is bad to the birder. Good to the farmer (cropland down to the water's edge) is bad to the scuba diver. Conflicts and more conflicts. Who gets the most standing in how a lake is "managed"? That's often every bit as variable as the lake ecology itself.

Read more about land use at the end of the chapter.

We're getting into the dangerous ground of politics. Darkness, in more ways than one, may descend if we don't keep moving.

≈ ≈ ≈ ≈ ≈

Spider Lake connects into 139-acre Stone Lake, maximum depth 43 feet. An eagle's nest is located on an island near houses in Stone Lake.

An eagle's nest near private homes is not all that unusual these days. Bald eagles have acclimated remarkably well to human disturbance. Still, wildlife biologists and lakeshore owners worry that development could alter the presence of eagles in northern Wisconsin.

To that end, in 1997, the WDNR undertook a study to examine the impact of shoreline development on breeding bald eagles. The researchers reasonably hypothesized that bald eagle reproduction would be reduced near lakes with a high rate of shoreline development when compared to eagles nesting near lakes with little development.

They mapped the approximately 170 occupied bald eagle territories in Vilas and Oneida counties, which comprises 24 percent of all the eagle nests in Wisconsin. The researchers then randomly selected 90 of those nest sites, all of which had been active in at least four years between 1980 and 1996. They measured three parameters: the number of young produced per year when the nest was active (productivity rate); the number of years that a particular nest was active in succession during 1980 to 1996 (nest intensity value); and the number of years that a specific nest tree within a given territory was active divided by the number of years that the entire territory was occupied (nest tree fidelity).

They then analyzed the data to determine several things. First, they looked to see if there were differences

Bald eagles and nest

in the productivity rate relative to the percent of surface water and the percent of development (the number of houses, human population, house and human density per square kilometer). They found that none of these landscape characteristics explained productivity variations in the 90 nests.

Second, they grouped the nests into categories of risk for human disturbance or development, based on property ownership. From this, they could determine the number of nests that were at risk from human disturbance. They found 36 percent of the eagle territories to be situated in areas at a high risk for development, and another 20 percent at a moderate risk. Despite these differences in risk, they found no differences relative to development that impacted productivity, nest intensity, or nest tree fidelity.

The upshot? Bald eagle breeding success appears unaffected by the current patterns of lakeshore development in northern Wisconsin, a very heartening result.

Before you take that to the bank, however, two factors may be affecting the results of this research. One, the density of human settlement on the lakes in 1996 was relatively low compared to the high density we will undoubtedly see in the next 15 years. And two, while bald eagles appear to be habituating to increased human disturbance, there is evidence that individual eagle pairs abandon nest trees and entire territories when a certain threshold of disturbance is exceeded. Continued monitoring is essential to ensure that increasing development doesn't stress breeding eagle pairs to the point that they abandon nest territories for quieter waters elsewhere.

≈ ≈ ≈ ≈ ≈

Stone Lake empties to the west into the Manitowish River channel.

Paddle due north and Stone Lake narrows into 74-acre Fawn Lake, a shallow pan of water with a mean depth of 7 feet. A narrow channel leads from Fawn Lake into 555-acre Clear Lake and the end of the chain to the north. Clear Lake's 45-foot maximum depth and 16-foot mean depth help keep its water cooler than other lakes in the chain.

Still, warmwater fish like muskie, northern pike, and walleye can be found in good numbers here, but a mercury warning influences many anglers to practice catch and release.

Some 341 lakes in Wisconsin are subject to consumption advisories due to mercury levels. From 1980 to 1990, roughly one-in-three lakes tested in northern Wisconsin ended up on the mercury advisory list. As a result, numerous research studies were undertaken to identify the source of mercury and its deposition rates. Researchers learned that mercury was being deposited at a rate three to five times higher than historically.

Latex paint was fingered as the major source of mercury in the 1980s, leading to state and federal laws banning mercury in paints, although vendors were allowed to use up their remaining stocks. In 1993, mercury fungicides were banned from use on golf courses in the state. Mercury was also banned from frivolous uses in such items as sneakers with lights in their heels that were activated by a mercury switch, and from maze games in which mercury was used as the ball. The state also eliminated mercury from batteries, ultimately leaving the combustion of coal as the major source of mercury in our atmosphere.

The late 1990s brought good news. Since 1995, data show that mercury deposits, as measured by rain samples, have declined 30 to 50 percent in northern Wisconsin. This decline coincided with the 1995 closure of the White Pine copper smelter in White Pine, Michigan. This smelter was the major source of a large number of air pollutants, including mercury, in the Upper Midwest.

The mercury decline has taken place so recently that researchers are unable to say what the effect has been on fish contamination. But certainly the trend is promising. Read more about mercury contamination at the end of the chapter.

Mercury contamination has also been a serious issue of concern relative to breeding loons. Situated at the top of the food chain, loons receive high doses of bio-accumulated mercury. Of all species on the Manitowish Chain, loons are surely the most ardently appreciated and vigorously protected species. Their presence allows us to hear their extraordinary calling, an experience usually described as something akin to holy.

Shoreline development has been of great concern to loon researchers. Loon numbers are stable, if not rising. But what does the future bode, given our development and our ever-increasing recreational impacts? In a 1997 study, the WDNR attempted to assess the relationship between lakeshore development and common loon reproduction. However, they found loon reproductive success to be more complex than a simple causative relationship with development. Concurrent studies found two additional factors that strongly influenced loon reproductive success. One factor was water chemistry. Researchers found that loon productivity was lower on acidic lakes (low pH), and that chicks raised on low pH lakes received less food from adults than on higher pH lakes. Given that acidic lakes typically have an altered fish composition, fewer fish, and smaller fish, it appears an acidic lake may not have enough available food to sustain a loon family.

Mercury levels were also higher in chicks and adults on acidic lakes. In turn, chick production was lower on lakes where the chicks had elevated mercury levels.

The other factor that must be put into play is where nests were placed. Loons nesting on islands or artificial nesting platforms had a higher rate of reproductive success than did loons nesting on shorelands.

Hatching rates were nearly twice as great on artificial platform nests than on shoreline nests, and were substantially higher on platforms than on island nests. Fledging rates were twice as high on platform nests compared to shoreline nests, and significantly higher on platforms than on island nests.

So, where does shoreline development fit in? Previous studies in 1973 in Alberta, 1983 in central Ontario, and 1986 in Michigan found that shoreline housing development was associated with a decline in common loon reproduction success.

But the WDNR realized its study as structured could not sort out shoreline development from the many other variables. The researchers then revamped their study to include monitoring the reproductive performance of at least 75 loon pairs on lakes in Vilas, Iron, Oneida, and Forest counties for at least three years. They measured lake water chemistry, lake size, shoreline development, and location of nest sites on these study lakes through the summer of 1999. They are currently using multivariate analysis to tease out and evaluate the effect of each of these factors on loon productivity. Results will be announced sometime in 2001.

≈ ≈ ≈ ≈ ≈

For an alternate route that leads into the Trout River, paddle due south out of Spider Lake, under the bridge on Highway 51, until 496-acre Manitowish Lake beckons with its 61-foot maximum depth and 23-foot mean depth. An active eagle's nest is located southwest of Manitowish Lake in section 27, nearly a half mile off-water.

Little Star Lake, an exceptionally clear and beautiful 245-acre lake, lies to the west and has no exit. A combined public beach and picnic ground is on its far eastern shore.

Continue paddling southeast to enter Trout River. Alder Lake at 274 acres and a 33-foot maximum depth, and Wild Rice Lake at 379 acres and a 26-foot maximum depth are next up the line. An eagle's nest is located off-water on the eastern shore of Alder Lake.

Where the Trout River enters Wild Rice Lake, an eagle's nest has been active off and on in a big pine since the early 1960s. The nest declined and was abandoned in the 1960s as the effects of DDT hit, but eagles returned in the mid-1980s. They chose the very same tree that had been used before, demonstrating that some trees have just the right size, configuration, and location to be THE site for an eagle's nest. WDNR eagle researchers who have flown over and counted eagle nests for several decades, say that they can spot likely nest trees from miles away. It seems that some trees are just made for eagle nests!

The nest on Wild Rice Lake was blown out in a big storm in July of 1999. The long-standing nest tree simply snapped in two. But by that autumn, the eagles had chosen another nearby tree, and were hard at work re-establishing their nest.

Alder and Wild Rice Lakes border the east side of the 12,000-acre Powell Marsh Wildlife Area. A portion of this end of the Powell has been converted into commercial cranberry marshes.

Wild cranberries, of course, provided the inspiration and genetics for the development of the commercial strains of cranberries. Henry Schoolcraft explored much of the Upper Midwest in the early 1800s and wrote extensively of his experiences. He had this to say about cranberries while portaging in northern Minnesota in July, 1820:

"We found ourselves at the edge of a horrible swamp, covered with water or mud, in which we sunk to our knees at almost every step. The traveling was more difficult on account of the trees that had been blown down in great numbers by a violent wind. Over these we were obliged to climb, sometimes to a great height and not infrequently at the risk of our necks.

"We succeeded after a painful struggle, in getting through this swamp about the middle of the afternoon, but it was succeeded by another much worse.

"This was a kind called Tamarack Swamp, from the timber that grows in it, tho' we found very few trees of this kind here. These swamps are covered with water as the others, on which lays a thick moss, so tender that it will not bear the weight of a man. Consequently at every step we took we were entangled in the moss, and often prostrated headlong in the water . . .

"In crossing the swamp we found the cranberry (*oxycoccus macrocarpus*) in great abundance . . . The agreeable taste of this berry was a grateful treat, at a time when we were much fatigued by traveling for many miles over an elastic bog where no drink-water could be procured."

Whether in northern Minnesota or along the Trout River, wetlands were an absolute curse for portaging in those days, as they are now if you care to try your hand at historical re-creation. Put 100 pounds or more on your back—a voyageur often carried two packs over long portages, each weighing nearly 90 pounds—and see how you do. Remember to wear a hat so searchers can find the spot where you went through the bog mat.

Wild cranberries grow in bogs, not marshes or swamps. The flowers emerge in June and look very much like the upland flower called shooting star. Deliciously tart fruits form in August and are usually ready for picking in late September.

Cranberries are one of only a handful of fruits native to North America that are now major commercial crops. The blueberry and Concord grape are the other two grown in our region. Wisconsin produces more cranberries than any other state in the nation, about 240 million pounds annually, in 20 different counties. Economically, cranberries contribute more than 334 million dollars to

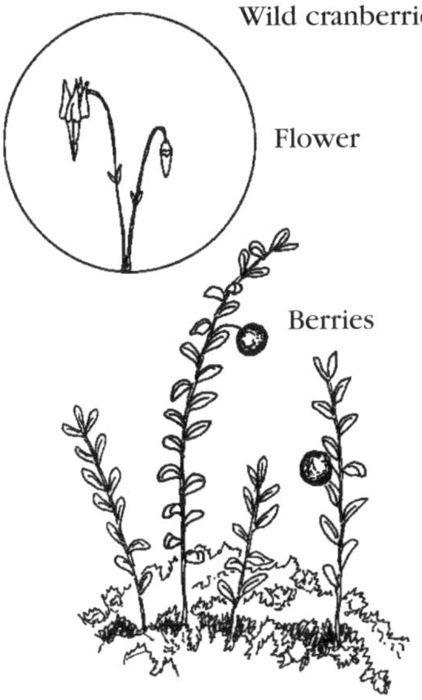

Wisconsin's economy, and provide thousands of jobs.

But as with most commercial agricultural endeavors, there's a downside. Commercial cranberry growing dramatically alters wetlands. In the late 1990s, Wisconsin's commercial cranberry bogs occupied more than 150,000 acres, of which only 15,500 acres actually grew cranberries. The balance of the land consists of man-made wetlands, woodlands, dikes, ditches, and roads. Ditching and diking the area allows the companies to control water flow to the beds based on the seasonal or daily needs of the cranberries.

Cranberry growers convert wetlands to cranberry beds by scalping off the top 2 feet of soil and placing a layer of sand on the bed. The sand serves as the root zone for the cranberries.

The cranberry beds themselves are a monoculture, except for undesired weeds, which are usually controlled with herbicides. Monocultures, as a general rule, defy the maxim that ecological health is measured through diversity—the more diverse a landscape, the healthier it is.

The question arises: What is the actual effect of cranberry culture on wildlife? A 1991 study on the impact of cranberry growing on Wisconsin's avian life found that cranberry beds by themselves failed to serve as a nesting site for any species, although they did provide some food for some species.

However, there is a difference between the monoculture of cornfields and that of cranberry beds, the difference lying in the overall footprint each type of field imposes on the landscape. Cornfields often occupy vast acreages on a farm. The fields are planted to the maximum extent possible, and typically exclude hedgerows and woodlots, which once provided floristic diversity on the farm.

Cranberry farms usually take a very different approach. The 1991 study discovered that only 7 percent of the total acreage owned by cranberry owners was in cranberry beds. Another 3 percent was in ditches, dikes and roads; 21 percent in reservoirs; 29 percent in wet marshlands; 22 percent in shrub and wooded wetlands; and 17 percent in wooded uplands.

The dikes and ditches proved to be more valuable to wildlife than the cranberry beds. Their usefulness as nesting and cover is nevertheless limited because they have steep sides, are mowed, and are often used as roads. However, the study showed that reservoirs constructed for the cranberry beds by impounding streams and wetlands provided habitat for large number of waterfowl. So one must look at the wildlife value of all the operations on all the areas owned by the companies.

The 1991 study sampled three commercial bogs in three different counties using transect surveys for bird sightings and calls, and also mist-netting at all the sites. The researchers found an average of 36 species at each site, and a total of 63 different species.

The data showed that the fewest number of birds species used the cranberry beds—only 11 species. The adjacent areas of wetland and the disturbed areas had the greatest number of species, likely because of edge effects. Researchers found 23 species in disturbed areas, while observing 26 species in adjacent wetlands.

The study actually found a greater total number of bird species in cranberry wetland systems than on sedge meadows of comparable size. This isn't surprising, however, given the amount of edge and shallow open-water habitat found in commercial cranberry systems.

Before we strike up the band in celebration of cranberry farms, three important concerns must be addressed. One, the use of pesticides on the cranberry beds can and does kill birds. Two, this study took place in 1991. Since then, expansion of commercial cranberry bogs has taken place, the result being an increase in monoculture and a decrease in habitat diversity. Three, attaining the maximum number of species is not what wildlife management is supposed to be all about. If that were the case, we would always work to create the greatest amount of edge habitat, since

Locations of Birds Observed in Three Wisconsin Cranberry Operations in Monroe, Jackson and Burnett Counties in August 1989 (Source: Hollands et al. 1990)

	Habitat species observed in:					
Species	**Cranberry Beds**	**Reservoir**	**Ditches/ Dikes**	**Disturbed Areas**	**Adjacent Wetland**	**Adjacent Upland**
Common Loon		X				
Double-crested Cormorant		X				
American Bittern			X			
Great Blue Heron		X	X			
Green-backed Heron			X			
Wood Duck		X				
Green-winged Teal			X			
Mallard		X				X
Bald Eagle		X				X
Accipiter Species	X			X		
Red-tailed Hawk	X		X	X	X	X
Merlin	X					
Sora			X			
Sandhill Crane			X			
Killdeer			X			
Lesser Yellowlegs			X			
Solitary Sandpiper		X	X			
Spotted Sandpiper			X			

Continued

Locations of birds observed in three Wisconsin cranberry operations in Monroe, Jackson and Burnett Counties in August 1989 (Continued)

Species	Cranberry Beds	Reservoir	Ditches/ Dikes	Disturbed Areas	Adjacent Wetland	Adjacent Upland
	Habitat species observed in:					
Common Snipe			X			
Great Horned Owl					X	
Barred Owl					X	
Ruby-throated Hummingbird					X	
Belted Kingfisner		X	X		X	
Downy Woodpecker					X	
Northern Flicker					X	X
Pileated Woodpecker						X
Eastern Wood Pewee						X
Least Flycatcher				X	X	
Eastern Phoebe				X	X	X
Eastern Kingbird		X		X		
Tree Swallow	X	X				
Barn Swallow	X	X				
Blue Jay					X	X
American Crow	X		X			X
Common Raven		X				
Black-capped Chickadee				X		X
White-breasted Nuthatch						X
Grey-cheeked Thrush				X		X
American Robin		X				X
Gray Catbird				X	X	
Cedar Waxwing	X	X	X	X	X	X
Red-eyed Vireo				X		X
Golden-winged Warbler					X	
Tennessee Warbler					X	
Nashville Warbler					X	
Yellow Warbler					X	
Chestnut-sided Warbler				X		
Pine Warbler				X	X	
Black-and-white Warbler				X	X	
American Redstart				X	X	
Ovenbird				X		X
Northern Waterthrush			X		X	
Wilson's Warbler				X		
Common Yellowthroat				X	X	
Scarlet Tanager				X		
Rose-breasted Grosbeak				X	X	
Indigo Bunting				X	X	
Rufous-sided Towhee				X		
Field Sparrow	X		X	X		
Song Sparrow						
Red-winged Blackbird	X				X	
American Goldfinch	X		X		X	
Total	11	15	18	23	26	16

the greatest number of species lives there. Wildlife management has more to do with preserving natural systems and biodiversity than with inducing big species numbers. Intact large sedge meadows are the habitat most often transformed into cranberry farms. The meadows now comprise only 30,000 acres in Wisconsin, down from an estimated pre-settlement presence of 1,135,000 acres. Only 2.6 percent of Wisconsin's original sedge meadow remains, a habitat rich in its own right.

The Army Corp of Engineers has historically viewed reservoir construction as a process that destroys one type of habitat suitable for some species, while creating habitat useful for others. Certainly that's one way to look at it. The question remains, is the trade-off worth it?

≈ ≈ ≈ ≈ ≈

Back on Spider and Stone lakes, and on into Rest Lake, the issue of good shoreline conservation and stewardship raises its complex head. The variety of ways people attempt to stem the erosion of their shoreline, the amount and type of aquatic vegetation left alone in shallow water, the size of docks, the type and amount of vegetation on shore, the placement of boathouses and homes, the amount of lawn along the water, the type and amount of shrubs on the shorelands, the amount of woody debris left in the water, and the amount and placement of pavement leading to the water all personify the ecological understandings and values of each shoreline owner.

The Chain provides a clear window through which to view the development that northern lakes have experienced in recent decades. Of northern Wisconsin's 12,000 miles of lakeshore frontage, over 80 percent is in private ownership. In the past 30 years (approximately 1970 to 2000), the development of lake shorelands has equaled or surpassed that of the previous 100 years, a trend that will continue with an estimated doubling of development in the next 15 years (2000 to 2015). The lakeshore frontage along the Manitowish Chain is roughly over 90 percent privately owned, and thus will likely be developed to that full extent some day, if present practices are any indication.

The Wisconsin's Shoreland Management Program (WHSP) establishes certain standards for cutting shoreland vegetation (Chapter NR 115) "in order to protect natural beauty, control erosion, and reduce the flow of effluents, sediments, and nutrients from the shoreland area; and to protect fish and aquatic life."

The WDNR undertook a series of studies in 1997 (some of which have been discussed earlier) to measure the extent to which these regulations were working. One study examined the vegetative structure of the buffer zone on 12 undeveloped lakes and 16 developed lakes in Vilas and Oneida counties. As one would expect, researchers found much greater tree canopy cover, subcanopy cover, and shrub cover on undeveloped lakes.

The most significant impact of development was the near-elimination of shrub cover and dryland coarse woody debris on developed shoreland sites. The WDNR researchers also found the proportion of the shallow water area covered by floating, emer-

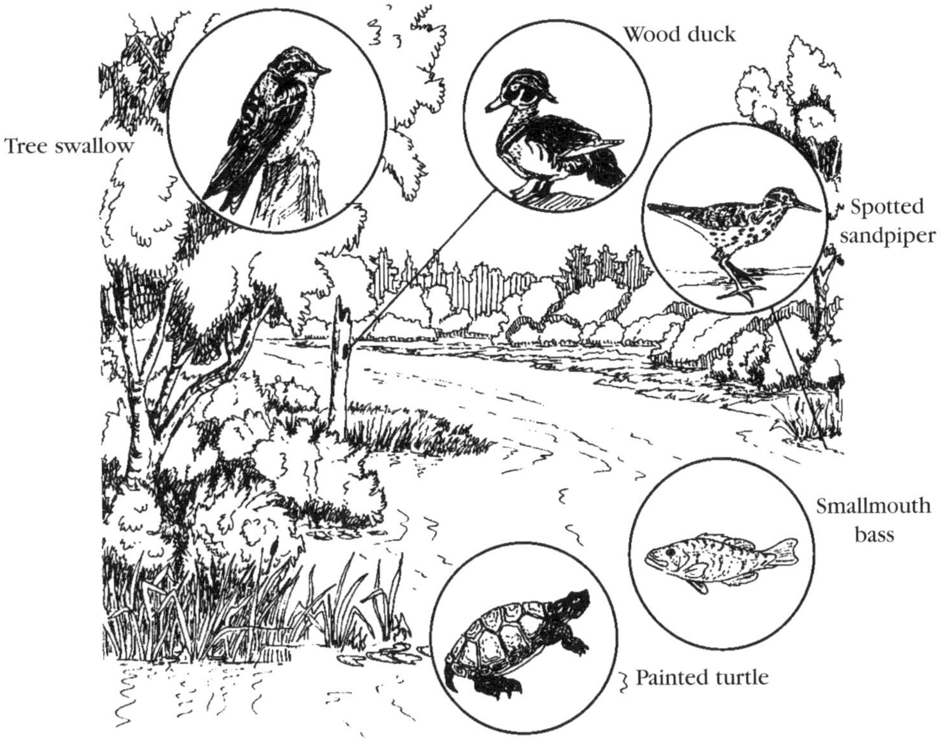

Wildlife associated with buffer zones

gent, or submerged vegetation was dramatically reduced on developed lakes, as was the coarse woody debris in the water.

The connection that must be made from this study is that the loss of vegetative structure to the extent seen on these lakes results in a serious reduction in wildlife habitat quantity and quality, including habitat available for spawning fish, juvenile fish, and fry.

How could this be if we have a shoreland management program that was designed to protect lake water quality and the ecological integrity of lakes? Let's look at the regulations. The WSMP says that in the buffer strip of land extending inland for 35 feet from the Ordinary High Water Mark (OHWM), clearcutting of trees and shrubbery must not occur for more than 30 feet in any 100 feet of width. Basically, about two-thirds of a 35-foot-deep buffer zone must not be clearcut, This is a good policy, because this buffer zone traps sediment from overland runoff, stabilizes streambanks, protects natural beauty, and prevents excessive sediments from covering up spawning gravels and degrading fish habitat. The buffer zone also provides essential habitat for riparian species like amphibians and reptiles, which use both land and water during their life cycles.

Within the shallow water zone, leaf fall and toppled trees and limbs (coarse woody debris) provide organic matter utilized by aquatic invertebrates, and also provide long-term cover for fish. On land, decaying trees and limbs provide food and microclimates for numerous insects, reptiles, and amphibians, as well as shoreline-dependent species like bald eagles, loons, waterfowl, kingfishers and mink.

The study sought to quantify just what effect the buffer zone policy had in practice. The researchers found the following in the 35-foot terrestrial buffer zone:

Average tree canopy cover	
undeveloped lakes	57%
developed lakes	38%
Subcanopy cover	
undeveloped lakes	18%
developed lakes	8%
Shrub layer cover	
undeveloped lakes	34%
developed lakes	7%

Researchers found no difference in the percentage of ground cover, but observed that the species composition changed. They also found much less coarse woody debris in the buffer zone on developed lakes. Basically, many of the developed buffer zones were little more than a canopy of big pines over a ground cover of bluegrass.

The vegetative cover right at the shoreline reinforced the story:

% of shoreline with tree cover	
undeveloped lakes	35%
developed lakes	22%
% of shoreline with shrub cover	
undeveloped lakes	64%
developed lakes	16%

And as one would expect, they also found much less coarse woody debris along the shorelines.

In the shallow water zone stretching out from shore, the story again repeated its themes:

Area covered by floating vegetation	
undeveloped lakes	13%
developed lakes	<1%
Emergent vegetation	
undeveloped lakes	18%
developed lakes	3%
Submerged vegetation	
undeveloped lakes	5%
developed lakes	<1%

The proportion of unvegetated lake bottom in the shallow waters of undeveloped lakes was 65 percent. On developed lakes, 95 percent of the lake bottom was unvegetated—a veritable desert. And again, significantly less coarse woody debris was found on the bottom of developed lakes.

The researchers fed the data and current regulations into a computer model, which took these habitat features and predicted the effects on a fully developed lake (none of the study lakes were fully developed). The model predicted that tree canopy coverage will be reduced 93 percent, shoreline shrub layer will be reduced 100 percent, and coarse woody debris will be completely absent from all zones.

Thus the current Wisconsin Shoreland Management Program fails to protect the key habitat features that are vital to wildlife and fish populations. It also fails to protect water quality and ecosystem health.

A very important point must be made, however. The study results don't necessarily indicate that the management program is worthless. Rather, these results are strongly influenced by the effects of limited enforcement

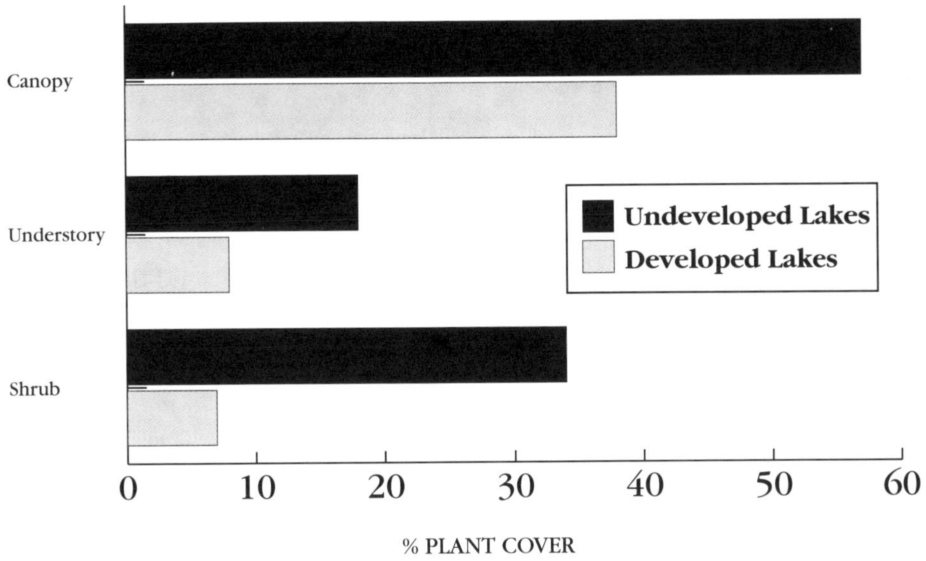

of the current regulations and the ambiguity of some of the regulations.

What to do? The study recommends the following:

•Let the vegetation grow back. They found that where vegetation was not mowed, a surprising number of native species had returned. This suggests that a native seed bank remains if we simply let the land and shallow water recover.

•Leave the shrub layer undisturbed. It's well-known that more wildlife species exist in structurally diverse habitats than in simple habitats. Yet the first thing many homeowners do is to simplify the vegetation around their homes by removing shrubs and planting grass.

•Staff and fund the county zoning offices adequately, so they can enforce existing laws.

•Review the language of the laws to remove the ambiguity. What exactly does "clearcut" mean, for example? Does leaving one tree mean that the area is not clearcut?

•Support new lake classification systems.

•Investigate options for increasing the width and amount of protected buffer zone.

•Find a way to get existing information on landscaping for wildlife into the hands of shoreline owners, and help them understand it.

These recommendations offer a very positive approach to change that doesn't require enormous expenditures of money or a doctorate in lake ecology. Protecting shorelines along lakes and rivers can be done, and therein lies the hope for our northern waters.

≈ ≈ ≈ ≈ ≈

We normally provide for our suburban streets and our interstate highways what we so seldom accord our waterways: setbacks for buildings to keep the roadsides green and attractive, public walkways along them, broad highway rights-of-way that preserve not only scenic beauty but considerable wildlife habitat as well.

— John Kaufmann

Wild Lakes Acquisition Criteria

Lakes with no artificial structures or other forms of cultural disturbance on their shores are termed "wild lakes." In 1995, in order for future generations to have the opportunity to experience a wild lake ecosystem, the state established a wild shoreland acquisition committee to select the lakes of highest ecological value and to recommend their purchase. Unfortunately, limited funds and the skyrocketing cost of lakeshore property confined the WDNR's current purchasing power to approximately five new lakes through the year 2000, provided there were willing sellers.

The Wisconsin Stewardship Program (reauthorized in 2000) includes the purchase of wild lakes and shorelands. Other funding being pursued includes the Land and Water Conservation funds (LAWCON) held by the federal government. Private nonprofits like the Nature Conservancy and land trusts will hopefully also play a major role.

Lakes are rated for acquisition according to the following criteria:

	3 points	**2 points**	**1 point**
Size	50+ acres	10-49 acres	<10 acres
Current development	none	3 or fewer dwellings/mile	stretches of 1/4 mile with no dev.
Development potential	high	medium	low
Endangered resource	high	medium	low
Fish	good	average	poor
Game	good	average	poor
Other (wild rice beds, geological, historical sig.)	good	averge	poor
Cluster lakes	many in good quality in group	3-4 lakes average quality	2 average quality

A maximum total of 24 points is possible. A rating released in late 1999 listed only three lakes in Vilas County at 20 points or more. All of the other lakes in Oneida and Vilas Counties were rated at 17 points or fewer, indicating that either the rating system was impossibly high, or that we have very few lake systems left in pristine condition that can be acquired.

If shoreland owners want to preserve the ecological integrity of the shorefront they worked so hard to purchase, they must be aware of how their actions affect the water and land. Educated choices can then be made; hopefully, these choices will be made in the best interest of the lakes' flora and fauna. More than all of our laws put together, the voluntary efforts of private lakeshore owners will determine what all of us experience as we travel these lakes.

The heart of the matter is how each person perceives "their" lake. People differ in how they value water. It's their perception of what a lake is that drives their values equation. Are lakes perceived simply as beautiful scenery? As a surface for running boats and skis? As a site to harvest maximum numbers of fish? As a haven to watch wildlife? As a place to swim and cool off? As a combination of these things or others?

The most damaging perception may be when lakes are viewed like a ballfield. Ballfields are open expanses of cleared land created expressly for human recreation. No user cares, or should care, what species of plants grow on or around a ballfield, or which animals might utilize the field for food, nesting or denning, or shelter. No significant plant or animal community is disturbed by the play that occurs here, since most native plants and animals were driven out long ago from these sites. Ballfields are for human play, period.

To many people, particularly visitors, lakes have the appearance of aquatic ballfields. Lakes are open expanses of water with a native plant and animal community that is often difficult to see—or if seen, is difficult to understand, and therefore appreciate. The hidden loon or grebe nests along a shoreline swamped by a boat wake don't typically enter into the consciousness of those who do the swamping. Or if they do flicker in their consciousness, their value rates far below the thrill that speed produces. It's a playground they say, Wally's World, Wisconsin Dells, urban pleasures in rural areas.

Don't get me wrong. Lake recreation can be lots of fun. I thoroughly enjoyed water skiing as a youth, though, importantly, I don't recall ever giving a thought to the effect of all those waves we were generating. Doubtless, I would have enjoyed jet-skiing in my youth as well.

Enjoyment isn't the issue. The problem arises when one person's fun directly or indirectly destroys a plant or animal that has no perceived value and no legal standing. More often than not, the plant or animal is not even understood to exist, much less regarded as valuable.

In a society that often loudly proclaims the inviolability of individual rights, slowing down is seen by some as an assault upon their Constitutional rights.

The juiced-up speedboats have a hard time even idling at a speed that produces "no wake," the signed admonition that greets all boaters as they enter the Manitowish River channel between the lakes. What good is 150 horsepower if you can't let the horses run? What value is sleek aerodynamic design if you can't go fast enough to create a wind-in-your-teeth?

A lake is a commons, as is a river, meaning both are owned in common by the public. Commonly owned land and water have been the source of dispute forever. It was the subject of a story called "The Tragedy of the Commons" written in 1833 by a mathematician named William Lloyd. Lloyd described a pasture open to all, where every herdsman was free to keep and graze his animals. As long as the population of herdsmen and their animals remained low, conflict was minimized. There was room enough for all. However, each herdsman eventually sought to maximize his gain, recognizing that the positive side of keeping additional animals on the pasture meant a gain of +1 to him for every additional animal he added. The negative side, that of overgrazing, was shared by all the herdsmen. As such, it was perceived as a much smaller loss than the gain offered by grazing additional animals.

Each herdsman sharing the commons, of course, reached the same conclusion. Each man wanted to increase his herd without any limits placed upon him in a world, a commons, that was limited. Because each herdsman pursued his own best interest, the commons was ruined; the moral of the story being that total freedom in a commons brings ruin upon all.

Reaching critical mass is another way to look at it. The straw that breaks the camels back. When enough is enough, and anymore is too much.

The story is about the natural limits inherent to the land and water, and the legislated or "matters of conscience" limits that we place upon our human behaviors. The issue is clear, however unclear the answers are. Limited resources require limited usage. A state forest, a national park, a lake, a river, have boundaries that limit how many people can enjoy them, unless we are willing to allow the values that we seek in those places to be eroded and eventually destroyed. It's fine, and desirable, for a place to be all things to all people, as long as the people don't exceed the carrying capacity of the site.

The answer is found somewhere in the knowledge that we must limit the use of every commons or they will be of no value to anyone.

There are many partial solutions, all of which are objectionable. We can sell the commons off as private property, and wash our hands of the mess. We could limit access to the commons on the basis of wealth, or through an auction or lottery, or by some agreed-upon standard of merit. We could utilize a first-come, first-served system based on who gets in line first. Or we can allocate time usage—you can use it from 10-3, I'll use it from 3-8, and the other folks can have it from dark till 10 a.m.

No regulation is most desirable, but regulate we must or the tragedy of the commons will inevitably occur.

Pollution works in the opposite way. Rather than taking too much out of the commons, it's a matter of putting too much in that the rest find undesirable. Pollution laws were finally implemented when it was crystal clear that polluters would continue to rationalize their own gain as more important than the community's loss of clean water, air, soil, health, and beauty.

We hear today the clamor of property rights people demanding their rights to do as they please with their property, wrapping themselves up in what they believe is their Constitutional right to place their individual gain/pleasure/etc. above the needs of the community, beyond the integrity of the commons. That argument only works when there's no public around to be harmed by their action(s). It's a matter of tempering our actions to fit the system we're in and the time during which we're performing the action. We do this all the time, recognizing when we're in a frontier and can do much as we please, or when we're surrounded by others and must act socially responsible.

We must treat the commons, and thus the community, ethically. We can educate and defer to the conscience of every man and woman to do what is right, or we can legislate appropriate behavior within the commons, something we do every day with traffic laws to criminal laws.

Writer Garrett Hardin posed the essential question in an essay in 1968: "How do we legislate temperance?"

He answered by saying that we must enact "mutual coercion mutually agreed upon." The statement sure doesn't have a pleasant ring to it. But this is what those shelves of lawbooks in every lawyer's office are all about. We provide legal incentives or penalties as a means of coercing people into acting in a manner that is regarded as responsible by the majority of society.

The best example may be taxes. We all dislike taxes, but we all agree they are necessary for the social good. We make taxes compulsory because 98 percent of society wouldn't pay taxes if given the choice, but they still would want the benefits derived from taxes. Hardin says we support taxes and other coercive devices to "escape the horror of the commons."

With people stacked on top of each other in lots 100 feet apart along many river edges and lakeshores, the tragedy of the commons is at-hand. I believe that for every person who cries that his or her freedom and rights are being infringed upon by zoning laws and recreational restrictions, there are many who are screaming that their rights to peace and quiet, observations of wildlife, or fishing are being curtailed. For every ATV running full bore down a trail, nearby, a far larger number of walkers are swearing and making plans to go somewhere else the next time.

We're seeking mutual coexistence here, but our numbers make the job difficult.

Hardin writes, "What does freedom mean? When men mutually agreed to pass laws against robbing, mankind became more free, not less so. Individuals locked into the logic of the commons are free only to bring on universal ruin; once they see the necessity of mutual coercion, they become free to pursue other goals. I believe it was Hegel who said, 'Freedom is the recognition of necessity.'"

≈ ≈ ≈ ≈ ≈

Given all the varying perspectives on lakeshore development and water quality protection, is there one solution that will keep everyone happy? If you have one, I suggest you run for some high office immediately. Solutions tend to be like holding water in your hands.

We employ now the multiple-use concept, which permits virtually all uses, with some limitations. This con-

Lake Associations/Districts/Self-Help—Making A Difference

Lakes are under stress from erosion, invasion of exotic species, loss of native plants and animals, contamination, nutrient enrichment, overcrowding and use conflicts, acid rain, mercury, and more. How does one go about organizing property owners, area residents, and other lake users to protect and improve lake quality and the quality of life along the lake? Two of the more common ways are forming a lake association or a lake district.

A lake association is a voluntary "friends" group with formal membership comprised of those who own lakeshore frontage, use the lake, or want to better these public resources. The goals of a lake association may include improving water quality, improving the fishery, resolving recreational conflicts, or simply understanding more about the lake's ecology. Associations often raise money through dues or fund-raisers to finance lake improvements. Voluntary membership means that not all lakeshore owners may be members of the association, but the more inclusive the membership is, the more effective the association can be.

A lake district is a formal special-purpose unit of government. Property owners within the district pay fees, usually as part of their property tax bill, which finance the district's activities. Lake district property owners elect commissioners to approve budgets and contracts and elect district officers. A city or village must consent to have its lakefront properties included in a lake district. Lake districts work to regulate difficult issues like boating laws on the waters, recreational use restrictions and conditions, and others.

Organizations create power through numbers, but nothing can substitute for an individual's thorough knowledge of how a lake "works." Understanding the whys and whens and wheres and hows of your lake makes all the difference in resolving issues the lake, and the lakeshore property owners, face. In Wisconsin, three groups—the Wisconsin DNR, the University of Wisconsin Extension, and local lake people—formed the "Lake Partnerships" model to help ensure healthy aquatic systems. This nationally acclaimed model draws on the expertise of each group. The DNR provides the technical expertise and regulatory authority; the University of Wisconsin Extension provides educational materials and programs, and helps to bring a diverse community of lakeshore owners together; and local lake people provide the elbow grease to get the job done. Today, over 650 lake organizations and thousands of volunteers help to monitor the well-being of Wisconsin lakes through this partnership.

Continued

Lake Associations/Districts/Self-Help—Making A Difference (Continued)

The Wisconsin Association of Lakes (WAL) provides the umbrella under which lake organizations and governmental organizations gather. WAL's a major player in statewide and local politics, advocating for the conservation of our waterways, and providing or supporting a host of educational programs and opportunities. Contact WAL at:

Wisconsin Association of Lakes
One Point Place, Suite 101
Madison, WI 53719
(800) 542-5253
wilakes@execpc.com
<www.nalms.org/wal>

cept assumes that the land and water can adapt to all uses, and expects all users to tolerate one another.

Multiple-use fails on two major accounts, however. The first, and most important, is that regulations based on multiple-use are made without giving the natural lake or river community the highest standing. Human priorities supercede the needs of wildlife or plant life. The kingfisher's and cedar waxwing's need for lakeshore perch trees isn't figured into the equation. Neither is the green frog's need for unfragmented, undeveloped lakeshore, nor the need of the dragonfly or loon for no wakes.

The second failure results from the preconception that all uses and users are compatible. They're not, as much as we would like to believe they are. Multiple-use compatibility grinds to a halt as the density and frequency of users grows and the seriousness of technological impacts increases. When will hovercraft become common on the lakes? And what will follow them? A lake buzzing with jet-skis is fundamentally incompatible with any quiet-use recreation.

We've tried to resolve some of these conflicts by instituting times-of-use rules. In some townships, jet-skiers are only allowed on the water from 10 a.m. to 5 p.m. Time-of-use helps, but it fails the most important test: How well do our rules address the protection of the natural lake community? The wave that swamps the loon's nest does its swamping equally well at 9 a.m. or 3 p.m. The aquatic plantbed doesn't care what time of day it was when the propeller ripped it up.

One caveat must be added. Many species of wildlife have found ways to adapt to human pressures. They have learned either to acclimate to human presence, or to alter their habits to fit within "time-of-use" pressures. Beavers feed predominately at night when things quiet down; birds forage in the early morning and then retreat to safer havens as the lakes "wake up" to human use.

All is not lost. The sky isn't falling. These are not the ravings of a placard-carrying "the world is going to end" zealot. Still, a 1993 mail questionnaire completed by 2,334 of 14,000 subscribers to Lake Tides, a University of Wisconsin Extension Lake Management Program newsletter, revealed that 78 percent "enjoy Wisconsin's lakes mostly for their peace, quiet, and natural beauty." Given that statistic, I believe these ravings are instead the legitimate and reasonable concerns expressed by many, if not most, lakeshore owners and users.

So we return to the issue of what to do. Frankly, there is no absolute resolution—there's only compromise. How then to compromise?

None of us likes anybody telling us what to do. But necessity dictates that a body of laws must be created in order that the commons may retain some semblance of its values. These are the very same values that brought most of us here in the first place. We may not like the laws, but we'll like their results.

I believe we should zone lakes for either recreation or wildlife, with a possible additional zone in between that permits limited motor use (less than 10 horsepower). This would create a simple three-zone classification system. Property owners would know exactly what they're getting into, enforcement efforts could be concentrated, and user conflicts would be cut markedly.

We would, of course, be sacrificing the natural community on some lakes so we could preserve it on others. That would be an unhappy choice, but it is one that we have already made on many lakes. Zoning choices must be made now on the remaining lakes blessed with little development, or these lakes will end up in full recreational use at the expense of the natural community. Inaction, of course, is action.

The WDNR is pushing for a lake classification system in each county, recognizing that it's impossible for state or local governments to individually manage all 15,000 lakes in Wisconsin. The WDNR also realizes that all lakes can't be managed alike, because of the profound differences in hydrology, size, shape, depth, locations, water quality, level of current development, unique flora or fauna, and so forth. One size doesn't fit all in the lake business any more than it does in the other businesses.

Lake classification came into being as a middle-ground compromise, an attempt to place lakes into broad categories. The intent is for each category to have a specific cadre of management strategies best suited to its lake type.

Nearly half of all Wisconsin counties have adopted, or are considering adopting, lake classification systems. Each county must consider the biological and physical characteristics of its lakes, gauge all of the lakes' susceptibility to environmental problems due to shoreland development and lake use, and then place each lake into a category that receives appropriate management.

Typically, a lake classification system develops three classes of lakes, which look like this:

Natural or Wild Lakes: Lakes with unique wildlife values and excellent water quality making them unsuitable for significant development or high recreational use. These lakes receive the most restrictive development standards.

Intermediate Lakes: Lakes that are relatively intact as natural resources. Development is typically light or moderate, and management strategies are implemented to keep the lake from becoming more developed.

Developed Lakes: Lakes that have a high level of existing development and recreational use. These lakes are so developed that restrictions on shoreland development would be ineffective.

As you might imagine, the politics played out in each county on how to manage its lakes can reach a fever pitch. The politically compromised result is often a watered-down classification system that fails to truly protect the lakes.

I've always liked the title of one of Canadian musician Bruce Cockburn's songs, "The Trouble With Normal Is It Always Gets Worse." Normal use will continue to get worse. Our perpetual challenge will be to define an acceptable "normal."

The earth is beautiful because of water. — John Jerome

≈ ≈ ≈ ≈ ≈

Riprap and seawalls are used extensively along the Manitowish Chain to try to prevent erosion of shorelines. (A note to music lovers: Riprap, hip-hop, and be-bop have nothing in common.) Riprap is made from gravel and larger stones and boulders, and is the most common structure used for erosion control along Wisconsin lakes. Seawalls, built of stone, concrete, wood, or metal, are less common. Boathouses and homes built right at the water's edge effectively serve as seawalls as well.

While riprap and seawalls do stem erosion, here's a case of good intentions failing ultimately to produce good works, particularly in the case of seawalls. If the entire river ecosystem is taken into account, a series of seawalls can make life very difficult for the individual just downstream, who now receives the deflected waves and increased currents created by the seawalls, causing greater erosion on his/her property. Riprap and seawalls also destroy plants that grow at the water's edge, like sedges, rushes, bulrushes, spikerushes, and pondweeds, and thus remove habitat for invertebrates like snails and aquatic insects that are eaten by ducks, frogs, and fish.

Vertical seawalls also often prevent ducklings, frogs, turtles, aquatic insects, and mammals from easily entering and exiting the water along the shoreline, and remove habitat for basking snakes and turtles.

Hydrologically, a vertical seawall can create an undertow from breaking waves that scours the lake bed, destroying aquatic plants.

And because seawalls usually are the final extension of a large lawn, seawalls remove coarse woody debris from the water's edge, reducing a very important characteristic of prime wildlife habitat. Nutrient enrichment and siltation typically increase as well, fueling algal blooms.

Basically, a concrete wall makes the river into a gutter, preventing the natural meeting of land and water, and profoundly altering the riparian zone where so much wildlife movement occurs.

Piers are another intrusion to the natural shoreline. They jut into the Manitowish River and lakes with great regularity. They too create a mixed blessing, and excessively large piers are now becoming a major issue along many waterways. Wisconsin law distinguishes the rights of owners of riparian areas from public rights. There's one simple reason for such discrimination—Wisconsin laws hold navigable waters in trust for all citizens. The state is charged with the "affirmative duty" of keeping navigable waters safe, open to public hunting and fishing, and open to other recreational uses such as the viewing of scenic beauty. The Wisconsin Supreme Court has further instructed the DNR to consider the cumulative impacts of gradual intrusions into navigable waters.

A hornet's nest of contention has arisen around what constitutes reasonable use of a shoreline. Definitions of such words and concepts as "cumulative impact" and "scenic beauty" hover indistinctly over the legal landscape, and thus over the natural landscape. In the case of piers, how long and wide and close together should they be? How many piers should a shoreland owner be allowed to build? When does the cumulative impact of one more pier impair the public's interest in the lake? We even have current litigation in Wisconsin courts regarding the legality of selling "pier slip condominiums." In effect, this involves selling the right to own a piece of water where you can dock your boat, even though all public waterways belong to the people of Wisconsin. How can someone sell what he or she doesn't own?

Seasonal and permanent piers in Wisconsin are regulated by the DNR (ss. 30.12-13, Wisconsin Statutes, if you want some light bedtime reading). The bottom line is that shoreland owners can build a pier if the structure does not exceed "reasonable use" of the property. Piers can't obstruct navigation, or damage spawning fishes, waterfowl nests, or beneficial plants. They cannot extend offshore beyond the line of navigation, a distance usually limited to a water depth of 3 feet.

Ah, fertile grounds indeed for lawsuits.

Let's set the legalities on shore for a moment and look at the ecological impact of piers, which is somewhat murky as well. The negatives of adding piers to lakeshores are as follows:

- They destroy plant growth simply by their construction.
- They shade out underwater foliage.
- Coarse woody debris is typically removed from all around a pier, reducing important in-water and onshore habitat.
- They typically fragment continuous natural habitat along a shoreline, reducing amphibian, bird, and mammal habitat.
- They typically increase boat pressure. Most piers are obviously built with docking boats in mind, and this damages plants.
- Because piers are usually a site for swimming, aquatic plants are removed from around the pier to make a "clean" swimming area.
- Piers continually protruding into the water can be an aesthetic blight, turning waterfronts into parking lots.

On the other hand, adding piers to lakeshores can:

Three desirable lakescaping options

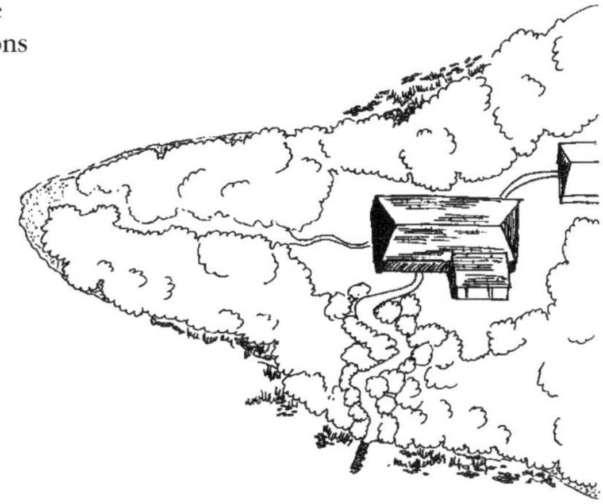

•Improve plant growth on wave-washed shorelines. Sediment can collect on the lee side of the pier, allowing aquatic plants to take root.

•Provide refuge habitat for fish in the darker waters under the piers. Prey fish hidden under piers can see predators up to 2.7 times farther than the predators can see the prey.

•Provide important recreational benefits like a site for a boat lift, a site for sunbathing and swimming, and a site for watching sunsets and Northern Lights.

All of the above brings us back to the continual themes of balance, determining how much is too much, and finding a way to avoid loving our lakes to death. How does one minimally impact the shoreline while still having access to swimming and boating? Can every one of us have all that we want on every lake, and still have a healthy lake ecosystem? Is it possible to have your cake and eat it too?

The answer is a qualified yes, if one does it right. "Doing it right," however, is a value-laden phrase. It is ultimately a matter of what you believe is right for your property and what uses and activities are allowed by local ordinances.

Here are the issues and questions that need to be considered:

•Erosion control. How can you stabilize shoreline banks? How can you prevent overland erosion from occurring on your property?

•Buffer zones. How big should they be? What species should they contain? What do you want your buffer zone to provide—noise reduction, privacy, wildlife habitat, beauty, etc.?

•What are "good" plants to keep? How many? How do you transplant each species to maximize survival and growth?

•What are your needs and desires? They might include swimming, a pier, boating, a view, viewing wildlife, good fishing, more privacy, maximum quiet, a lawn, etc.

•What can you do to improve water quality?

•How can you understand your site? What type of soil do you have? How much does wind and wave action affect the shoreline? What plants are helping you right now? How does your lot drain? How much sun and shade reach the forest floor? What is the nearshore like? How deep is the water at the edge, and how fast does it deepen?

•How do you draw up a plan, and what should it look like?

While these questions may add up to what appears to be a relatively complex study, all of them get answered either consciously or unconsciously over time. The trick is to think about them as a whole, and early on, rather than addressing them piecemeal over years.

The answers may not always be clear, but exploring these questions will help you gain a deeper awareness of what is around you and your place within it all. The largest payoff may be that your enjoyment of where you live inevitably increases as you know more about its ecology.

At the heart of the matter is shoreline and in-water vegetation—what is often called the buffer zone. But just how much vegetation is the right amount? And what species are "best"? A vegetative buffer zone needs to be long enough and deep enough to be more than just a fragment of vegetation trying to cover a dozen other sins. Size often counts in the natural world, and here, more is better.

In its book *Landscaping for Wildlife and Water Quality*, the Minnesota DNR says an ideal buffer zone should include at least 50 percent and preferably up to 75 percent of the shoreline frontage. The zone should extend from 25 to 100 feet from the water's edge onto land, and 25 to 50 feet from the edge into the water. As a general rule, deeper buffer zones create better protection than narrow ones.

The buffer zone obviously has two components: aquatic plants in the water and plants above the water line. However, if the lake bottom drops off rapidly into deep water, emergent vegetation may never have existed along the shoreline, so restoration of aquatic plants would be impossible.

Since no two lakes or rivers are alike, the only appropriate answer to the questions of which species and how much to plant of each should be: Whatever belongs there. Paddle up and down the shoreline, find where natural vegetation has been left intact, and model your shoreline along those lines. If you are unsure of species identification, call your local University Extension and see if they can offer some help. Many offices in lake country now employ individuals to work on lakeshore restoration and management.

Please note that you may not own the lake or river bottom in front of your property, and you may need a special permit to plant within the water. Consult your local natural resources agency for information.

Creation of the buffer zone is the key to lakescaping. The buffer zone will probably reduce your lawn by one-third to one-half, also reducing the attendant fertilizer, herbicide, and pesticide inputs. The vegetation will create a zone that Canada geese won't cross to reach the lawn, a major plus for those of you desiring to keep goose droppings off your property. Designing and creating a lakeshore landscape can be every bit as enjoyable as gardening, America's number one outdoor pursuit. Give it a go—the results will give you pride and pleasure on many fronts.

≈ ≈ ≈ ≈ ≈

Numerous excellent printed resources exist to help with restoring shoreline. Here are a few of the best:

Lakescaping for Wildlife and Water Quality, Henderson, C., Dindorf, C., Rozumalski, F. Minnesota DNR. To order Minnesota DNR publications, call (800) 657-3757 or (651) 297-3000.

"Becoming a Lake-Front Property Owner: Tips on Buying Lake-Front Property" DNR PUBL-WR171 99 Rev.

"Life on the Edge . . . Owning Waterfront Property" UW Extension.

"Shoreline Plants and Landscaping" UWEX GWQ014.

To order UWEX materials, call (608) 262-3346, or write Cooperative Extension Publications, Room 245, 30 N. Murray St., Madison, WI 53715.

Rest Lake is the last lake in the Manitowish Chain. Privately owned Fox Island lies near the middle of the lake. A quick look at a water depth map shows very shallow water between the east side of the island and the main shore. Given the 9 foot or so rise in lake levels from the Rest Lake Dam, it's probable that Fox Island was once the high point of a peninsula attached to the mainland.

In 1902, a group of four burial mounds was reported on Fox Island by James Albright in *The Wisconsin Archaeologist.* An exploration of one mound unearthed iron axes, pieces of gun barrels, a hammer stone, hexagonal spectacle glass, glass beads, a tin pail, and other articles. The specimens were placed in the Milwaukee Public Museum.

Ninety-two years later, an intensive archaeological land and underwater survey of the western tip of the island failed to find any features indicating man-made mounds or burials. Some artifacts were recovered in the area during surveys in 1992 and 1993.

Archaeologists theorize that over the century either all traces of the mounds have disappeared into the lake, or that Albright really only found one isolated burial mound.

Most of us arm-chair history buffs tend to associate mound-building with southern parts of the state. However, mounds existed commonly in the northern lakes area too, though not to the same extent as in other parts of the state. Vetal Winn, in the April, 1924 *Wisconsin Archaeologist*, discusses mound sites on dozens of area lakes: Flambeau, Pokegama, Crawling Stone, Gunlock, Squirrel, Squaw, Minocqua, Tomahawk, Kawauguesaga, Big St. Germain, Trout, Bass, Little Muskellunge, Boulder, Wild Cat, Spider, Stone, and Rest, as well as other sites. Most sites have long since been disturbed, and are on private land.

I give these general locations to encourage intellectual exploration only. Please refer to the sidebar on the Archaeological Protection Act if you're feeling an itch to find a shovel. If you would like to legally participate in a professionally led dig in the northern lakes area, contact the Nicolet-Chequamegon National Forest to join their exceptional Passport in Time program. This program utilizes summer volunteers who are supervised and trained in how to excavate sites—I recommend it highly.

≈ ≈ ≈ ≈ ≈

The Rest Lake Dam is the last stop on this section of river. A pull-out on the left side of the dam into tiny Kohler Park offers a small parking lot, grassy area, and picnic tables. A short walk away are the restaurants and shops of the Manitowish Waters business area.

The Rest Lake Dam has some fascinating history.

"Charles H. Henry, his assigns, or legal representatives are hereby authorized to improve the north fork of the Flambeau River for log driving purpose by building, maintaining and operating on land owned or controlled by him or them, in town forty-two, range five east, in Oneida county, one or more dams across said stream, and by building maintaining and operating in and along the said north fork of Flambeau such piers, booms and other structures as he or they may deem advisable or necessary to carry out the purposes of this act . . . Such improvements shall

Passport in Time Program

The Passport in Time (PIT) is a classic win/win deal. With volunteer help, the National Forest Service can accomplish historical research or restorations impossible within their limited budget, and volunteers get the training, structure, and leadership necessary to participate in a fascinating activity usually reserved only for highly trained professionals.

This common sense approach came about when the Archaeological Resources Protection Act was revised in 1988 and directed the National Forest Service, among others, to develop cultural awareness programs. The NFS responded with an initiative called "Windows on the Past," which urged each National Forest to develop its own cultural resources programs. The Nicolet and Chequamegon national forests in Wisconsin, and the Superior and Chippewa national forests in Minnesota, worked cooperatively to begin a program called "Passport in Time," an effort so successful that it is now sponsored in 75 national forests across the country.

Through Passport-sponsored programs, 15 sites on the Nicolet and Chequamegon forests have been investigated, using hundreds of volunteers. In addition, thousands of visitors have observed the excavations in progress at archaeological open houses that are held at each site. Artifacts dating as far back as 8,000 years have been uncovered in these studies.

Volunteers typically spend one to two weeks at a site, working full-time under the supervision of an archaeologist. Cost is free, as is registration. Campsites are often provided for free as well.

The *Partners in Time Traveler*, announcing the current season's opportunities, is published twice a year, in March and September.

For more information, contact:

Jill Osborn
PIT National Coordinator
USDA Forest Service
1249 S. Vinnell Way
Boise, ID 83709
Call (800) 281-9176
Or call the national forest you wish to work in.

be operated for the use and benefit of all persons desiring to navigate said stream with saw logs without discrimination."

So reads Chapter 449 of the *Laws of Wisconsin*, published on April 27, 1887, which authorized the construction of the Rest Lake Dam.

On February 5, 1997, 110 years later, the Federal Energy Regulatory Commission (FERC) issued an order requiring the current owner of the dam, the Chippewa and Flambeau Improvement Company (CFIC), to file an application for licensing the Rest Lake and Turtle-Flambeau dams. Neither dam had ever required a license before. A license requires an Environmental Impact Statement (EIS) to determine the effects of the dam on the natural river system.

In 1998, after a rehearing request by the CFIC, the FERC determined that the Rest Lake Dam "is neither used and useful nor necessary or appropriate to maintain or operate the downstream projects . . . and conclude that we have no jurisdiction over this reservoir." As a result, no license is deemed necessary to operate the Rest Lake Dam, no EIS needs to be done, and thus no one knows what the impacts of the dam are on the Manitowish River.

The FERC rehearing order gives some background on the area affected by the dam: "The drainage area of the reservoir is about 243 square miles, or 12.9 percent of the Flambeau River drainage. It impounds nine natural lakes with a total surface area of about 4,200 acres and a usable storage capacity of about 15,150 acre-feet under current operating conditions."

Historically, the dam is said to have held back even more water. Long-time residents and local historians believe the dam backed up a head of water all the way into Stepping Stones Lake near Little Star Lake, and into Statehouse Lake, across what is today Highway W. If true, that would have constituted an enormous amount of water. Considering the shallow nature of the Manitowish, a head of water this large would likely have been necessary to carry logs all the way to the mills down on the Chippewa River.

The logging companies were only temporary forces. Everyone knew they would only be there for 10 years or so. By then, the timber would be played out, and the companies would move on. But behind them, the companies would leave many legacies, one of which was a long string of dams along our waterways. The dams created reservoirs that were seen as "normal" by the new settlers who arrived after they had been built. Property was purchased with the understanding that lake and riverfronts were stable, and would remain that way.

To make a long story unfairly short, the Upper Midwest has many dams, like the Rest Lake Dam, that have had no significant purpose for 100 years. Certainly, they help regulate water levels and can smooth out flood surges, but if we've learned anything about rivers, it's not to build in flood plains, which are nature's system for regulating high waters. If we follow the simple maxim that warns us to leave flood plains alone, then what is the value in retaining many dams?

Dams in the Northwoods help control floods, holding back snowmelt runoff and storm surges in the spring, and releasing water in the summer to raise water flows for recreation. Some (none on the Manitowish) also produce "clean" and cheap electric power. This sounds good. Year-round stability of river flow and the creation of electricity to light homes seem like good things. But they're a major trade-off, and often represent a trade that should never have been made.

We have historically seen rivers more as conduits that carried water than as healthy, living ecosystems. But we now know that dams are often cataclysmic events for rivers and the life forms within them. I use "cataclysmic" with trepidation, because not all dams are the same. The mega-dams we built on large rivers like the Colorado and the Columbia rivers have profoundly altered those ecosystems, and "cataclysmic" describes the degree of change.

But with small dams on little rivers like the Manitowish, the changes are not always so apparent, nor anywhere near so extensive.

Here are the known impacts of dams on any river system, big or small:

•Dams fragment habitat. Dams stop fish, mussels, and clams, and other aquatic organisms from moving up or down streams, affecting the ability of organisms to reach their spawning grounds and to migrate to optimal habitats based on seasonal needs.

•Dams change the water temperature. Increased temperatures play a particularly negative role in cold headwater streams. Water temperatures are also typically colder in winter.

•Dams alter the transport of sediment, woody debris, and food. They block the flow of sediments and logs, the raw materials of river structure. The river downstream from a dam is starved of the materials that build riverbeds and gravel bars. Without logjams, for instance, a river gradually loses deep-pool cover, refugia for organisms, and an important substrate for the food chain. Water still goes downstream, but without its texture or load. In other words, without the materials that renew the river channel.

•Dams alter the dissolved oxygen content of the water. Warmer, slower waters contain less dissolved oxygen.

•Dams add to the amount of suspended solids in the water, increasing turbidity and creating better habitat for bottom feeders that stir up the bottom even further. The reservoirs behind dams are warm, slow and muddy artificial lakes—often in stark contrast with the clear, cold, fast-flowing waters that formed them.

•Reservoirs are nutrient sinks, collecting and concentrating nutrients in sediments and plants. Reservoirs eventually fill up with sediments and may require dredging.

•Dams reduce biological diversity.

•Dams increase the amount of water lost to evaporation by pooling the water into reservoirs, an exceptionally important issue in dry climates.

•Dams increase plant productivity by warming the water and slowing it down.

One question that often arises about human-made dams is how they differ from beaver dams. Actually, beaver dams create many of the same impacts as well, but to a far lesser degree, and over a transitory period of time. Beaver dams cause small ponding along a river or stream, but nothing like a huge reservoir. Fish and insects can often get by a beaver dam, but they can't get over 5 or more feet

Floods—What Causes Them?

Floods have always existed, but they appear to do more damage now than historically. When we live on flood plains, drain wetlands, channelize waterways, farm up to the edges of rivers, pave thousands of acres of land with asphalt, we see more floods and think they're happening more often these days. But that's not the case. The real problem is that society expects rivers to fit into "economic progress" visions, but rivers don't read code books well, or pay enough attention when variances are given to people who want to build where the river has flowed off and on for 10,000 years.

Rivers exist primarily to do one thing—drain water to the next lowest point. But people often don't have enough personal history in a place to know what to expect from a river's drainage. Or they don't pay enough attention until the river is in their basement. Or they want to think the problems have been fixed through technology. When we humans move frequently, we bring with us no experience with the natural history of the place. Then we blunder ecologically, often at great cost.

Fortunately, from an ecological standpoint, floods will always be with us (see chapter 3 on the benefits of floods). We literally have no choices relative to the existence of floods—they will always occur. Instead of trying to prevent them, we must learn to understand and respect the processes that lead to flooding in order to minimize the economic damage to human society.

So what do we need to know? Let's start with terminology. Luna Leopold, former chief hydrologist for the U.S. Geological Survey, refers to discharges large enough to overtop riverbanks and flow over the flood plain as "overbank flows." Now I doubt "overbank flow" will soon become part of our everyday lexicon, but what Leopold is trying to do is point out the relative frequency of such flows without invoking a value-laden word like "flood," which conjures up raging calamity. Floods, or overbank flows, are relatively common occurrences on rivers, and their impact is typically far more beneficial than harmful.

Leopold states that a river channel will have only a moderate or small amount of water flowing in it on most days. On a few days in any given year rain or snowmelt will fill a channel to its peak, but not over its banks. And only approximately twice a year will heavy rains or snowmelt produce enough surface runoff to fill a channel to the top of its banks and overflow onto its flood plain. The depth of this floodwater typically will be equal to the average depth of the water in the channel during normal flow.

Continued

Floods—What Causes Them? (Continued)

A great, catastrophic flood typically occurs only once in several generations, or about 50 years. Some readers may remember the enormous floods in the Upper Mississippi River Valley in 1993—the rainfall that fell during this flood will likely not be repeated for another 1,000 years.

So, "overbank flows" occur every year on rivers, but how do they come about? In small drainage basins like the Manitowish, floods are more likely to occur when the soil is already wet prior to a major rainfall. The infiltration rate of the rainwater into the soil is significantly decreased when the soil pores are already partially filled with water. Water can't penetrate as well, so much of it runs off. On rare occasions, the soil becomes completely saturated like a full sponge, and all ensuing rainfall becomes runoff.

Frozen ground also prevents infiltration. Early spring snowmelts offer meltwater nowhere to go because the ground is still frozen, so water runs and pools in depressions everywhere. Driveways become tidal ponds, and rivers often layer meltwater on top of snow and ice.

The intensity of a rainfall also helps determine peak runoff. An inch of rain over 24 hours is a very different animal than an inch of rain in one hour.

Human activities greatly increase runoff rates, and thus the likelihood of flooding. Deforestation, surface paving, leaving farm fields barren, grazing and soil compaction by animals, and other activities all decrease the rate of water infiltration into the soil. Roof downspouts, street gutters, and storm sewers help speed runoff downhill and into rivers, causing higher peak volumes of water.

On a river like the Manitowish, which has some 90 percent of its shoreline still intact and natural, runoff is seldom a major issue. The Manitowish also still has its flood plains intact. These wetlands soak up overbank flows when the natural capacity of the drainage basin soil is overwhelmed. So, the river floods nearly every spring after snowmelt. The difference is no one has built on the flood plains. We gaze on the overbank flows with curiosity, wondering where the ducks and geese are, rather than gazing with fear, wondering if we will survive the flood.

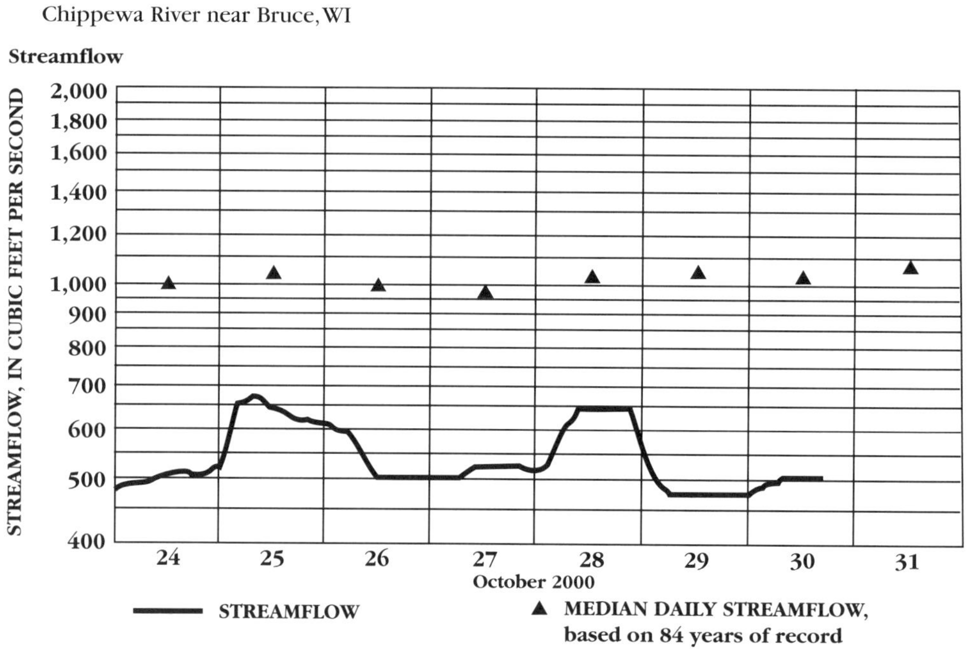

of concrete. Beaver dams come and go with the rise and fall of beaver populations. They're momentary rather than permanent structures on a stream.

Beaver dams also typically wash out in flood conditions. I've endeavored in an earlier sidebar to put floods in the positive light they ecologically deserve. Look at a hydrograph sometime, showing the flow variations of a river over time periods as long as decades, or as short as hours. Valerie Rapp, author of *What the River Reveals*, aptly describes hydrographs as cardiograms for a river, illustrating the rise and fall in riverflow. One can trace the river's history by comparing the pulses over many years. If the pulses are gone, Rapp writes, there will only be a faint beat left—and eventually a flat line. Then we've destroyed the heartbeat of the river.

We've tried technological fixes to resolve a dam's technological problems. Agencies operating dams have attempted many expensive solutions to induce what the river used to do for free. They've installed screens for the turbines, bypasses to keep fish away from the turbines, spillways to reduce the fall that fish experience, fish ladders to try to allow fish to migrate. But problems have arisen. Fish ladders are reasonably successful for upstream passage, but are another story for downstream passage. Many fish pass through turbines where they are injured or stunned by the tremendous force of the water, or are trapped in the backroll below the dam, or are injured in the drop over the spillway.

Hatcheries have produced fry to replace the fish that fail to survive the other "fixes," but sometimes the hatchery fish haven't gone through the specific natural genetic selection needed for that particular river. They die, or they fail to reproduce. Sometimes the fish have carried hatchery diseases that have caused additional problems rather than alleviating the existing ones. On the Manitowish to date, however, hatchery fish have not caused any discernible problems.

When we say that renewable hydroelectric power is cleaner than air-polluting coal plants, have we factored in the potential cost to the river ecosystems? Usually not, in large part because attaching monetary value to factors like increased siltation and turbidity confuses bankers and government officials. These are vague notions compared to numbers on an electric bill.

Sometimes the changes to a river ecosystem may be seen as positive, confusing the issue further. Warmer water causes shifts in fish populations from coldwater species to warmwater species like walleye, crappies, catfish, northern pike, bass, bluegill, yellow perch, and others. While dams break the continuous ranges of many of these fish species into isolated populations often cut off from their spawning areas, they seem to be doing fine in some places, like the Manitowish Chain. Walleyes draw huge numbers of anglers to the Northwoods, and to the Manitowish Chain, so the dams have been seen as producers of income and desired resources.

Today, the Rest Lake Dam's main purposes are to draw down water levels in the fall, so those who live along the Chain of Lakes don't have to remove their docks; and to make power boating between the lakes possible. This is not an inspiring mission.

I can hear my friends and neighbors protesting this line of thought. "Why not leave well enough alone? The dam's not hurting anything." So goes the normal reaction.

The problem is, we don't know what the effects of the dam are on the natural system, and our ignorance may not be bliss.

The dam clearly stops fish migration. One of the more remarkable historical aspects of the Rest Lake Dam was the installation of the Barr Fishway in 1931, a unique system that moved spawning fish from the bottom of the Rest Lake Dam up into Rest Lake. In 1931, during one spring month 1,181 fish went through the fishway. In 1932, 2,099 fish were counted using the fishway. In spring of 1933, 3,282 spawning fish were lifted into Rest Lake. Most of these fish were suckers, however, a fish not well-regarded by the fishing community. Clearly though, without the fishway, the dam halted the migration of spawning fish.

Read more about the Barr Fishway at the end of the chapter.

Of specific interest is the apparent failure of any natural reproduction of lake sturgeon in the Manitowish River over the last 75 years. No one has been able to pinpoint the cause of their reproductive failure during this time period, but the coincidental building of the Turtle-Flambeau Dam in 1926 and the fluctuating water levels created by the Rest Lake Dam are major suspects. Without an EIS that addresses this issue, the connection between sturgeon failure and the dams remains unknown.

But, let's cut to the chase—to the inevitable question: Should the Rest Lake Dam be removed? From a sociological perspective, probably not. Dam removal would have significant repercussions on many lakeshore properties, and would affect recreational opportunities by altering the necessary depth for motorboating between lakes on the chain. The ensuing legal mess would make all of our heads spin.

From an ecological perspective? Well, without an EIS, the jury is still out. The answer, though, for all the reasons we have already discussed, will likely be "yes."

Like the earlier discussion on the Fishtrap Dam, the ultimate question to be asked is this: Do the perceived gains outweigh the perceived losses? And, as before, it depends on whose eyes are doing the perceiving. Who has the highest standing, the natural lake and river floral and faunal community, or the human residents and users of the natural lake community? Is there a middle ground where the human residents learn to see themselves as simple equals in the natural lake community, and derive their pleasures from that perspective?

In a meeting on June 5, 2000, the CFIC requested that the Township of Manitowish Waters purchase the dam from them, because the company has no use for it whatsoever. The town, however, isn't sure it wants the dam due to excessive liability questions. So what to do?

These are very tough questions, not just for Rest Lake, but for lakes and rivers throughout Wisconsin that are unnaturally altered by unneeded dams. No single answer is the correct one. Keeping a dam may or may not be appropriate. Each lake or river has its own story. The questions for humans to ponder are all about how and whether we wish to rewrite those stories.

Spring Equinox

In my canoe I scan the sky,
then the riveredges.
I peer under the protective ledges of ice
overhanging the banks.
I am seeking the leaders of the spring migration,
the air pioneers.

But they've yet to arrive.
They must be somewhere in fleeting transit,
or waiting,
waiting until the moment to launch
like a wind rising.

I notice then the first spring insects.
They're hatching and crawling
across the ice ledges to leap into the river.
The ice drips too, goosebumping the water,
dankly chiming,
and with the insects,
vanishes into the push downstream.

Trumpeter swans arrived three days ago,
clouds given flesh upon their descent.
Crows returned a week ago,
radiating their blackness from the worn limbs
of an old pine along the banks;
the same old pine that was flooded one too many springs
and now stands all limbs like a skinny school girl.

Chickadees, bless the chickadees,
lend their encouragement as I paddle on.
They whistle, "The river is open, open, open."

All else is oddly quiet for a spring equinox,
as the world changes course
and hemispheres exchange dreams.

Original Pre-Settlement Vegetation

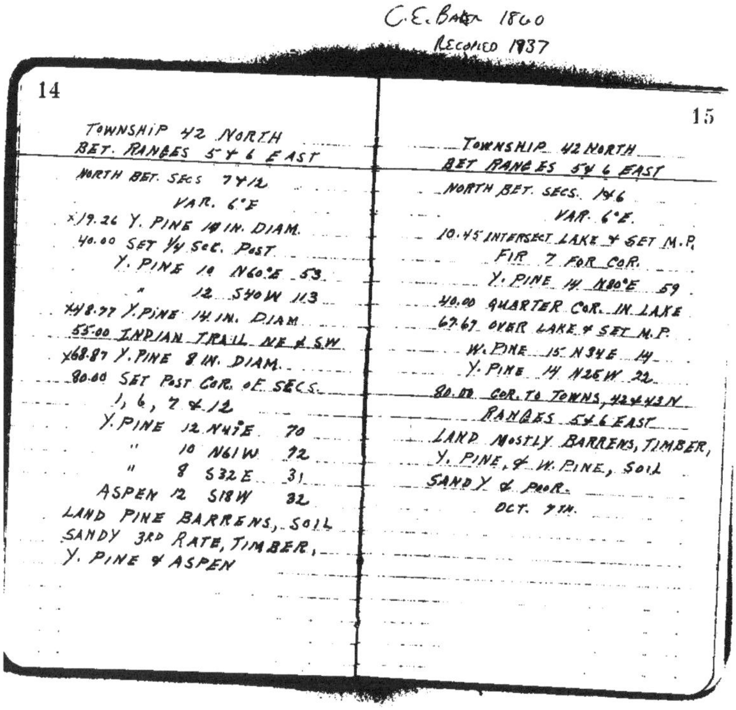

C. E. Baker 1860
Recopied 1937

14

TOWNSHIP 42 NORTH
BET. RANGES 5 & 6 EAST
NORTH BET. SECS 7 & 12
VAR. 6°E
x19.26 Y. PINE 14 IN. DIAM.
40.00 SET 1/4 SEC. POST
Y. PINE 10 N60°E 53
" 12 S40W 113
x48.77 Y. PINE 14 IN. DIAM
55.00 INDIAN TRAIL NE & SW
x68.87 Y. PINE 8 IN. DIAM.
80.00 SET POST COR. OF SECS.
1, 6, 7 & 12
Y. PINE 12 N47E 70
" 10 N61W 72
" 8 S32E 31
ASPEN 12 S18W 32
LAND PINE BARRENS, SOIL
SANDY 3RD RATE, TIMBER,
Y. PINE & ASPEN

15

TOWNSHIP 42 NORTH
BET RANGES 5 & 6 EAST
NORTH BET. SECS. 1 & 6
VAR. 6°E.
10.45 INTERSECT LAKE & SET M.P.
FIR 7 FOR COR.
Y. PINE 14 N80°E 59
40.00 QUARTER COR. IN LAKE
67.67 OVER LAKE & SET M.P.
W. PINE 15 N34E 14
Y. PINE 14 N25W 22
80.00 COR. TO TOWNS, 42 & 43 N
RANGES 5 & 6 EAST
LAND MOSTLY BARRENS, TIMBER,
Y. PINE, & W. PINE, SOIL
SANDY & POOR.
OCT. 7TH.

Though we have no actual detailed record of pre-settlement vegetation, much can be inferred from the original surveyor records taken in the mid-1860s, some 15 years or so after the removal of the Ojibwa to reservations. The surveyors were required to lay out a grid system of rectangular coordinates that were spaced 6 miles apart in north-south and east-west directions. These blocks were called "townships," and within each township, 36 square sections, one mile on a side, were also laid out. At every half-mile interval in each direction along the section lines, a corner point was established. Several trees were blazed to mark the corner points in order to help settlers find the exact location of their claims.

The surveyors listed each tree and its diameter at these corners. They also listed what they saw along their section lines, and described the understory vegetation. Some surveyors were more descriptive than others in summarizing the vegetation, no doubt due to a combination of basic personality and day-to-day environmental factors. If the mosquitoes were killing them, I suspect they wrote less.

Continued

Original Pre-Settlement Vegetation (Continued)

From these records, researchers have reconstructed a picture of what the original Wisconsin landscape looked like. Geographer Robert W. Finley created a map in 1951 of the original vegetation of Wisconsin—copies can be purchased from the University of Wisconsin Press in Madison, Wisconsin.

A great deal is made of pre-settlement "original vegetation." We perceive it as pristine old growth, untrammeled, untouched, unbroken. That's a myth. Pre-settlement vegetation had been altered by human activity far earlier than when the Europeans arrived. The American Indian people conditioned the Northwoods landscape to suit their needs just as they did in the prairie regions to the south, though certainly not to as great an extent.

The "original" vegetation was also a result of a complex interplay of climate, soils, topography, disturbances, and the web of interactions among the organisms that lived there. Forests only a mile apart could be, and often were, dramatically different. There was little static or primeval about the Northwoods. While vegetation may appear to be stable in the time scale of one human life, it's highly dynamic. What the surveyors saw was simply one snapshot in time—not "THE SNAPSHOT." To try to get back to a re-creation of that one moment, while admirable and exciting, is to deny the continual role of disturbance on the landscape by wind, fire, insects, disease, and native people.

This means there's no real single, stable "balance of nature." Too much flexibility and chance is involved. Instead we have a continual balancing of nature, requiring us to understand the mechanisms of change rather than to think in just one model or perspective.

Still, it's fascinating to try to picture what the landscape may have looked like prior to settlement. To go way back, our primary source of information is fossil pollen preserved in lake, marsh, and bog sediments. Radiocarbon dating of these microscopic grains of pollen has been analyzed to provide dates that were used to reconstruct the chronology of our vegetation over time. It's not a perfect picture—the view can be skewed by the nature of pollen production, transport, deposition, and preservation. Most plants can only be resolved to generic or family levels. But despite these shortcomings, we still can infer the general character of this region.

The last glacier retreated some 10,000 years ago from north-central Wisconsin. Sparse spruce woodlands first invaded the open ground. Jack pine, red pine, white birch, and oaks followed onto the tundra-like habitat, beginning about 9,500 B.P. (Before Present). White pine, sugar

Continued

Original Pre-Settlement Vegetation (Continued)

maple, and yellow birch appeared on the scene about 7,500 B.P. as the climate continued to warm.

From 10,000 B.P. until 6,000 B.P., temperatures are thought to have risen until the warmest and driest conditions occurred. Precipitation is believed to have decreased by 10 to 25 percent over much of the Midwest during this time, while July temperatures were likely 1 to 4°F higher than today. A return to moister and cooler conditions began sometime around 6,000 B.P., and led to the return of more boreal species like balsam fir and white spruce, which show up in the pollen record again about 3,500 B.P. With them came the first arrival of hemlock.

Much later, a brief cold period occurred. Called "The Little Ice Age," temperatures declined between 1450 to 1850 A.D., creating a cooler, wetter, and stormier climate than present. The climatic conditions during this time period were responsible in part for the creation of the vast white pine and hemlock-hardwood stands of our region. If this time period had been warmer and drier, our Northwoods forests may have been more grassland than forest.

The overall concept to remember from pollen analysis of Wisconsin vegetation is that the forests were engaged in constant change.

Today in the Northern Highlands/American Legion State Forest, and in particular along the Manitowish River, we mostly have poor sand and sandy/loam soils, which best support pines, along with a mixture of aspen, white birch, red maple, and red oak. The original pre-settlement forest vegetation was dominated by large stands of white and red pines with a mixture of northern hardwoods. The NH/AL was at the heart of an area that is commonly referred to as "the pinery." Pineries were not found continuously across the Northwoods, but grew in scattered large blocks on poor soils, in large part because pines can tolerate nutrient-poor, water-deficient soils better than nearly all other tree species. Where northern soils were better, pines were a small component of forests that were more often dominated by sugar maple, yellow birch, hemlock, and basswood.

Today, aspens prevail over the former sites of the pinery in the NH/AL, accounting for 33 percent of the forested uplands. Their dominant existence is a product of extensive clearcutting practices.

Color maps of the pre-settlement and current vegetation in the NH/AL can be found in "Shaping the Future: Master Planning for the Northern Highlands/American Legion State Forest." Contact the DNR offices in Woodruff or Rhinelander, Wisconsin, for a copy.

Bands of the Lac du Flambeau District

Six individual bands of Ojibwa are thought to have inhabited the Lac du Flambeau District: The Lac du Flambeau, Pelican Lake, Lac Vieux Desert, Turtle Portage, Trout Lake, and Wisconsin River bands. Specific locations and histories for each band are difficult to pin down given the incomplete historic records that exist.

Lac du Flambeau Band

The Lac du Flambeau Band has the best historic documentation of the six bands. Lac du Flambeau was the central village of the district, most likely due to its location at the head of the Bear River, which provided the most centralized access to the other bands and to Lake Superior. Two fur-trading posts, the American Fur Company and the XY Company, were built on Flambeau Lake to take advantage of the trading opportunities. The actual sites of the fur posts are unclear and remain the object of continued investigation. The main Ojibwa village was situated at the source of the Bear River on Flambeau Lake.

Members of the band lived not only in the village but also spread out on the many lakes that surrounded the village. Strawberry Island on the west side of Flambeau Lake was the site of 5 acres of gardens, and was used as a summer residence, likely because of the safety of living on an island. A major battle was fought here between the Ojibwa and Dakota Sioux tribes.

The Bear River provided extensive wild rice in September, and the many lakes provided not only a good fishery, but also a longer growing season due to the fact that lakes function as a heat sink to help prevent early frosts.

The 1854 treaty consolidated all the other bands onto the designated reservation in Lac du Flambeau.

Pelican Lake Band

The Pelican Lake Band was well inland from Lac du Flambeau, nearly five days travel. The location of the band on Pelican Lake is unknown, although the peninsula in the northeast corner of the lake was noted by 1880 white settlers as a village of Potawatomi Indians. From Pelican Lake, only a 3.5-mile portage was required to reach the Wolf River, crossing the watershed divide and permitting access to the Green Bay through the Fox River.

The Pelican Lake Band was very mobile, likely traveling farther than the other bands of the Lac du Flambeau District in their pursuit of seasonal natural resources. The Mole Lake Band of Ojibwa were the closest

Continued

Bands of the Lac du Flambeau District (Continued)

neighbors to the Pelican Lake band, and they likely traded, hunted, and traveled together to a significant extent.

Pelican Lake once had extensive rice fields until a dam increased the water levels and destroyed the crop. The volume of water in this very large lake would have protected crops from early frosts as well. A sugar camp was located at the likely village site in the northeast corner of the lake.

William Warren wrote that after 1848, diseases like smallpox nearly destroyed the Pelican Lake Band.

The Lac Vieux Desert Band

Two French translations exist for Lac Vieux Desert: "Lake of the Desert" or "old planting ground." Old planting ground would seem to work best given that excellent crops were grown on the islands, the most important of which was potatoes.

However, Lac Vieux Desert existed as an important American Indian site not for its potato potential, but rather for its location. Lac Vieux Desert was located at the intersection of five different routes of travel:

•the Wisconsin River itself, which rises in Lac Vieux Desert and flows over 400 miles to the Mississippi.

•the 15-mile-long Laura Lake portage trail, which provided access to the Manitowish River system and the Chippewa River.

•the north-flowing Ontonagon River, which was less than two days travel away.

•the Brule/Menominee River system, which led into Green Bay and Lake Michigan and was less than two days away.

•the L'Anse overland trail from L'Anse on Lake Superior, which ended at Lac Vieux Desert.

Lac Vieux Desert was the earliest known inland American Indian site within the Lac du Flambeau District. The French first encountered Ottawa Indians at Lac Vieux Desert in the 1660s. The Ojibwa are thought to have arrived in the early to mid-1700s.

The traditional village was located on South, or Cow, Island. The Ojibwa moved to the northeast corner of the lake in the 1880s, returning home to Lac Vieux Desert after being removed to the L'Anse reservation in 1854. By 1887, only one building was still inhabited on what was known as "Indian Point."

Trout Lake Band

The Trout Lake Band built a seasonal village at the source of the Trout River on Trout Lake. This site appears to have been used in the

Continued

Bands of the Lac du Flambeau District (Continued)

mid-1600s, before establishment of the traditional villages at Lac du Flambeau, Pelican Lake, Lac Vieux Desert, and Wisconsin River. Many journals and maps document the existence of this site.

The 1854 treaty intended to remove the Trout Lake Band to Lac du Flambeau, but the Band refused to leave their traditional summer and fall village, remaining until the late 1800s at the source of the Trout River.

Turtle Portage Band

The Turtle Portage Band's summer village was located in present-day Mercer next to a portage trail between Echo Lake (Turtle Lake) and Tank Lake (Grand Portage Lake). The site apparently was on the west side of Grand Portage Lake, likely where the town park (Carow Park) is today. The Flambeau Trail led from Lake Superior to Long Lake and then down-river to the Turtle Portage. Many travelers, often hungry after the long Flambeau Trail portage, passed through this area. The present town of Mercer means "merchant" or "trade."

From here, travelers could continue down the Turtle River to the Flambeau River and into the Chippewa River, or they could head into the Manitowish River system by way of a 3.5 mile portage from Mercer Lake (Sugar Camp Lake) to the Manitowish River, just a little ways west of the town of Manitowish.

Norwood wrote that potatoes and corn were raised at the village. The gardens were located on the narrow strip of land between Grand Portage and Turtle lakes, now bisected by County Road J. Sugar camps were undoubtedly utilized in the area, as evidenced by the naming of nearby "Sugar Camp Lake." While no historical documents indicate the presence of wild rice nearby, Wild Rice Lake is just west of Mercer and part of the Turtle River, the name certainly indicating that rice was once present.

Wisconsin River Band

The Wisconsin River Band, or the "Ouisconsaint" as they were called by the fur traders, apparently traveled extensively along the Wisconsin during the summer and fall. The site of the village remains controversial, some suggesting Lake Tomahawk as the site, others saying current day Minocqua was the spot. Local historian Jim Bokern refers to Schoolcraft's hand-drawn map from circa 1832, to show that the Wisconsin River Band were probably located above the outlet of Little St. Germain Lake. It's also quite possible that the band was fluid enough to occupy both Minocqua and Little St. Germain Lake.

Land-Use—The Key to Water Quality

I've always liked the old Fram Oil commercial where the mechanic looks at the car in his shop, then at the owner who will destroy his engine if he doesn't change the filter, and says, "You can pay me now, or you can pay me later." That's just how it is with land-use. We can manage our lands, and therefore our waters, with foresight right now, and pay a minimum price. Or we can suffer the consequences of having the land and water "engine" break down, have to spend exorbitant money to rebuild them, and then still have to implement the changes we should have been doing all along.

Many land-use issues relative to the biological integrity of rivers aren't matters of rocket science. We still permit things such as untreated storm drains to pour directly into streams. As a boy, I used to dump all kinds of interesting things down our city storm drains, from paint to antifreeze to oil. We would never have dreamed of pouring this stuff into the lake where we vacationed, but we never made the connection that storm sewers are like rivers beneath our feet, carrying away the rain that falls on our cities. "Away" is a place we tend not to think about, but it's usually the nearest river or stream.

Storm sewers exist because pavement has replaced soil, an exchange made in hell for rivers. Ever watch water pouring in a gutter after a hard rain? It roils at breakneck speed, with little to sponge it up, slow it down, divert it, and absorb it, the opposite of what occurs in a functioning forest. The pulses of water that hit rivers after heavy rains alter the normal flow regime, creating a bipolar feast or famine condition, spate or dribble. Green space in cities isn't just about aesthetics and shade. It's about controlling water flow and protecting rivers, too.

We allow other things, too, things most of us wouldn't think possible anymore in the 21st century. Things like permitting farmers to plant fields right down to the shoreline of a river, rather than by law requiring simple buffer zones to catch and filter runoff. We have 5 million acres of corn planted every year in Wisconsin. Soil erosion averages about 5 tons per acre per year on flatter lands, and up to 90 tons per acre per year in southwest Wisconsin on steeper slopes. Wisconsin farmers apply 16 million pounds of pesticides to their field corn every year in Wisconsin, some of which washes into unbuffered streams. Forty percent of all fish kills in Wisconsin are attributed to pesticides.

We still permit farm animals access to streams (though in priority watersheds, this practice is no longer allowed). A dairy cow produces 50 pounds of manure a day, and we have 3 million cows in Wisconsin. Cows also break down the stream banks, caving soil into streams and making them wider and slower. If farmers pasture animals next to a stream, law should require them to fence the animals out of the stream.

Continued

Land-Use—The Key to Water Quality (Continued)

These are obvious examples. Let's put land-use issues into a framework of four factors, which determine the health of our waters, and offer some examples of simple choices we can make.

1- Flow Regime - We need land-use practices that allow rain and meltwater to be absorbed and slowly released over time, rather than in pulses that often produce intense flooding during hard rains. How?

•minimize hard surfaces (roads, etc.) near streams in particular, and throughout populated areas.

•protect or restore wetlands that help filter and recharge water.

•demand by law riparian buffer zones along shorelands.

•end all stream channelization (guttering).

•require practical, inexpensive soil conservation practices like contour planting, buffer zones along streams, and fencing out of animals.

2- Habitat Structure - We need to keep all the natural parts of the plant community, and not assume all plants are created equal and provide the same benefits. How?

•minimize cutting in riparian habitat.

•maintain shrub habitat as well as trees.

•control erosion along shorelands and from uplands.

•encourage native plant diversity along streambanks.

3- Water Quality - We need to maintain the right balance of nutrients and dissolved oxygen in our water, and to keep out what doesn't belong. How?

•encourage farmers to use minimum tillage, contour plowing, and strip cropping.

•keep fields in winter cover so snowmelt doesn't carry the soil away in the spring.

•improve management of animal wastes like manure storage or spreading.

•use fertilizers/pesticides in minimum amounts at appropriate times on necessary crops (wheat is necessary, lawns aren't).

4- Energy Relationships - We need to understand how energy moves through a river system, who the players are, what their roles in the drama are, and keep the energy transfers balanced. How?

•keep shoreline vegetation intact to provide shading and cooling of streams.

•practice simple soil conservation to reduce sediment-loading and the resultant warming of waters.

•let leaves and woody debris return to the soil or water system.

Mercury Contamination

Mercury contamination seems almost impossible given the extraordinarily tiny amounts that are found in the air and surface water. Take one droplet of mercury, or about the same amount in a standard mercury thermometer—that's the same amount that is deposited annually on a lake in northern Wisconsin with a surface area of 27 acres.

Even in these minute amounts, mercury becomes a toxicological problem through bioaccumulation and biomagnification. Here's how these processes work, and a little background.

The nutrients for plant growth are found in very low concentrations in most natural waters. Phytoplankton must collect these nutrients from a relatively large volume of water in order to obtain enough nitrogen and phosphorous for growth.

They also inadvertently end up collecting human-made chemicals in the process, even though these chemicals may be in such low concentrations as to be virtually unmeasurable. Over time, the chemicals biologically accumulate (or "bioaccumulate") in the organism at a much higher level than is found in the open water. Certain chemicals, like mercury, DDT and PCBs in particular, break down very slowly in the water, and thus end up accumulating in the fatty tissues or muscle tissues of higher level organisms.

Zooplankton and small fish eat vast quantities of phytoplankton, concentrating toxic chemicals in their bodies, a process repeated at each step in the food chain. The process of increasing concentrations of a chemical through the food chain is known as biomagnification.

The top predators at the end of the food chain, such as walleye, musky, eagles, osprey, loons, and humans, may accumulate high enough concentrations of a toxic chemical to cause serious deformities or death. The concentration of chemicals in the tissues of top predators can be millions of times higher than the concentration in open water.

The eggs of aquatic birds often have the highest concentrations of toxic chemicals because they are at the end of the food chain, and because egg yolk is rich in fatty material. Thus, chicks often show the first signs of a toxic chemical problem.

Humans typically do not achieve as much exposure as an osprey might because we eat a varied diet. Osprey, on the other hand, are 98 percent fish eaters. Nevertheless, low-level exposures in humans may still have adverse effects, and must be taken seriously.

Fish consumption advisories focus on pregnant mothers, unborn children, and pre-teens because the developing fetus and child are most

Continued

Mercury Contamination (Continued)

susceptible to chemical exposure. Recent research suggest that prenatal effects occur at intake levels 5-10 times lower that that of adults.

Mercury contamination comes to us by way of coal combustion power plants, waste incineration, metal processing, and chlorine alkali processing. Increased emissions worldwide have made mercury contamination a global problem, as well as a problem in the U.S.—33 states have issued fish consumption advisories due to mercury contamination as of the year 2000. Best estimates to date suggest human activities have doubled or tripled the amount of mercury in the atmosphere, and we continue to increase atmospheric mercury at a rate of 1.5 percent per year.

Natural mercury occurs in the environment, but studies of sediment cores in northern Wisconsin lakes show the younger sediments deposited since industrialization have mercury concentrations that are 3-5 times higher than those of historical sediments.

Mercury cycling in water is complex; one form of mercury can be converted to another, in particular into methylmercury, the most toxic form to higher level animals. Scientists currently think that lakes with higher acidity (low pH) and higher dissolved organic carbon enhance the likelihood of methylmercury entering the food chain, though many details of the aquatic mercury cycle remain unknown.

Mild mercury poisoning affects the central nervous system and can result in reduction of motor skills and dulled senses of touch, taste, and sight. Severe poisoning, such as occurred in Iraq in 1971, can kill hundreds of people and injure thousands of others.

Lakes that have never seen any development and have minimal human use can still have high mercury levels and fish consumption advisories. The task is to reduce and/or eliminate mercury emissions to the atmosphere, a task that bills in the Wisconsin legislature have tried to bring about, but have failed to pass due to lobbying efforts of the utility companies who say there is no existing or cost-effective technology to reduce the emissions.

Malhiot's Journal

STATEMENT OF THE GOODS SENT TO THE OUISECONSAINT CONFIDED TO THE CARE AND CHARGE OF J. BT. BAZINET AND J. Q. RAICICOT BY FR. VT. MALHIOT

Lac du Flambeau, October 4th 1804

October 4th, 1804							Plus
3 Pieces common Cloth, blue			@	40	Plus the piece		120
3 Brasses	do	H. B.[26]	@	4	do	the brasse	12
4	do	do Scarlet	@	6	do	do	24
4	do	Calico	@	2	do	do	8
2 Blankets	3-	points[27]	@	4	do	each	8
11	do 2-1/2	do	@	3	do	do	33
6	do 2	do	@	3	do	do	33
6	do 1-1/2	do	@	2	do	do	12
6	do 1	do	@	1	do	do	6
3 Capots	4	ells	@	3	do	do	12
3	do 3-1/2	do	@	3-1/2	do	do	10-1/2
3	do 8	do	@	3	do	do	9
4	do 2-1/2	do	@	2-1/2	do	do	10
1	do 2-1/2	do	@	1-1/2	do	do	1-1/2
4	do 1	do	@	1	do	do	4
8 Rolls of braid			@	2	plus	each	16
3 skeins of wool			@	2	do	do	6
2 Laced caps			@	2	do	do	4
1 Chief's coat							8
1 Chief's shirt							2
2 Hats			@	2	plus	each	4
1 Plume for hat							1
3 Small children's shirt			@	1	do	do	3
2 Black silk handkerchiefs			@	2	do	do	4
3 Packages of White porcelain beads			@	4	do	do	12
1 Dozen large knives			@	4	for one plus		3
6 Fine knives			@	1/2	a plus each		3
1 Dozen of Steels for striking fire			@	6	for a plus		2
2 Dozen Awls			@	1	dozen for do		2
3 Dozen Wormers[28]			@		do	do	3
1 Dozen horn combs			@	6	for	do	2
6 Box-wood Combs			@	3	for	do	2
1/2 Roll of wire for snares							3
3 Packs of cards			@	1	plus	each	3
2 Boxes with burning glass			@	2	do	do	4
2 pieces of ribbon			@	3	do	do	6

Continued

Malhiot's Journal (Continued)

3 Looking-glasses	@	1	do	do	3
3 Steel boxes	@	1	do	do	3
30 Needles	@	25	for 1 plus		2

Return

May 21, 1805					Plus
69 large bear skins	@	2	plus each		138
16 small do do					
47 Deer Skins	@	2	for a plus		23-1/2
327 Musk-rat skins	@	10	do	do	32-3/4
68 Beaver skins, making					58
3 Lynx skins	@	2	plus	each	6
20 Otter skins	@		do	do	40
5 Fisher skins					5
100 Marten skins	@	2	for a plus		50
1/2 a Moose skin					1

Goods Brought Back

1 Capot of 3-1/2 ells	3-1/2

Silverware

3 Large double crosses	3
8 Pairs of earrings	1-1/2
30 Small brooches for the hair	1

Utensils

1 Large brass kettle		7
1 Small tin	do	3
2 Large axes		4

[26] Probably a kind of cloth manufactured especially for the Hudson's Bay Company and their trade. -Ed.

[27] For the explanation of this term, see Wis. Hist. Colls. xvi. p. 400, note 2. -Ed.

[28] A wormer was a small coil of iron or steel, used in cleaning a gun. -Ed.

Barr Fishway

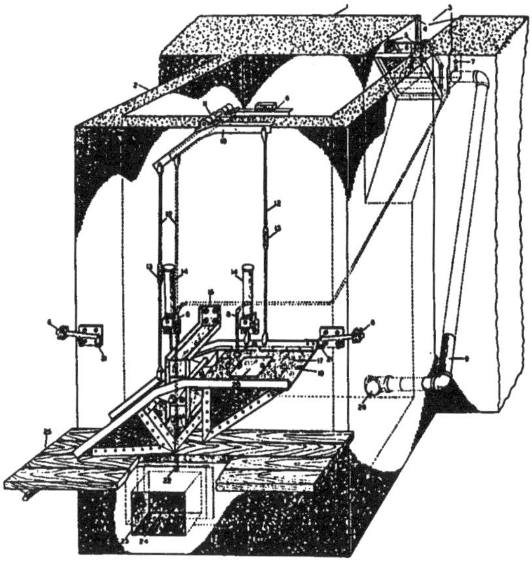

At the 1931 convention of the American Fisheries Society held at Hot Springs, Arkansas, B. O. Webster, the long-time Superintendent of Fisheries for the State of Wisconsin, wrote:

"Unless it be water pollution, there is probably no other one factor which has done so much to deplete the numbers of game fish in American waters as the construction of innumerable dams across the thousands of rivers and streams of North America . . .

"Of the inland water species of fish, the only one which could adapt itself in any measure to the ladder type of fishway was the trout . . . In the north central part of the United States, and particularly in the Lake States, this problem has become very important. The species of fish most directly concerned are the wall-eyed pike, or pike perch, the pickerel or great northern pike, the bass, and in Wisconsin, the muskellunge and sturgeon. None of these species will use any type of fishway that requires leaping as does the fish ladder type; also, each of these species is entirely too cautious and wary a fish to enter any fish wheel. . .

"Since 1912, the fisheries division has been experimenting with every type of fishway that seemed practical. However, none of them, until 1931, showed any promise of being satisfactory . . .

"Radically different in type from any fishway suggested before, the Barr type is really a fish lock or fish elevator rather than a ladder or wheel. In operation it is quite similar to the locks which elevate boats from one level to another . . .

Continued

Barr Fishway (Continued)

"The lock is extremely simple in operation. It is filled by means of the inlet tube from the lake above to the bottom of the concrete box. Water enters the box at a considerable pressure which results in a constant swirling, which is an attraction to the fish.

"The fish are first attracted to the entrance to the fish lock by the great rush of water which results from the emptying of the box. After they pass through the inlet, the swirling of the water in the box acts as a further attraction. In the corner of the box immediately below the entrance to the egress tube, there is a break in the wall which slants down from the lake level more than halfway to the bottom of the box. As the fish swim around in the box following the current, they find this opening. The second or third time they again find the opening and by this time the water is high enough so that the automatic trap has opened and they can pass out through the egress tube.

"The entire operation of the fish lock is automatic . . . During the record taking the first spring the fish lock operated at 40 minute intervals.

"The best test of a fish lock is found in the answer—does it work? The Barr fishway installed at the Rest Lake dam in the spring of 1931 did work . . . It worked so well that the Wisconsin Conservation commission which has been made skeptical of all fishways, adopted a very commendable statement about it at a meeting held July 25, 1931:

'The Conservation Commission recommends that fishways of the Barr type be installed in all dams in Wisconsin where it is considered that such installation be beneficial to fish life and practicable.'"

That first spring, from May 19 to June 18, 1931, a test was made at the Rest Lake dam. A total of 1,181 fish were counted that went through the fishway, even though the fish run for all these species was considered done for the year:

Pike	399
Bass	173
Suckers	552
Lawyers	6
Musky	32
Sunfish	19
Total:	1,181

Continued

Barr Fishway (Continued)

The following year, 1932, from April 21-May 31, 2,099 fish were counted:

Pike	543
Suckers	1,398
Perch	67
Musky	4
Lawyers	13
Ciscos	6
Rock Bass	59
Bluegills	2
Crappies	7
Total:	2,099

Subsequently, the Barr Fishway was installed at the Prairie du Sac dam on the Wisconsin River, and 5,170 fish were moved up river from April 15-May 26, 1932.

From April 25-May25, 1933, on the Rest Lake dam, 3,282 fish were counted using the Barr Fishway.

At the Cisco Lake Dam in Michigan, 30,081 fish were counted over the two-year span of 1936 and 1937, including thousands of rainbow trout.

The fishway was tested on numerous other lakes in the 1930s as well, with apparent success.

An apparent success story.

However, the story ends here! I'd love to tell you more, but that's all the information I can find. In an interview with current fish managers on the Manitowish Chain, they have no knowledge of the Barr Fishway, nor any records. What was so glowingly described by the Wisconsin Superintendent of Fisheries in 1931 fell out of favor sometime, somewhere, for some reason.

To date, the Rest Lake dam still has no means by which fish or other organisms can migrate beyond it, unless they wish to hurl over its edge downstream. However, the Manitowish Chain of Lakes remains known for its excellent fishery. Clearly most fish are finding sufficient spawning habitat on either side of the dam without having to get past the dam. It's an interesting question to ponder how many more would spawn successfully if the dam was passable.

Chapter 5
Our Course:
Rest Lake Dam to Hwy. 47/Manitowish

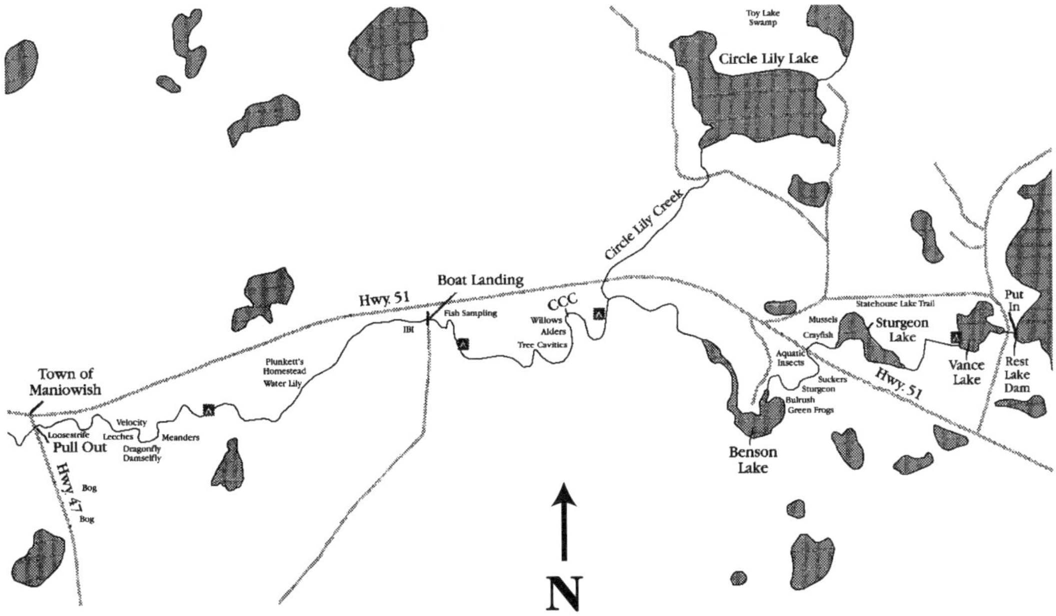

There is nothing softer and weaker than water, and yet there is nothing better for attacking hard and strong things.

—Tao Te Ching

Chapter 5

Of Mussels, Bulrush, Sturgeon, Meanders, and Dragonflies

A home river is that rarest of friends, the one who frequently surprises you with new elements of personality without ever seeming a stranger.
— Paul Schullery

The put-in is on river-left near the pier below the dam. Allow about four hours for this stretch; less for exercise maniacs, more for those who poke along and enjoy the pleasures of a quiet river.

Immediately below the dam, decaying wooden pilings project from the water, indicating the site of the old Rest Lake Fish Hatchery that was built by the Civilian Conservation Corps and operated from 1932 to 1964. Fish eggs, primarily muskie and walleye, were brought here, raised to fry, and then released. Small CCC-built fish hatcheries like this one were numerous at the time. Island Lake, Big Lake, Mercer, and Sayner all had hatcheries, most of which closed in the 1940s.

The river soon enters Vance Lake, which is more of a widening of the river than a lake. Vance Lake was named after Peter Vance, reportedly the first white man to settle in the Manitowish Waters area. Vance pioneered a site in 1884 on the north shore of the lake.

The river narrows down as it leaves Vance Lake. A large boulder on the north shore of the river is reminiscent of spirit stones that the Ojibwa singled out as places to pray.

A campsite lies just downriver of the boulder, marked by a DNR sign. Located on an old homesite, the openness and slow recolonization of the site shows the clear ecological signs of former habitation.

The river flow narrows down through two shallow rocky stretches that would have made excellent sites for native people to spear or net spawning sturgeon. Nothing high-tech was needed. They could have simply moved rocks to constrict the river's path and force the sturgeon through a narrow slot for easy capture.

Read more about the seasonal economic cycles of the Ojibwa at the end of the chapter.

An eagle's nest appears on river-left in a large white pine. On March 5, 2000, we observed an adult sitting on the nest, incubating eggs. The nest is not recorded historically in DNR records, so it was new as of the year 2000.

The river soon widens again into Sturgeon Lake, named for its obvious association with the lake sturgeon that are seen both downriver from here as well as upriver to the dam.

Freshwater mussels live on the sandy bottom all along the length of the Manitowish. From samples taken under the Highway 51 bridge and downstream near the Wisconsin Departament of Transportation wayside, the most common species in the Manitowish are the mucket (*Actinonaias ligamentina carinata*), the fat mucket (*Lampsilis radiata luteola*), the three-ridge (*Amblema plicata*), and the pocketbook (*Lampsilis ventricosa*). Six other species were found at these two sites, their abundance a sign of good water quality.

Mussels lead a rather remarkable life for all their seeming inaction. If you have ever wondered how a mussel mates (after all, they just don't seem to have the requisite parts), well, here's your answer. In midsummer, when water warms to a particular temperature, the female releases her eggs, which get lodged in her gills. Females of certain species can release incredible numbers of eggs. One individual female can generate some 30 million fertile eggs in five or six years!

The male meanwhile releases his spermatozoa, which hopefully float past the female and are drawn into her gills, fertilizing her eggs. The eggs develop rapidly, until they are released as minute larvae each less than one-fiftieth of an inch long. The larvae are grouped in packets of 50 to 100, called glochidia. The glochidia must attach within a few days to the scales or gills of a specific species of fish or they will die. They're often brightly colored and attractive, luring the fish to bite into them.

The glochidium parasitizes the fish, though without harming it. Each mussel species requires a particular fish or group of related fish to parasitize. Only a small fraction of the glochidia manage to attach to any fish, much less the right species. Once attached to the fish, the fish's tissue grows over the glochidia and encloses them. The glochidia will remain here for several weeks to a month until they reach the size of a pinhead and drop off into the current and are swept away. Eventually, they drop onto the lake or stream bottom, where over several years a slow-maturing process ensues until adulthood is reached.

Freshwater mussels are filter-feeders, drawing water in through their syphon-like mouths and absorbing nutrients and dissolved oxygen. Waste products exit through a second syphon opening. Flowing water is essential to bring a continuous supply of drifting nutrients and oxygen to them. Without water currents—for instance, when a dam is built and stops up the flow—the tiny mussels will die.

As a mollusk grows, its shell enlarges to accommodate its body growth, forming annual rings like a tree. Mussels grow rapidly at first, but like trees, the growth rate slows eventually to almost nothing. Old age for a mussel might be 50 years or more.

The annual layers can be teased apart in a laboratory, and analyzed chemically to tell the story of what was in the water in any given year of the mussel's life. By examining mussels at different sites along a river,

researchers can pinpoint specific sources of pollution. Mussels thus serve as long-term biological monitors.

Each shell has two valves, or halves. When the shell is open, a large foot and two syphons are visible. The mussel sticks out its foot to grip the substrate and slowly draws its shell along. They're not out for long walks. A mussel may move all of 100 feet in a lifetime. The shell mantle can sense touch and shadow, serving as a warning light to the mussel to close its shell.

Don't bother gathering the freshwater mussels in the Manitowish for a dinner—their taste leaves a whole lot to be desired. Given the extraordinary life cycle they must go through to reach adulthood, they've earned their immunity from our needs anyway. The muskrats, otter, and raccoons don't see it the same way. Their piles of shells on the shoreline illustrate how good mussels obviously taste to them.

Also residing along the bottom in the shallows are crayfish. Often many, many crayfish! The rusty crayfish (*Orconectes rusticus*), an exotic species, now dominates crayfish populations in northern lakes and rivers, and has dramatically impacted aquatic plant beds throughout the Manitowish chain. According to work done by researchers at the University of Wisconsin Trout Lake Center for Limnology, the Manitowish Chain of Lakes has a significantly lower number of aquatic plant species, almost certainly due to the high abundance of rusty crayfish.

The rusty crayfish, native to Ohio, Indiana, and Kentucky, began to appear in northern Wisconsin lakes and rivers around 1960, likely introduced through its use as favorite bait

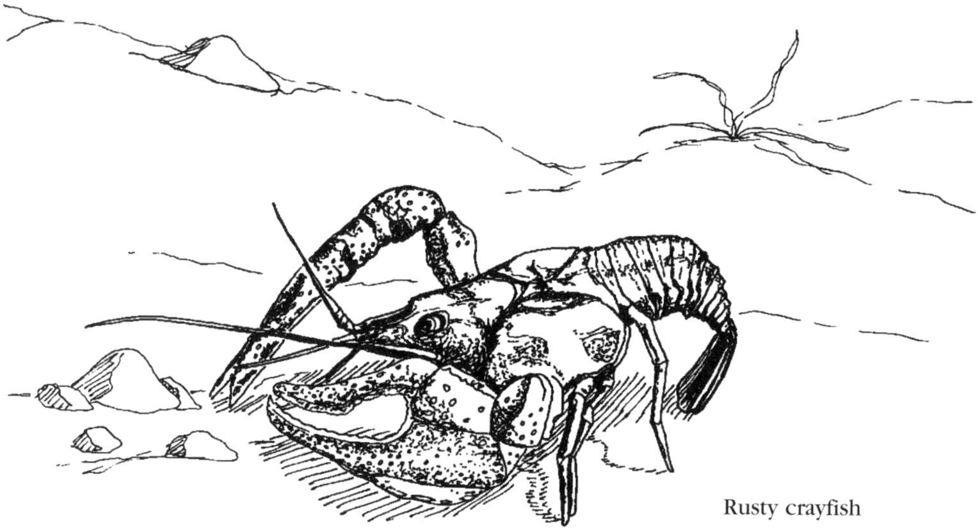

Rusty crayfish

by out-of-state anglers. Commercial crayfish trappers, pleased at the prospect of a larger and more prolific crayfish, added a helping hand by introducing the rusties into many of their favorite trapping lakes. No one foresaw the dramatic ecological disruption such actions would create, and the use of live crayfish for bait wasn't outlawed until 1983.

Sixty years ago, only one crayfish species, the fantail (*Orconectes virilis*), was native to northern Wisconsin waters. The blue crayfish (*Orconectes propinquus*) then began to appear in northern lakes. It flourished, but without causing any significant ecological issues or imbalances. Both species acted as benign members of the ecological community, playing out their roles as omnivores, eating fish eggs, carrion, insect larvae, and plant foods, and serving as prey for large fish, wading and fishing birds, and aquatic mammals.

The rusties changed all that. While all three species have similar life cycles, the rusties have a competitive advantage in nearly all aspects of their life cycle. Consider the following:

- The rusties' eggs hatch earlier in the spring than the blue and fantail eggs, and grow faster.
- The rusty is considerably larger than the blue and slightly larger than the fantail, conferring an advantage in territorial battles and in warding off predators.
- The rusty is the most aggressive of the three species. It's even willing to take on research divers by facing the divers and raising and lowering its pincers in a threat display.
- Fantails and blues beat a retreat when approached by a rusty, often hiding in their burrows under the edge of a rock or sunken log. Rusties follow them into their burrows and oust them. Not only do the fantails and blues lose their favored feeding territory, but they are much more exposed to predation by predatorial fish.

•The rusty is so large and aggressive in its defensive displays that few fish—including the main predator of crayfish, the smallmouth bass—will attack them.

•Rusty males frequently mate with blue and fantail females, creating a hybrid. But blue and fantail males aren't known to mate with rusty females. So rusties reduce the reproductive success of their competitors.

•The rusty feeds for longer periods and on a greater variety of substrate types than the blue or fantail.

The long and short of it is that rusty populations sometimes attain densities of 15 adults per square meter, which wouldn't be a problem if the rusty didn't do to aquatic vegetation what clearcutting does to upland forests. Aquatic plant beds that once thrived on a host of lakes have been decimated by the voracious rusties.

A study on Sparkling Lake in Vilas County compared the effects of the fantail and the rusty crayfish on large aquatic plants. The researchers built 12 crayfish corrals in a rich plant community on the bottom of Sparkling Lake. Four enclosures held rusties, four contained fantails, four had no crayfish. Both species ate a similar amount of vegetation, indicating that the heavy impact of rusties is probably due more to their increased density and their ability to ward off predation.

The researchers then compared the growth of plants in a crayfish-free enclosure with that of an enclosure containing 10 rusty crayfish per square yard. By the end of summer, the only plant remaining in the enclosure with the crayfish was water celery, while a dense mixture of four to five plant species thrived in the crayfish-free corral.

Another study, conducted on Trout Lake in Vilas County, looked at low densities of rusties to determine whether smaller numbers still had such a profound effect. They found that at a density of one crayfish per square yard, the biomass of aquatic plants was reduced by 40 percent. At a density of 10 per square yard, not a single plant stem remained after 12 weeks.

The issue is larger than just the survival of plant species alone, since the plants serve as cover, grazing, and egg-laying sites for a host of aquatic invertebrates. Snails and insects that utilize the plants are displaced, or eaten directly by the rusties, reducing their availability as prey for fish predators like bass, walleye, and muskie.

Of 107 lakes and 50 stream reaches surveyed in northern Wisconsin and the upper peninsula of Michigan, a region where the fantail was the only common crayfish in the first half of the 20th century, the fantail now occurs in only 44 percent and 38 percent of the lakes and streams respectively.

What does the future hold? No one's quite sure. Rusties have clearly altered the ecology of many pristine northern lakes and rivers, but at what point will other members of the aquatic community adjust to their presence and learn to exploit them, or at least find a way to tolerate them? At some point, natural selection will catch up to the rusties and relegate them to merely integral members of the lake and stream ecology. When this will happen is the 64,000 dollar question.

≈ ≈ ≈ ≈ ≈

A river—with its attendant cascades, eddies, boils, and whirlpools—is the most expressive aspect of a natural landscape, for nothing else moves so far, so broadly, so unceasingly, so demonstrably, and nothing else is so susceptible to personification and so much at the heart of our notions about life and death. Across generations and around the globe, humans, we double-footed jugs of seventy percent water, have seen rivers as both our source and the way out of this world.

— William Least Heat-Moon

The Statehouse Lake Trail, the only hiking trail along the river, runs for a mile or so along the north shore of Sturgeon Lake and the river. The trail connects into a series of upland loops that lead to the North Lakeland Discovery Center on Statehouse Lake. This nonprofit environmental education center organizes a diverse schedule of outdoor activities.

Sturgeon Lake attracts a small but consistent number of migrating waterfowl in spring and fall. Its quiet, shallow water and healthy aquatic vegetation provide good layover habitat for birds on their way north to breeding grounds, or on their way south to wintering grounds.

The river soon goes under the Highway 51 bridge, the site of a diesel-fuel spill in January 2000. A semi-truck ran off the bridge here, killing the driver, and spilling 175 gallons of diesel fuel into the river. Chemical or oil spills usually bring to mind foundering ocean-going tankers or leaks from large factories along major rivers, but here the context was much smaller, both in terms of the spill and the river volume affected.

The spill had no apparent long-term effects on the river, but probably couldn't have occurred at a worse spot. The small rapids just downriver and around the corner from the bridge is the major spawning site for lake sturgeon and a number of other fish.

This site is rich in other ways, too. I found an Eastern spiny softshell turtle here in 1994, a species still not listed as present in Vilas County. Unfortunately, without a photograph or specimen proving their presence, my hand-held specimen doesn't count.

Eagles fish this site, often in large numbers; we've seen up to 10 gathered here at a time when the fish are spawning. Kingfishers commonly fish from perches hanging over the river. We've watched deer swim the rapids. And once when I was putting-in with a group of paddlers here, two enormous snapping turtles drifted by very slowly, locked in a mating embrace that was still in progress as they rounded the river bend.

It's a wonderful spot, despite the noise of the traffic crossing the bridge, the activity of canoeists putting-in, and the presence of anglers trying their luck.

This area provides excellent opportunities for sampling aquatic insects due to the variety of stream habitats. Rocky riffles, quiet pools, and natural shorelines clothed with overhanging vegetation—each provides different habitat characteristics. I've sampled here for aquatic insects, and found a diverse array of stoneflies, mayflies, dragonflies, damselflies, and others.

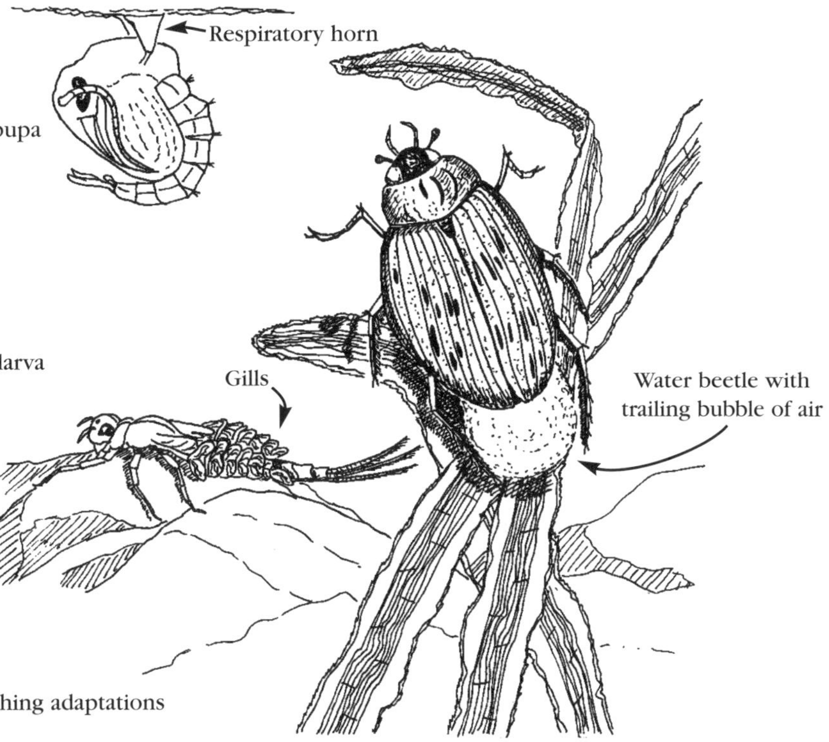

Insect breathing adaptations

As numerous as aquatic insects are here, only about 10 percent of all insects are aquatic. Living in water requires a host of specialized adaptations. Aquatic insects must know how to breathe, how to stay in one place when the river current wants to deliver them to the ocean, how to transition from a larval aquatic stage of life to an adult terrestrial life, how to find food without becoming food for something else, and how to survive a northern winter.

Let's briefly look at the array of aquatic insect adaptations for breathing. Oxygen is available directly from the atmosphere, or in a dissolved form in water. Terrestrial insects generally utilize a system of tubes, called a tracheal system, which distributes oxygen throughout the body. The largest tubes usually occur as external openings, known as spiracles, along the body. Surface-dwelling aquatic insects and semiaquatic insects typically possess these tracheal tubes and spiracles.

But the respiratory systems of submerged insects are quite different and amazingly diverse, utilizing a host of structural, physiological, and behavioral adaptations.

Aquatic insects can be divided into two major categories based on their respiratory systems: aeropneustic (air-o-NEW-stick) aquatic insects, which use atmospheric oxygen even though they are submerged or partially submerged, and hydropneustic (hi-dro-NEW-stick) insects, which are submerged and use dissolved oxygen in the water.

Mosquito larvae and pupae are likely the most recognizable of aeropneustic insects. They typically use a breathing tube/syphon or respiratory horn extended to the surface of the water to obtain atmospheric oxygen. The tips of the tubes have water-repellent hairs that allow them to remain suspended at the water surface.

Other aeropneustic insects live a submerged life, but must come to the surface now and again to obtain air. Called air-storage breathers or periodic-contact breathers, these insects carry a supply of air under water with them, much like scuba divers do. They capture air at the surface, and then carry the air under water for gradual use. Many water bugs and adult water beetles capture an air bubble on a part of their bodies or in an underwing chamber, then breathe from it until it's depleted. Dense microscopic hairs or scales hold the bubble of air against the insect's body where spiracles are located. The bubble of air keeps the spiracles dry so the insect doesn't drown.

The air bubble doesn't deplete as rapidly as one might expect, because it acts as a gill, permitting dissolved oxygen in the water to diffuse into the bubble, replenishing it. In human terms, this would be like a scuba diver whose tanks continually refill by absorbing oxygen from the surrounding water.

A few other aeropneustic insects have evolved to pierce the stems of submerged plants and take oxygen from them, since aquatic plants must pump air down to their roots in order to survive.

The hydropneustic insects obtain oxygen through cutaneous respiration, meaning absorption of oxygen directly through the body. In their larval stages, most mayflies, dragonflies, damselflies, caddisflies, alderflies, and stoneflies utilize gills or other membranous appendages for external respiration. Gills appear typically as either flat, plate-like structures or as filamentous, bushy structures.

Many of the submerged insects beat their gills (some mayfly larvae), undulate their abdomens (caddisflies within their tubecases), push themselves up and down (some stoneflies), or repeatedly take in and expel water (many dragonfly larvae) in order to increase the movement of water around them, thus increasing the availability of dissolved oxygen. This ventilation is much like opening a window to get a good circulation of fresh air in the house.

≈ ≈ ≈ ≈ ≈

I've driven more than a million miles over American highways, but I don't recall loving, for itself, even one road. How can you love an unmoving, stone-cold strip of concrete, ever the same except for its aging, its attrition? But a river comes into existence moving, and it grows as it moves, and like a great mother carries within itself lives too varied and multitudinous for our myriad sciences even yet wholly to number and name.

— William Least Heat-Moon

The river hooks around to the right and offers the only genuine Class 1 rapids in its entire 44-mile length. In high water, the riffles bobble canoes and kayaks over most of the rocks. In low water, it's pinball time, bounce-and-scrape, but only for 75 yards or so.

Mary and I have paddled this little stretch in late May and have literally had fish bouncing off our hull. Significant numbers of suckers and redhorse spawn among these rocks, as do much smaller numbers of lake sturgeon. The sucker and redhorse spawning runs attract a dozen or more eagles that must positively salivate at all the action in the shallow waters. Another shorter rocky run just downstream also attracts spawners and their associated predators.

Read more about suckers at the end of the chapter.

Fish biologists are considering signposting these riffles to educate the public about the sturgeon spawn. Sturgeon appear to congregate in Benson Lake, about one half mile downstream from here, and in early to mid-May when water temperatures are just right, about 58°F, they move up along the rocky shores to spawn.

We're not talking big numbers of adult sturgeon. Only four to five females on average may spawn in any given spring. A big year might involve 15. While the males appear at the spawning sites every spring, the females show up only on a three to five-year rotation.

The females are ripe for about 12 to 18 hours, spawn quickly, and then head back downriver. Females mature in 22 to 26 years, males in 15 to 18 years. Both genders grow quickly, reaching 30 inches or more in the first five years and 40 inches or more in 10 years.

Historically, the Manitowish held good populations of sturgeon, but the population has dwindled to almost entirely adults in the 75-year and older age class. Sturgeon on the North Fork of the Flambeau River below the Turtle-Flambeau Flowage (TFF) Dam continue to thrive, but after the dam was built in 1926, sturgeon between the Rest Lake and the TFF dams have apparently failed to successfully reproduce. DNR biologists have found virtually no fish that would have been produced in the Manitowish River system since 1926.

In 1993, in an attempt to gain information that might shed some light on this reproductive failure, WDNR researchers fitted six females with radio transmitters to monitor their movements. All were longer than 50 inches: The largest sturgeon monitored weighed more than 140 pounds and was 74 inches long. Another 24 sturgeon were fitted with aluminum tags in their dorsal fins. Unfortunately, the transmitter technology then was only good enough to last for two years, and a minimum of information was gained.

In 1994 and 1998, DNR fish biologists also captured a few spawning females, and stripped their eggs for artificial incubation in the Wild Rose Hatchery downstate. A single female produces around 100,000 eggs. In 1994, after incubation of the eggs, 76,000 one-inch-long fry were released in these rapids and at the state wayside farther downstream. Another 9,724 five-inch-long fingerlings were later released at four sites between the Rest Lake Dam and Benson Lake. In July of 1998, 11,800 2.5-inch-long fingerlings were released in these rapids and at the state wayside. Another 6,018 five-inch-long fingerlings were released later at the same sites.

The biologists don't know if these fish have survived or not. Sturgeon

Talking the Same Language of Difficulty

Since my concept of rapids as "hard" may be your epitome of "easy," the International Scale of River Difficulty was devised to categorize rapids worldwide into six classes, from easy to extreme. Be very aware that ratings are derived from average conditions. When seasonal or daily weather changes occur, water depths may rise or fall, with water velocities following suit. A river paddled in early May after the snowmelt or after a heavy spring rain can be an entirely different animal from the bucolic little stream that barely floats a canoe in August. When water is high or particularly cold, a river's rating should be considered at least one class more difficult.

Class 1: Fast, unobstructed water with small waves, requiring little maneuvering and involving only slight risk.

Class 2: Fast water with sizable, often irregular waves, and other river features requiring river-reading and maneuvering abilities; suitable for trained paddlers.

Class 3: Very difficult rapids with large, irregular waves, drops of 3 feet or more, requiring difficult maneuvers and precise control; suitable only for experienced whitewater paddlers.

Class 4: Intense, powerful, turbulent, unpredictable, dangerous; requires highly skilled paddlers with strong rescue skills. Portage required for anyone else.

Class 5: Extremely dangerous rapids, featuring big drops, many obstructions, "keeper" holes; grave danger with little chance for rescue. Teams of experts only in optimal conditions. Mandatory portage for everyone else.

Class 6: Experts using the best equipment and with all their experience still risk their lives; suicide likely for the rest of us.

In the absence of any rating, one may assume the river is entirely quietwater, but the old saying about assuming should likely be remembered (assume is spelled a-s-s-u-m-e, and u and me are asses if we assume anything).

The Manitowish has two rapids that are rated Class 1. Both are rocky with small riffles, and are usually easily run with adequate water. When water levels drop in summer, canoes may have to be dragged through the rocks, or a pinball run can be made by bouncing off the many rocks that are now above or just below the surface. Both sections can be easily portaged if one wishes.

fingerlings are vulnerable to larger fish like suckers, and to other animals like crayfish and otter. Little evidence of fry or fingerling survival surfaced until three 26- to 30-inch juveniles were caught by the DNR on the Turtle-Flambeau Flowage in 2000. Given their size, these three were likely in the batch of fish released in 1994. Two of these juveniles were implanted with radio transmitters so they can be tracked in the future. Unfortunately, the fish biologists have precious little other evidence to say whether the releases in 1994 and 1998 have made a difference or not.

The WDNR implanted 12 adult sturgeon, six in 1999 and six in 2000, with tracking devices designed to last a minimum of 16 years. These should provide long-term data on the movements of sturgeon in the Manitowish system.

Fishery biologists speculate that the abundant populations of rusty crayfish and greater redhorse are the most likely factors limiting the sturgeon reproductive success, though the rusties didn't become a factor until their introduction in the 1960s. Biologists also suspect that the fluctuating water temperatures, water velocities, and water levels produced by the upstream Rest Lake Dam may significantly impact success. A release or withholding of water during the critical two week period when sturgeon stage and spawn could critically affect them.

In summer, watch for sturgeon breaching like freshwater whales in Benson Lake. Sturgeon will leap completely out of the water until they appear to be standing on their tails, and then land with a whack. Why they behave so curiously is unknown, though in areas with lamprey eels, the sturgeon may be trying to shake off attached lampreys. There are no lamprey, however, in the Manitowish. Benson appears to have a resident population of sturgeon—several monitored fish haven't moved out of Benson for two years.

Incongruously, a fall fishing season currently exists for river sturgeon on the Manitowish. Given the exceedingly low numbers of sturgeon in the river, a moratorium on taking any adults would seem to be an easy call.

Benson Lake, and the river channel just above it, also attract moderate numbers of migrating waterfowl in spring. This is one of the first places Mary and I go as soon as the ice is off, in order to see hooded and common mergansers, bufflehead, goldeneye, and scaup.

Benson also provides excellent shoreline habitat for breeding frogs, including the creatively named green frog (*Rana clamitans*). Male green frogs defend their shoreline territories and attract females by "singing." I put singing in quotation marks because, just as beauty is in the eyes of the beholder, song is in the ears of the hearer. Green frogs sound like the twanging of an untuned banjo string, a common sound heard around the edges of northern lakes usually from June through mid-July.

I wrote earlier of a series of five WDNR studies that was done to determine the impact of shoreline development on northern lakes. One of these studies looked at the effect of lakeshore development on green frog distribution and abundance in northern Wisconsin. The study found that green frog numbers were greatly reduced on developed lakes.

The decrease was directly related to the reduction of optimal green frog habitat, and to the fragmentation of their habitat. Fragmentation refers to the break up of large and continuous habitat into a patchwork quilt of smaller areas surrounded by disturbed land.

At this point, a fair question might be, "Who needs green frogs?" Like the proverbial canary in the coal mine, the green frog can serve as an ecosystem indicator, reflecting the health of aquatic life in northern Wisconsin lakes.

Why use green frogs and not other frog species? Because they're shoreline dependent and habitat generalists, inhabiting nearly all types of permanent water in Wisconsin. The territorial adult males will defend their shoreline throughout the summer from other males. The males call throughout the breeding season, with peak calling at night, making their detection by researchers relatively simple.

The disappearance of green frogs is serious, for reasons ranging from the loss to the food chain to the aesthetic loss of living on a lake sans the chorus of frogs. But while the disappearance of green frogs as a species is an important issue, it's not the issue. The study concludes that current zoning regulations on lakes and rivers are not adequate to protect green frog populations on northern Wisconsin lakes. The implications for other aquatic life forms are grave.

Here's the bottom line. By clearing our shorelines and shallow waters of vegetation, we get a better view. But of what? We get views of more erosion, greener lakes, and significantly reduced wildlife in the water and along the shoreline, a trade-off no knowledgeable shoreline owner wants.

Read more about the decline and fall of green frogs at the end of the chapter.

≈ ≈ ≈ ≈ ≈

In the aquatic plantbeds along Benson's shoreline live a host of bulrushes. Through their Bible study, most people know bulrushes as the hiding place of baby Moses. Botanists get testy about such attributions when, as one wetland plant expert writes, Moses was most likely hidden in papyrus sedge (*Cyperus rostrata*). Biblical scholars had to be expert interpreters of many things, but botany may not have been within their collective expertise.

Like most shallow-water, emergent species, bulrush typically grows in densely colonial patches. Colonial growth makes sense for aquatic plants. Thick stands of plants help absorb and withstand the buffeting of waves, the disrupting power of currents, and by dint of sheer numbers, the appetites of mammals like muskrats. Perennial horizontal rhizomes in the sediments keep the colonies expanding by sprouting new clonal stems every spring.

Bulrushes do yeoman work of pioneering shorelines and building land. The thick growth buffers waves to help reduce erosion, while the rhizomes hold together sediments that would otherwise continually be resuspended in the water with the churning of every wave.

Break off a stem and note the fine

Bulrush species

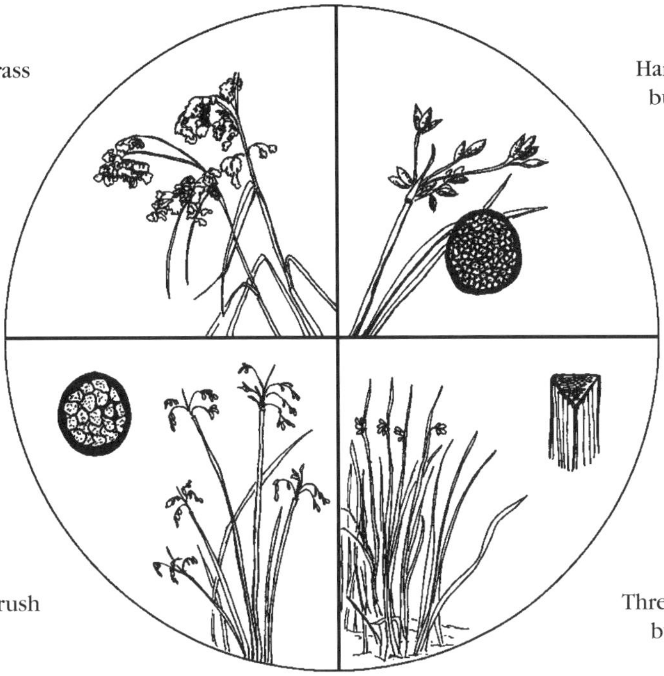

partitions that allow air to flow down to the buried roots. Bulrush, like most emergent plants, has evolved stems that pump air through those channels while providing architectural support.

Bulrush stems are often the ladders upon which dragonfly nymphs climb to escape the water and transform into flying adults. The brittle husks of dragonfly exoskeletons can typically be found still clinging to bulrush stems.

The standing stems also provide nesting sites and cover for birds like marsh wrens and red-winged blackbirds, while the broken stems may be gathered to make a floating platform for birds like black terns. Muskrats eat the stems (as well as the tubers and rhizomes), and gather them for use in building their lodges.

Waterfowl, shorebirds, and certain marshbirds commonly eat the hard-coated seeds, known as achenes. The appetites of coots, mallards, blue-winged teal, black ducks, shovelers, Canada geese, trumpeter swans, sora rails, and a host of other birds make bulrush a highly important food source.

A bulrush stem can be soaked in fat or wax and be used for a candle. One author even claims that you can bind the stems together to serve as an emergency raft, a la Thor Heyerdahl's Kontiki. Along the Manitowish, it would be easier to just walk out to the road.

Unfortunately, bulrush usually performs all this useful labor in anonymity. As is true for most wetland plants, the bulrush provides a multitude of values, yet is unlikely to be identified by one person in a thousand.

Hopefully, the word will spread about its good deeds.

Read more about bulrush at the end of the chapter.

≈ ≈ ≈ ≈ ≈

Heading downstream out of Benson Lake, the river briefly braids into two channels separated by a narrow island. Both channels usually convey enough water to carry a canoe, so take either side.

Braided channels typically change as the river carves new configurations. Sediments fill in parts of the braids, and islands form when vegetated spurs jutting into the river are breached and cut off. The braids live an ephemeral life, changing over time as the river seeks new identity.

The relatively quiet flow of the Manitowish minimizes this chameleon existence, but high waters during snowmelt have often changed portions of the channel. It's the nature of rivers to be amorphous.

We've seen this firsthand. We live today in Mary's grandparent's home in Manitowish. Mary's mother Alyce was raised here. Mary often recalls going swimming as a child in a certain pool along the river where the highway came close to the channel. Today that pool no longer exists, and the site looks little different than any other stretch of the river. Now other pools formed on the outside edge of meanders make for good swimming. The inside curves have built up sand deposits sufficient to make tiny beaches perfect for sunbathing, or for otters to eat a meal.

Soon Circle Lily Creek enters from the right, draining the many thousands of acres of bog and wetland to the north, an area known to locals as the Great Circle Lily Marsh. This wetland complex is wild country. Wisconsin's Natural Heritage Inventory Program calls part of this area the "Toy Lake Cedar and Ash Swamp," an apt description that helps keep the riffraff out. Hardwood swamp, white cedar swamp, and alder thicket make this 3,473-acre area a tangle. If you were fleeing from UFOs in a B movie, this would be a great place to run into the deepest swamp to shake off the aliens. I recommend this area to low-budget filmmakers and refuge-seekers alike.

But I'd also recommend it to birders and wildflower enthusiasts. Boreal birds are well-represented here. A winter snowshoe hike can yield boreal chickadee, spruce grouse, gray jay, and black-backed woodpecker.

The size of the area renders it worthy of consideration for protection. Fragmentation has occurred in so much of our Wisconsin landscape that finding an intact ecological community of several thousand acres is rare.

Circle Lily Creek seldom holds enough water to float a boat, except possibly in spring. But if you can find passage, the likelihood of seeing another person is nil.

Circle Lily is a traditional put-in and take-out site for paddlers. Manitowish Access Road runs to a now-defunct bridge that crosses the creek just before it enters the Manitowish. You can park next to the road near the river.

A canoe/kayak/tubing outfitter has just begun business here. Hopefully, ecotourism will not display its uglier side and lead to numbers that exceed the river's capacity for absorbing travel.

Morning on the River

When dawn rinses
the black heavens away,
and satin pink
runs in on the suntide,
an eagle sits perched in the flagged pine
beside the river.

Below the eagle
the marsh resonates with primal calls,
slowly goldening and shaking out the reflected stars
from its blackwater.
A blue luster creeps between the cattails,
tapping shoulders with the gold light
on the river.

Songs rise on the morning curtain,
each in its appointed thrush time
according to sparrow light,
warbler heat,
eagle hunger.
All carry in the wind and flow
over the river.

The sun backlights the eagle.
The bird's silhouette falls
from the naked limb,
then glides black on the uplifting hands
of rising heat
above the river.

The songs die in an instant
as the shadow plunges,
scraping the river's surface
with blazing yellow talons,
and emerges . . .
empty.

Reluctant to give
this one time,
the river pushes on with no celebration
no gain, no loss.
Just blue with sky, white with clouds.

Many a southern Wisconsin river has been overrun by paddlers and tubers who in the context of seeing so many others on the river, tend to devalue the river and degrade it with litter and roguish behavior.

With its rich history, excellent wildlife values, and peaceful nature, the Manitowish deserves respectful, quiet travel. The challenge for ecobusinesses is to make a living without significantly diminishing the resource they utilize.

≈ ≈ ≈ ≈ ≈

Keep your eyes on the trees just downstream from the entry of Circle Lily Creek; eagles often perch along this stretch.

Three campsites are conveniently located on river-right between here and the town of Manitowish, all on upland banks under tall pines.

Just past the first campsite, a series of large willow trees lean out over the river (river-right). During the entire paddle from High Lake to here, very few willow trees have poked their heads from the shoreline—and for good reason, because willow trees are poorly adapted to northern winters. Five species of willow trees may be seen in northern Wisconsin, but only one native willow tree, black willow (*Salix nigra*), barely extends its range this far north. The nonnative willow most likely to be seen along northern rivers is the crack willow (*Salix fragilus*), which sometimes escapes onto disturbed sites.

However, thousands of native willow shrubs clothe the wetland shores, often associating with their brother-in-arms, the tag alder. Willow and alder shrubs provide superb habitat cover for a host of birds that find sanctuary in the tangle of their limbs and foliage (read more about the ultimate tangler, tag alder, at the end of the chapter).

The sudden intrusion of a highly unusual species like these willow trees is often a red flag, one typically waved from human hands, indicating that some disturbance occurred on a site. Scramble up the high sandbank and there stands a large field partially invaded by trees. A CCC (Civilian Conservation Corps) camp, housing several hundred men, thrived here in the 1930s. These men replanted many forests, fought fires, built bridges and roads, and overall, made life in the Northwoods significantly easier.

Only a few odds and ends of the camp remain, but the disturbance to the land still speaks clearly. Scrubby, scattered trees have worked to take hold here, but 65 years later, the open-field plant community still predominates. In most of the soils of the Northern Highlands, plant succession after a clearcut usually leads to the rapid invasion of big-tooth and quaking aspen, better known as "popples." If a fire followed the cut, jack pine, red pine, or white birch may move in instead. But on sites that are heavily trampled, where the sandy soils become compressed, recolonization proceeds very slowly, providing long-term evidence of human habitation. Old fields with nonnative willow trees are a dead giveaway of human presence long since gone.

High sandbanks also mark the CCC site. We've observed over the years what appears to be a fox den in the south-facing banks. This location offers a reminder of winter, the ultimate limiting factor that determines the presence of plants and animals in the Northwoods. The advantages of building a den facing south on a high bank with an open wetland across from it have not been lost on the fox clan. Where better to collect the warming rays of sun on cold winter mornings? If you can't alter the macroclimate, learn to take advantage of the microclimate. The physics of passive solar heat collection have been the study of many northern animals, their research leading not to scientific papers and book titles, but to survival and young-of-the-year.

Along these same lines, note the number of cavity holes in dead trees that face the river, and which usually look south as well. While woodpeckers knock their heads against the wall all day, they've not knocked loose their common sense. Where better to lay your eggs in May, a month that believes in the equal opportunity weather of 75°F and southerly winds one day, and northerlies with snow the next? The entrances of woodpeckers' holes most often face in a southerly direction, increasing solar exposure.

If birds could speak, we would hear them saying, "Leave the dead!" Old dead trees are not only heirlooms; they are also full-convenience motels. Here, birds and mammals can find accommodations, food, rest, and a window, albeit perpetually open, from which to more safely observe the world. When the window is hammered out on the south side, the room comes complete with heat, at least when the sun shines.

Some 32 species of birds in the Upper Midwest need cavity trees for nesting, from chickadees to nuthatches, bluebirds, wood ducks, and various owls. Dead and dying trees also provide a host of benefits beyond nesting cavities. Called "residual trees," or better yet, "wildlife trees," these snags, standing or down, serve all of the following avian needs:

- Foraging sites for bark probers and gleaners.
- Cavities for roosting and nesting for primary excavators (woodpeckers).
- Cavities for roosting and nesting for secondary users (ducks, owls, some passerines).
- Hunting perches.
- Song perches.
- Shelter for passerines.

Standing trees are only part of the snag equation. Downed, dead woody debris provides niches for a host of small invertebrates, fungi, lichens, mosses, and microoganisms that exploit this pool of nutrients. Downed logs also provide sites for seedling establishment, particularly the seeds of yellow birch and hemlock. They're also used for display and drumming sites by ruffed grouse, and they provide feeding and nesting places for ground-foraging birds and small mammals.

I remember walking in several different-aged clearcut stands and being impressed at the array of birdlife, particularly in those stands that had left a series of standing snag trees. Those snag trees were like magnets, particularly for singing males proclaiming their virility.

My anecdotal observations are supported by studies indicating that species richness and avian density increase when wildlife trees are present in an early regenerating stand. One Superior National Forest study measured the density of 26 breeding bird species. It ranged from only 3.9 territorial males/hectare in the least complex habitat to 8.6 territorial males in the most complex.

Guidelines for maintaining wildlife trees (snags) in cutover areas have been around since the late 1970s when the U.S. Forest Service published a detailed manual for western forests and also summarized habitat characteristics in the Midwest. However, state DNRs and the U.S. Forest Service have different guidelines for snags. The Minnesota DNR wrote its guidelines in 1994, recommending that seven to 15 trees greater than 6 inches in diameter should be left on each acre.

Wildlife trees can be scattered throughout a timber sale, or left in reserve islands as live residuals. The problem with snags is that they are temporary, because of decay and blow-down. Therefore, enough trees must be left to provide habitat through the next cut, which may be 40 years or more in the future.

There's a trade-off involved in leaving single, dispersed trees. Habitat fragmentation increases, providing opportunities for predation by cowbirds, and others. But kestrels, bluebirds, and swallows prefer cavities in the open, so as with most things in nature, some species benefit, while others suffer.

Woodpeckers work as the primary excavators in our northern area, including yellow-bellied sapsuckers, hairy woodpeckers, flickers, and pileated woodpeckers. In southern ecoregions, red-bellied woodpeckers and red-headed woodpeckers are the primary excavators.

Larger woodpeckers like the pileated, red-headed, red-bellied, yellow-bellied sapsucker, and northern flicker work on larger diameter trees, generally 10 inches or more.

Cavity nesters
Barred owl family

Smaller birds, like chickadees, downy woodpeckers, and nuthatches, excavate their own cavities from soft snags in smaller and more rotten trees.

Cavity-nesting birds prefer live deciduous trees with heartwood decay, branch stubs, broken tops, and previously excavated cavities. Older aspen stands easily supply such trees, as long as some are allowed to stand past the normal 40 year cutting rotation.

≈ ≈ ≈ ≈ ≈

Another half hour downriver, more willow trees signify human alterations—in this case, a wayside park and boat landing. Picnic tables, outhouses, and an easy launch or pickup site await.

Snag-dependent Wildlife

F = food N = nest P = perch

Cavity Excavators

Black-backed Woodpecker	F	N	
Common Flicker	F	N	P
Downy Woodpecker	F	N	P
Hairy Woodpecker	F	N	P
Pileated Woodpecker	F	N	P
Red-bellied Woodpecker	F	N	P
Red-headed Woodpecker	F	N	P
Three-toed Woodpecker	F	N	
Yellow-bellied Sapsucker	F	N	

Hawks and Owls

American Kestrel		N	P
Bald Eagle		N	P
Barn Owl		N	P
Barred Owl		N	P
Merlin		N	P
Osprey		N	P
Red-tailed Hawl		N	P
Saw-whet Owl		N	P
Screech Owl		N	P

Open Farm and Meadow

Bewick's Wren		N	
Eastern Bluebird		N	P
Tree Swallow		N	

Residential Areas

Chimney Swift		N	
English Sparrow	F	N	P
House Wren		N	
Purple Martin		N	P
Starling	F	N	P

Water Birds

Belted Kingfisher			P
Black-crowned Night Heron		N	P
Bufflehead		N	
Common Goldeneye		N	
Common Merganser		N	P
Common or Great Egret		N	P
Double-crested Cormorant		N	P
Great Blue Heron		N	P
Hooded Merganser		N	P
Wood Duck		N	P

Woodland Birds

Black-capped Chickadee	F	N	P
Boreal Chickadee	F	N	P
Brown Creeper		N	
Carolina Wren		N	
Great-crested Flycatcher		N	P
Prothonotary Warbler		N	
Red-breasted Nuthatch	F	N	
Ruffed Grouse			P
Tufted Titmouse		N	
Turkey Vulture		N	P
White-breasted Nuthatch	F	N	
Winter Wren		N	

Reptiles and Amphibians

Most Salamanders		N	
Tree Frogs		N	

Mammals

Big Brown Bat		N	
Black Bear	F	N	
Bobcat		N	P
Deer Mouse	F	N	
Eastern Chipmunk			P
Eastern Pipistrelle Bat		N	
Fisher		N	
Fox Squirrel		N	P
Gray Fox		N	
Gray Squirrel		N	P
Hoary Bat		N	
Least Chipmunk		N	P
Little Brown Myotis Bat		N	
Mink		N	
Northern Flying Squirrel		N	P
Opossum		N	
Pine Marten		N	
Porcupine		N	
Raccoon		N	
Red Squirrel		N	P
Red Bat		N	
Silver-haired Bat		N	
Snowshoe Hare		N	
Southern Flying Squirrel		N	P
White-footed Mouse		N	

Courtesy of *Wisconsin Natural Resources Magazine*

On the other side of the river from the boat landing looms a concrete bridge abutment. This is the remains of a former highway, now called Sandy Beach Road. A bridge once crossed the river here, but now only a rutted dirt road leads back to Sandy Beach Lake.

In late July, 1990, the Wisconsin DNR conducted a fish and habitat survey from this boat landing. They first sampled a half-mile section of the river downstream for specific habitat parameters like stream width, depth, velocity, and substrate cover. They then assessed their data using the Wisconsin Warm Water Physical Habitat Rating System (see Appendix A for a sample form).

The researchers electroshocked fish along two segments of this stretch, stunning the fish long enough to measure and categorize them. They used the Index of Biotic Integrity to analyze the fish community, and then compared it to three other rivers they surveyed that same summer.

In the study section, the Manitowish consists entirely of runs, with no large pools or riffles. The average channel width was 83 feet, while the mean maximum "Thalweg" depth, which corresponds to the deepest part of the channel, was nearly 3 feet. Sand and silt dominated 85 percent of the substrate, while rock/cobble/gravel accounted for 15 percent—a number that's likely much higher than the average over the entire length of this sandy-bottomed river.

Available instream cover, where adult fish can find shelter or remain hidden, amounted to 16 percent of the river surface, a percentage considered "excellent." Instream cover includes woody debris, emergent and submergent vegetation, overhanging vegetation, and boulders.

Bank stability, measured as the percentage of the area not susceptible to erosion, presented little problem. The Manitowish received a 95 percent rating.

Bank vegetation within 5 meters of the stream channel was characterized as 55 percent lowland grasses, 32 percent lowland shrubs, 7 percent upland deciduous trees, and 4 percent open marsh.

Riparian vegetation from 5 to 100 meters from the stream channel was also estimated. Thirty-eight percent of this plant zone was found to be lowland shrubs, 31 percent upland deciduous trees, and 16 percent lowland grasses.

When researchers put all these values and a few others together in the WWPHRS, the Manitowish scored the highest relative to physical habitat of the four rivers studied. In layperson's terms, its score of 77, indicates good to excellent habitat. Stretches of the Wisconsin, Tomahawk, and Squirrel Rivers were also evaluated in the study.

When researchers applied the IBI to the fish community, they also rated the Manitowish the highest among the four rivers, with a score of 80. This score corresponds to a rating of excellent. Catch rates for all gamefish, including northern pike, walleye, burbot, and smallmouth bass, were generally low. In addition, no young-of-the-year sturgeon were found. However, overall species richness was excellent, and pugnose shiners and greater redhorse were found, both of which are threatened species in Wisconsin.

After all was said and done, the

Index of Biotic Integrity (IBI)

The IBI is similar to a battery of key medical tests given to determine an individual's health. The IBI supplements a chemical analysis of water, but doesn't replace it, because the biological integrity of a river requires many pieces of information. A channelized, lifeless river, or a clear acidified lake, could meet the chemical standards for clean water, given that their waters are free of contaminants. But they may also be lifeless and fail the test of biotic integrity.

The goal of the IBI is to determine the actual health of a river. And health in rivers, like in humans, is far more than the absence of chemicals or disease.

The IBI is based on a river's fish community, and requires a sample of all fish in the river. Researchers must asses the structure of the fish community, its abundance, and the condition of individual fish.

The piscivores, the top carnivores, are usually lower in number in a degraded stream. Instead, omnivores dominate in a degraded stream, because fish that eat anything usually tolerate higher levels of pollution and are habitat generalists.

Healthy rivers have high levels of biotic integrity, meaning they usually have a complete range of native fish for the region, with all ages represented from fry to grandpas. They also have a full range of intact biological processes, like energy production, reproduction, competition, predation, and adaptation. Healthy rivers have fish that are intolerant of pollution and that are in good physical condition. Good numbers of predatory fish typically exist at the top of the food chain in healthy rivers.

The IBI comes in another version to use with aquatic insect larvae (the Hilsenhoff Biotic Index), and is probably an even better index than utilizing fish species. Insect species have well-defined tolerances for pollution, and their presence or absence in a river tells a lot about a river's water quality. Some insects tolerate pollution, while others are intolerant. The proportion of grazers, scrapers, and predators provides valuable clues to the river's condition.

Whichever index is used in the end, they are both just tools. Ecologists must still translate the data and statistics into statements about the health of the river. Integrity means wholeness, completeness, and health. How does one truly measure these? Unfortunately, the subjective determination of a river's health will always be a matter of debate, much like the opinions of several doctors who differ in their prognosis of the same patient.

researchers recommended relatively little habitat improvement for the Manitowish. They suggested the possible use of habitat improvement structures like tree drops, half-logs, or boulder placements to improve cover for smallmouth bass and lake sturgeon reproduction. These suggestions make sense when you consider how little of the river runs over rocks and through upland forests. However, the study amply demonstrates that the Manitowish is doing fine without "improvements," so maybe we should leave well enough alone. The researchers also noted that additional surveys should be undertaken to arrive at a more complete picture of fish life in the Manitowish.

Most rivers are not so fortunate as to have had two experienced researchers survey them twice and write detailed reports about them. Even if reserachers receive funding for sampling a river, the research is typically limited to a few stretches of the river, which may represent only a tiny snapshot of the entire river. Thus the research data may be limited in its overall application. Given the state and nationwide lack of funding for professional research, the question becomes: What can you and your neighbors do to get a legitimate sense of a river's health?

Quite a lot, as it turns out. The University of Wisconsin Extension, in cooperation with the WDNR, has put together a "Water Action Volunteers" program, which offers an opportunity for volunteers to do stream monitoring. A series of "Volunteer Monitoring Factsheets" describes each of the parameters that volunteers measure on their chosen river: temperature, turbidity, dissolved oxygen, macroinvertebrates, and habitat assessment. The information is organized clearly, written well, and uses nontechnical terms without seriously diminishing the value of the research. It's an excellent program. See the resource guide at the end of the book for more information.

Read more about the fish sampling that was done on the Manitowish River at the end of the chapter.

Before continuing on from the landing, don't forget to pick up a DNR interpretive river trail guide to the next short section of the Manitowish. Numbered posts direct paddlers to stop for a moment and read the booklet.

≈ ≈ ≈ ≈ ≈

We've left all the lakes well behind and the river now begins to take on its wildest character.

We're now about an hour from the bridge on Highway 47, and the unincorporated town of Manitowish. At the landing, rocks briefly dominate the river bottom. The rocky substrate grinds many a canoe bottom in low water, but it also provides great cover habitat for aquatic insects and spawning habitat for walleye. The river bottom will soon revert to its sandy ways. In the next six hours of paddling between here and the Turtle Flambeau Flowage, rocky cobble will appear only a few more times.

A narrow island briefly divides the river into two braids, both of which normally have adequate water for passage.

A beaver lodge soon arises on river-right. This lodge has a checkered history of activity, fluctuating in its use over the last two decades. If the lodge structure still remains, check for recently peeled sticks on or around the lodge, and for signs of chewing on the many alder and willow shrubs along the shoreline.

An open field soon appears on river-right, the remains of the Plunkett homestead, cleared a century ago. The Plunkett brothers lived close to the river, just back from the high bank. Their idea of garbage disposal, as was the custom of many at the time, was to chuck their old bottles and cans over the edge of the bank and let the river carry the refuse away. The sandbank still yields a few rusted and broken remains of this early 1900s state-of-the-art landfill system.

I wrote a story about the Plunketts, which was published in a now out-of-print book called *Harvest Moon: A Wisconsin Outdoor Anthology*. The Plunketts' story speaks to the hardships, the changing landscape, and the personalities of the time. So does Plunkett Road, a slice of old Highway 51 that now bears the family name. I recommend having a picnic here. To appreciate history, it helps to sit with it for a while.

In the October chapter of *Sand County Almanac*, Aldo Leopold wrote of his visit to an abandoned farm:

"I try to read, from the age of the young jack pines marching across an old field, how long ago the luckless farmer found out that sand plains were meant to grow solitude, not corn. Jack pines tell tall tales to the unwary, for they put on several whorls of branches each year, instead of only one. I find a better chronometer in an elm seedling that now blocks the barn door. Its rings date back to the drought of 1930. Since that year no man has carried milk out of this barn.

"I wonder what this family thought about when their mortgage finally outgrew their crops, and thus gave the signal for their eviction. Many thoughts, like flying grouse, leave no trace of their passing, but some leave clues that outlast the decades. He who, in some unforgotten April, planted this lilac must have thought pleasantly of blooms for all the Aprils to come. She who used this washboard, its corrugations worn thin with many Mondays, may have wished for a cessation of all Mondays, and soon."

Leopold had the ability to read landscapes as if he were combing the personal journals of the occupants.

I've often wondered what Leopold would have read on the Plunkett homestead. I imagine he would first note the brittle remains of a giant willow tree standing back 30 yards from the river that Mary and I have observed, for at least the last two decades. He would wonder where its limbs have gone, and likely detect that they were chainsawed off by a firewood hunter. Mary and I watched this man work one day as we paddled by. He considered these limbs, which were once ideal perches for eagles, kingfishers, and insect-loving summer songbirds, more important for their BTUs, though willow throws off a weak fire compared to harder woods.

The size of the willow would say to Leopold that the land was in use by the turn of the 20th century. The nonnative willow was a shade tree of choice, planted quickly by new homesteaders who brought the seeds with them in the understanding of what a hard life was like under a hot summer sun.

The old field continues to hold its own against invading tree species like aspen, red pine, and jack pine. Leopold would likely say that no fires have run their course here. Why else the failure of the jacks, reds, and white birch to be thick upon the land? Instead, the soil resists the forest's advance, undoubtedly a testament to cows and horses that grazed this field over and again, trampling seedlings, compressing the soil, eating to the ground the most nutritious plant species, and leaving the rest.

The pine trees behind the field are mostly of similar height and diameter, except for a few big white pines. The plantation pines stand in testimony to the planting regime instituted by the DNR after it acquired the property. The Plunketts probably saved the big white pines as historical heirlooms to decorate the entrance to their homestead, museum pieces to remind them of the former regality of the Northwoods.

The foundation of a building lies among some of these young trees to the right of the path that leads to the road. The age of these trees indicates that the building was here as recently as 30 to 40 years ago, before it was torn down under its new public ownership.

No one can say why no descendants in the Plunkett family came forth to claim the land before it was sold to the state, but guesses come easily. Perhaps the sandy soils that grow pines and wild blueberries—but not two crops of hay or a field of corn—led them south to richer soil.

Possibly the depressed land prices made all their efforts seem fruitless. It would be another decade or more before the development boom in the 1980s would begin along the lake and river shorelines. The local economy, slow when at full steam, added to the economic hardships.

Maybe they grew weary of deep snows and long winters. Or tired of summer's bountiful mosquitoes.

Perhaps the strong memories of the land and departed family members were too hard to live with, or the memories too few to compel the next generation to take over the land.

Or there simply may have been no heirs to step forward, or no money to step forward with.

All that we can say is that the land was left to nature's reclamation some 30 to 40 years ago, and that the Plunketts' land use was hard enough to hold off most of the plant pioneers whose job is restoration, not civilization.

Reading a landscape requires knowledge of history, ecology, botany, economics, sociology, and a handful of other disciplines. Once successfully read, the landscape acquires a sense of place that becomes part of the reader's soul, a center point around which one's life revolves. This sense of place grows incrementally by living with the land over time. There's no shortcut, no virtual course, no one source of information that provides all the recipes for the slow cooking required of such an endeavor. Would that there was. The land would certainly be the better for it.

For a measure of your sense of place, take the test at the end of the chapter.

≈ ≈ ≈ ≈ ≈

Back on the river, and shortly beyond the Plunkett homestead, a lazy slough branches off to the right. Here is another historical text, the former main channel of the river that was cut off one spring by floodwaters and left to grow a new identity.

Sloughs are like waterfowl living rooms, places to relax out of the flow of the things and grab something to eat in peace. A quiet paddle back here may unveil wood ducks, mallards, blue-winged teals, painted turtles, green frogs, muskrats, or otters. Typically, an eagle perches in one of the big pines, overseeing, and adding constant tension to, the whole pastoral affair.

Water lilies provide a microhabitat within the sloughs, a veritable hatchery and restaurant for creatures of slow backwaters. White water lilies grace sloughs, acting as queens in the monarchial floral government that rules the backwaters. *Nymphaea odorata* stands as the Latin name of the most common species, a most appropriate title given that nymphs were mythological deities conceived of as beautiful maidens who inhabited the sea, rivers, woods, trees, and mountains. For its part, *odorata* speaks to the wonderfully fragrant odor of the blooms that reminds one of fresh oranges.

Thoreau had a different take on the fragrance, writing that it reminded him "of a young country maiden . . . wholesome as the odor of a cow." Faint praise indeed if you're a country maiden, but blissful words, I'm sure, to cows.

White water lily, along with its family member the yellow water lily (*Nuphar variegata*), cover the surface of shallow, quiet waters with their broad, flat leaves. These lily pads provide habitat for a host of critters that appreciate having the dry dock to lounge on. The topside of the leaves typically show the tracks and trails of the feeding insects, egg layers, cover-seekers, and loafers who have used and abused them. Feeding takes place on nearly every part of the water lily, even on the algae that often attach themselves to the stems and undersides of the leaves.

Dragonflies, damselflies, numerous frog species, and a variety of other less commonly known insects use the leaves for territorial perches. The undersides of the leaves often provide shelter for the egg masses of caddisflies, damselflies, whirligig beetles, and many other insects.

The flowers too provide habitat for many insects, so be sure not to inhale deeply from the sweet-smelling white water lily without first checking to see what you might be inhaling along with the perfume.

Northern fathead minnows spawn in water lily beds, and where there are minnows, there are northern pike. Green and mink frogs make a good living here, too. Once when stopped with a paddling group in a slough full of lily-pads, we watched a northern water snake grab a green frog from atop a lily pad and swim away with the frog in its mouth. The frog's legs were the only anatomical part still visible, and they were still kicking.

While water lilies are a relatively unimportant food source for waterfowl, muskrats and beavers absolutely relish them. In the summer, beaver feed almost entirely on aquatic vegetation rather than woody trees. They do a Johnny Appleseed-like job of planting water lilies all about by dropping rhizome fragments that root and form new plants. Beaver dams create back-

Winter Survival

In the deep snow
I labor in my snowshoes to stay afloat,
to swim over the surface
that wants to draw me down
and envelop me
as it has the spring wildflowers,
the blueberry stems, the lady ferns, the princess pines.
My dogs stay behind me, having learned
that to flounder in deep snow is to drown dry and cold.
I wonder how any animal can survive this.

At least the birds can remain in the trees.

The snow shallows
under the pines and balsams,
and tracks of squirrels and hares meander quickly
to tender buds and buried caches.
I hope they have planned well.

In March, when the sun is high but the snow still deep,
the buds and seeds will be gone,
and the energy lost in finding more food won't balance.
Death will come.
To die having made it to the sight of shore,
to die luckless when the thaw came sooner just two counties
south . . .
Well, I wish to die in the full flush of life,
not in its other promise of emaciation and despair.

So I snowshoe looking at more than tracks,
thinking of more than my own labor.

A red squirrel chatters,
a grouse flushes,
a flock of chickadees appraise me from branches so near
I could touch them.
"Hello, Hello," I say to them.
They don't respond,
but that is my expectation,
as it is theirs.
They understand I will soon retreat
and leave them their woods, their river.

water habitat ideal for water lily growth.

Our largest mammal in the Northwoods, the moose, feeds on water lilies extensively. White-tailed deer eat the plants, too, and the literature notes that even porcupines like water lilies. However, I would be quite amazed to see one lounging in a wetland eating lilies.

Read more about water lilies at the end of the chapter.

≈ ≈ ≈ ≈ ≈

Just past this slough, large white pines briefly clothe the point of the shoreline, providing attractive perch sites for bald eagles that fish the river. I've often found the oily black scat of otters under these pines. A slough rich with fish and crayfish and clams, and a dry upland from which to dine, comprise much of the whole world from an otter's perspective.

Matt Plunkett built a cabin on this point, its foundation still visible behind these old pines, and a home now to snakes and other baskers and burrowers.

On river-left, stands signpost #5, a post that originally directed attention to an eagle's nest in a large white pine across the intervening wetland. That nest blew down the day after the trail brochure was printed in 1994. Mary and I have kept track of the territorial nest in this area since 1980. The nest blew down once in the 80s, and was rebuilt in a neighboring tree that same year. After blowing down again in 1994, the nest reappeared in the year 2000, again in a different white pine. It's easily visible amongst the many pines that rim the wide wetland. Whether the wind now leaves it be—and for how long—is an unanswerable question.

A large beaver lodge appears on river-left a hundred yards past signpost #5. This lodge, like most lodges, has gone through a cycle of use and disuse, so it's anyone's guess if it will be active as you paddle by. A little farther downstream, another newer lodge has sprouted up on river-right (read more about beaver lodges at the end of the chapter).

The river now begins to take on its true meandering character, a slow, wandering personality it will exhibit until it empties into the Turtle-Flambeau Flowage. I've often wondered why rivers loop back upon themselves in an endless ribboning, when, as we all had drilled into us at a tender age, the shortest distance between two points is a straight line. Wouldn't the river want to be efficient and carry its load of sediments and loose ends directly to its destination, and be done with the job in the best tradition of UPS? Wouldn't a river simply follow a straight line downhill?

Nope.

Luna Leopold, renowned hydrologist, writes, "The typical meander shape is assumed because, in the absence of any other constraints, the sine-generated curve is the most probable path of a fixed length between two fixed points."

I'm pleased to know that there's a mathematical rule that explains meandering, but if your recall of trigonometry is equal to mine, this sentence means little to you. In fact, it may put you instantaneously to sleep in a Pavlovian reaction to high school math.

Stay awake—this math has meaning for rivers.

Leopold also wrote that rivers rarely flow in straight lines longer than 10 times their width. Applying this formula to the Manitowish, at an average width of 50 feet, the river should turn every 500 feet. And so it does on the whole.

Rivers meander. Meltwater running off glaciers meanders. Currents in the Gulf Stream meander. Rainwater trickling down your kitchen window or your car windshield meanders. Set up an inclined bed of sand in a laboratory, send a stream of water down it, and the water will almost immediately begin to meander. It may meander subtly at first, but eventually it will cut a true meandering channel. Wherever water flows, it meanders. This is the way of much of life, so why be surprised if water takes a similar journey?

Water follows the path of least resistance, turning away from obstacles and following variations in the terrain. But meandering occurs whether there are obstacles or not, significant topography or not. In fact, rivers in flatlands meander the most.

Leopold says, "River channels are curved, sinuous, or meandering because that is the natural and most probable form. It is the form that conserves energy and tends at the same time to make energy expenditure along the streamline most uniform. The physical forces act to promote a curvilinear form. A reach of stream that is straight tends to become curved, and in no known instance does a curved form become straight through any appreciable distance."

Once a river begins to meander, the curves grow via centrifugal force. The current speeds up on the outside of a curve, causing the outside bank to erode, while the slower water on the inside of the curve allows sediment to settle out and accumulate

The water deepens on the outside, or concave bend, and becomes more shallow on the inside or convex bend. The undercut banks on the outside of the curve offer deeper water, shade, and cover for fish. Wet a line here for the best chance of a fresh shoreline lunch.

Along the Manitowish, the deposited sediment on the inside bend, called a *point bar*, creates a fine basking

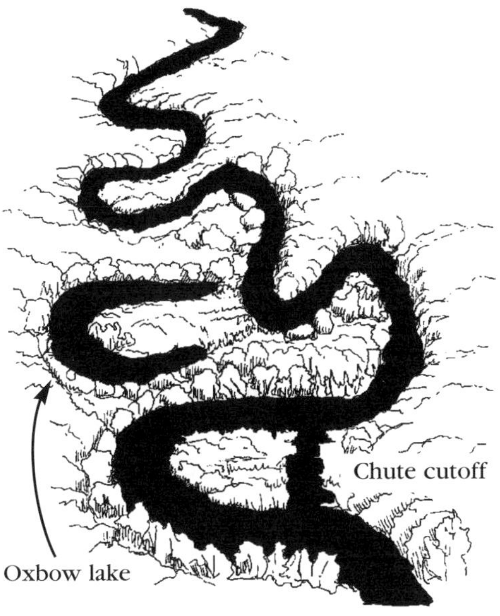

beach for ducks and turtles, and a dry table where otters can crunch a crayfish. Point bars sometimes form long, narrow points at the tip of a meander lobe, hence the name.

> *For the last word in procrastination, go travel with a river reluctant to lose its freedom in the sea.*
> — Aldo Leopold

The river-bends tighten as the combination of erosion and deposition continues. Eventually, the bends loop so radically that they become nearly circular. I've watched type A paddlers, driven loopy by all the looping, pick up their canoes and race across the short uplands between the meanders in order to make better time. At some point over the years, the river does the same thing, cutting a new, more direct route, called a *chute cutoff*. The river skirts the old loop, and leaves a crescent-shaped slough behind. Called an *oxbow lake* after the U-shaped collar that fits around a harnessed ox, these cutoff sloughs commonly provide peaceful habitat for ducks, turtles, muskrats, and the rest of the panoply of pond lovers.

Sometimes the neck of the loops becomes so narrow that the river channel simply expands and meets itself. The shortcut is then called a *neck cutoff*, but the result is the same—an oxbow lake, or slough, left behind.

The story's not over quite yet. While ducks and geese often use rivers as migration corridors and wayside rest stops, river channels often migrate downstream, too. Erosion occurs more rapidly on the downstream side of each bend of the river, and over time, a river may actually move downstream through its flood plain. An aerial photograph of a river valley often shows the previous locations of the river. The more rapid the current, and the softer the soil, the faster the river migrates.

Migrating rivers care little for political or property lines, much to the dismay of shoreline owners who are attracted to the river's beauty, but fail to understand its true character. On a large river like the Mississippi or Missouri, the river channel can move great distances. In the lower Mississippi, some of the meanders have lengthened nearly 3 miles over the last 500 years. Near the Missouri, a farmer who lives a mile from the main channel recently found in his soybean field the remains of a steamboat wrecked on Missouri River rocks in the 1880s.

Rivers are at work 24 hours a day—a long shift indeed. One of their

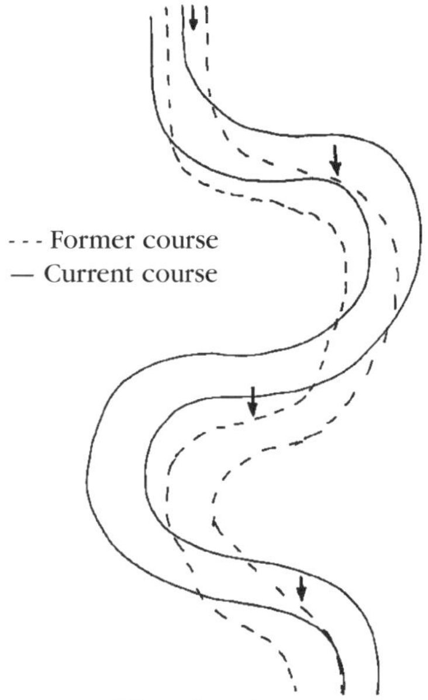

River changing course

jobs is to redistribute land. Think of rivers as conveyor belts, or freight trains, or excavators. They tear down, carry away, and build elsewhere. They are developers run amuck (pun intended), forever unsure if what they've built is in the right place, or if the design is exactly what they want. So they tear it down or tinker with it until it collapses, and then find a likely looking spot downstream to rebuild. They're like people who can't stop messing with the furniture. "Leave it alone, already!" you want to yell. But they just can't. Psychologists call it a compulsion. Rivers are compelled.

Rivers don't just move water and sediments around. They transport larger materials, too, consisting of everything from boulders down to pebbles, from whole trees to bits of their leaves.

A river behaves. It acts according to the terrain through which it flows. Which brings us back to the watershed, a topic we took up briefly when we first began this float on High Lake. Look to the land to understand the water. Are the slopes of the watershed steep or gentle? Does the river channel flow through solid rock or in loose soils? What is the vegetation like along the shorelines? How have people altered the watershed to increase the flow of materials into the river? Have they built storm sewers, removed trees, built hard-surface roads, planted corn, and used herbicides to keep the row "clean"?

Rivers running through hard rock tend to be crystal clear, while rivers carving their way through soft soils tend to be muddy. Rivers cutting through steep slopes tend to carry more sediments. Clearcut the hillside along a river channel; then with the first heavy rain, watch the hill slide into the river.

The sediment load is the material that enters the river from overland flow and also from erosion within the river's channel. The sediment itself helps to further erode a river's banks. Pure water can't erode anything, since a microscopically thin layer of water clings to anything solid, and the water actually slips across itself rather than on the solid material. Add some sediments—some sand, grit, or pebbles. Now, over time, the water can erode solid rock by abrading it, much like the effect of sandpaper.

How much sediment will a river carry? It depends on the speed and volume of the flow. The faster the current, the heavier the rocks it can move. The greater the volume of water moved by the stream, the more volume of rocks, or sand, or silt it can transport.

At its mouth, the Mississippi River unloads 312 million tons of sediment every year. However, China's Chang River (formerly called the Yangtze) wins the trophy for weight lifting. It picks up and carries to its mouth 500 million tons of soil annually. To picture this, line up 500 million one-ton pick-up trucks filled with sediments, and have them drop their cargo at a river mouth over the course of a year. Would there be a traffic jam backing up to the rivermouth? Absolutely! Here's the math: There are 3,600 seconds in an hour, 86,400 seconds in day, and 31,536,000 seconds in a year. In our scenario, about 16 trucks would have to empty their beds every second!

Rivers are the gutters down which flow the ruins of continents.
— Aldo Leopold

But, let's bring this down to scale, since the Manitowish seldom invites comparisons with the Chang. Using round figures, a current of 15 centimeters per second (one-third mph) is enough to move coarse sand; 50 cm/s (about 1 mph) shifts medium gravel; 2 meters per second (4.2 mph) will pick up medium cobble; 6 m/s (12.6 mph) will budge large boulders.

River velocity is measured another way—as discharge, or the amount of water that passes a specific point in a specific period of time. The rate is typically given in cubic feet per second (cfs). A small stream may only discharge 5 cfs, while the Colorado River rumbling through the Grand Canyon varies from 4,000 to 90,000 cfs. The Mississippi River varies from 1.5 to 12 million cfs, and the Amazon, the king of riverflow, has been recorded flowing at 52.5 million cfs. That's about the amount of water a building 1,000 feet high by 250 feet wide by 200 feet long could hold. Empty that building every second, and that's the volume of water the Amazon in flood can push by a single point every second.

How fast does the mighty Manitowish course along? Well, that depends on the season, of course. In spring, after snowmelt when the river typically overflows its banks, it may move at 300 cfs. In the summer, when the living is easy, the water levels are down and you're painting aluminum or fiberglass, or God forbid, KEVLAR, on the rocks, the flow is more like 40 cfs.

As water gets deeper, it tends to flow faster, so water speed increases the farther downstream one goes. This may seem counterintuitive, because rushing mountain streams appear to be traveling much faster than a large river sweeping around a bend. But put a current meter in both those waters, and the truth will be revealed. An average current in a mountain stream on a normal day may travel at 0.4 meters per second, while the current in the broad river much farther downstream may average 1 to 1.5 meters per second at low flow. In flood, a headwater torrent averages 1.5 to 2.5 meters per second, while on an average day on a large river like the Mississippi, water may be flowing at 2 to 3 meters per second.

Depth makes all the difference. When water moves downhill, it's slowed by friction when it rubs against the channel bed and banks. But as the water gets deeper farther downstream, the area of the channel

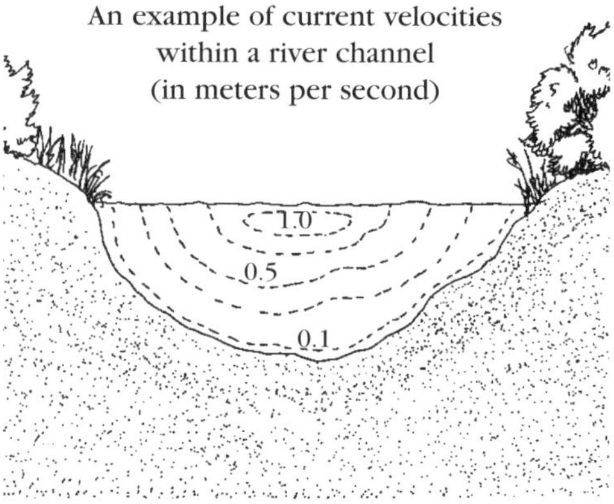

An example of current velocities within a river channel (in meters per second)

bed and banks doesn't significantly increase. So there's less friction, and thus more speed.

Velocity obviously increases with steeper gradients, but to complicate matters a bit further, velocity varies inversely with the roughness of the bottom. Highest current speeds are near the surface and near the center of the channel where friction is the least. We all know this, though few of us have likely thought about it, because of our observations of floating leaves. Leaves close to the banks move more slowly than those drifting down the center of the stream. Submerged, waterlogged leaves near the bottom of the stream trail behind those leaves above them.

Rivers with smooth, sandy bottoms typically flow faster than rivers with rocky bottoms, because as a general rule, rivers with sandy bottoms offer less resistance than rocky bottoms. This helps to explain why velocity usually increases along a stream's length. As channel depth increases, substrates typically become finer. The river transitions from a stony, steep, shallow stream to a deep river with sand. Less friction yields more speed.

In deeper rivers, the friction generated between the water surface and the atmosphere actually makes the velocity highest just below the surface, not at the surface.

This discussion isn't just academic, because everything living in a river has to adapt to the velocity of the current. The problem is that all these factors put together make it difficult to really determine what current velocity an organism is actually experiencing.

Read more about river velocity at the end of the chapter.

That's the big picture, muddled as it is by multiple factors. But the little picture may be even more important. Microhabitats exist in every river, and organisms are adept at finding and utilizing them. A classic microhabitat, known to all whitewater paddlers, is an eddy. A river current nearly stops in the eddy behind rocks. Organisms, including kayakers, can rest in these eddies even in powerful whitewater. Insects know that eddies are a good place to escape a current. Fish know that eddies are a good place to escape the current and eat insects. Good anglers know eddies are a good place to escape the current and catch fish.

Hiding in the spaces between rocks to get out of the current also makes sense for an aquatic insect, in the same way hiding behind boulders to get out of a strong wind makes

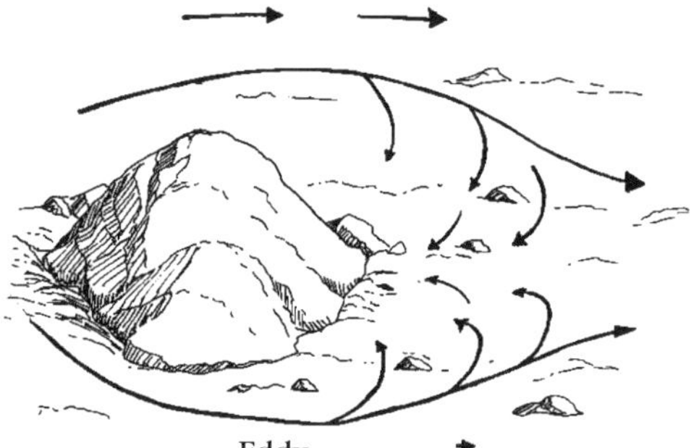

Eddy

sense for humans. Likewise, an aquatic insect hanging out in a plant bed can reduce its exposure to current flow, just like for people there's less exposure to wind in a forest than standing in a field.

Rivers with gravel or cobblestone bottoms usually support more life than sand or silt bottoms. Stones provide more surface area for an aquatic insect to attach itself to, more spaces for laying eggs, and more stillwater to hide in. Not only is current velocity reduced by friction, as discussed earlier, but a thin film of slow water occurs along the surface of rocks and along the bottom of the river. This boundary layer, or zone of stillness, may be seen by releasing a dye into a stream. In this zone of calm water, small organisms may carry on their lives in comparative ease, while unanchored organisms above them are being swept away in the current. In his book *The Bird in the Waterfall: A Natural History of Oceans, Rivers, and Lakes*, Jerry Dennis writes that it's like living where a constant hurricane roars overhead but can't touch you on the ground.

This boundary layer has traditionally been viewed by stream ecologists as a refuge area that is utilized by insects with dorsally (topside) compressed body shapes, like water penny beetles and a number of mayflies. However, this region of greatly reduced flow has been shown in recent studies to be quite thin. Consequently, only the smallest invertebrates and microorganisms could live within the greatly reduced flow. Larger invertebrates would have a difficult time finding sufficient shelter here. Still, the shape of these insects is

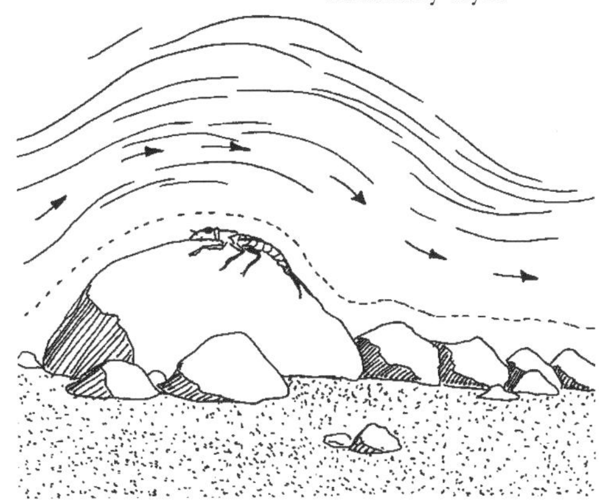

clearly more aerodynamic, and thus helps them stay put to some degree in a current.

Speaking of aquatic insects, literally thousands of damselflies live along this section of the river. The river clearly provides excellent habitat for these relatives of dragonflies. Both suborders, dragonflies and damselflies, are in the order Odonata. In the adults, they are easily differentiated by looking at how the wings are held. Both are unable to fold their four wings back over their abdomen when at rest as many other insects do.

Instead, a dragonfly typically rests with its wings spread out flat like a B-52, while a damselfly holds its wings folded up behind it. Damselflies are also generally smaller, and are comparatively weaker fliers, fluttering like butterflies. Dragonflies fly like they should be wearing Air Force insignia. One other thing to look for is that dragonflies have huge wraparound eyes. In many species, the eyes meet in a seam at the back of the head. A damselfly's eyes are more widely spaced.

The larvae are also easy to separate—the dragonfly's abdomen terminates in three short, wedge-shaped

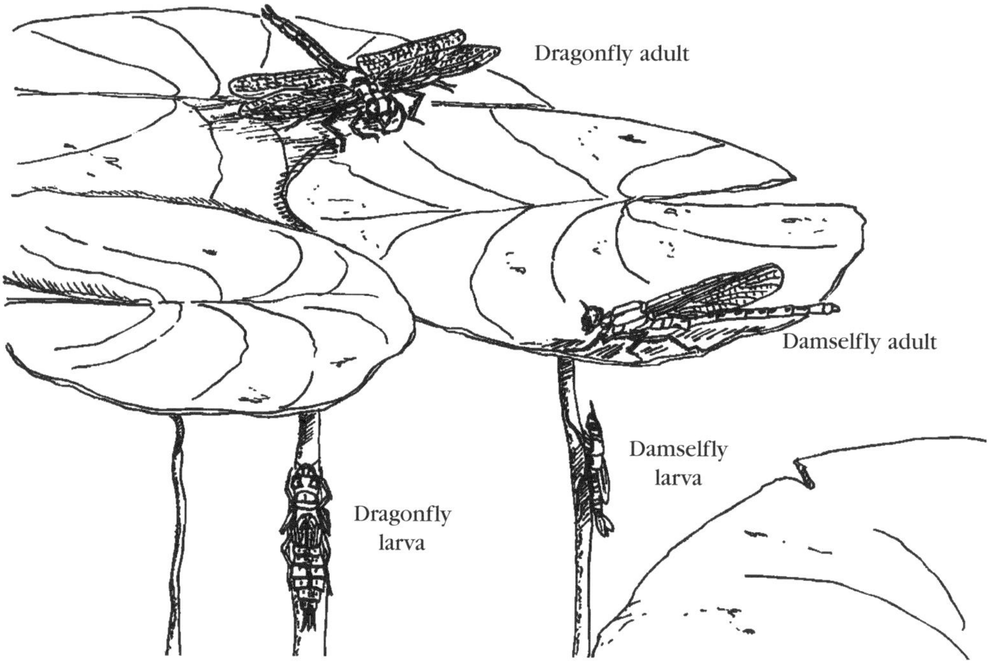

structures, while the damselfly's abdomen ends in three elongated, leaf-like "paddles."

Damselflies typically oviposit, or lay their eggs, in emergent or floating vegetation by puncturing the plant tissue with a knife-like ovipositor. Sometimes, while flying in tandem, the male anchors and guards the female as she lays her eggs. Other Odanata females may lay their eggs directly in the water or poke them into soft mud. Eggs hatch in 10 days or so into larvae, which in most river systems take two to three years to mature.

Dragonflies are often seen flying "in tandem," which is the polite way of saying they are most likely mating. Damselflies and dragonflies couple in a uniquely and highly flexible position that bears no resemblance to the missionary position, and is downright difficult to quite comprehend without an illustrated guide.

Male dragonflies produce sperm in an organ located near the end of their abdomen in their ninth segment. The problem is that their penis isn't there, but is instead at the top of the abdomen on the second segment. So before a male can mate, he must bend double, curl his abdomen under himself and transfer seminal fluid into an inner chamber behind his penis. The female must then curl her abdomen forward so that her genital opening, situated at the end of her abdomen in the ninth segment, meets the male's penis on his second segment, Together they form the familiar "wheel position." They appear to be tail to head, belly to tail.

The male's penis also serves another purpose besides insemination. The penis could be called a de-inseminator as well. The male makes regular undulating movements of his abdomen during copulation, but not in order to copulate. Rather, he's making an effort to mechanically remove the sperm of any other males that have previously mated with the

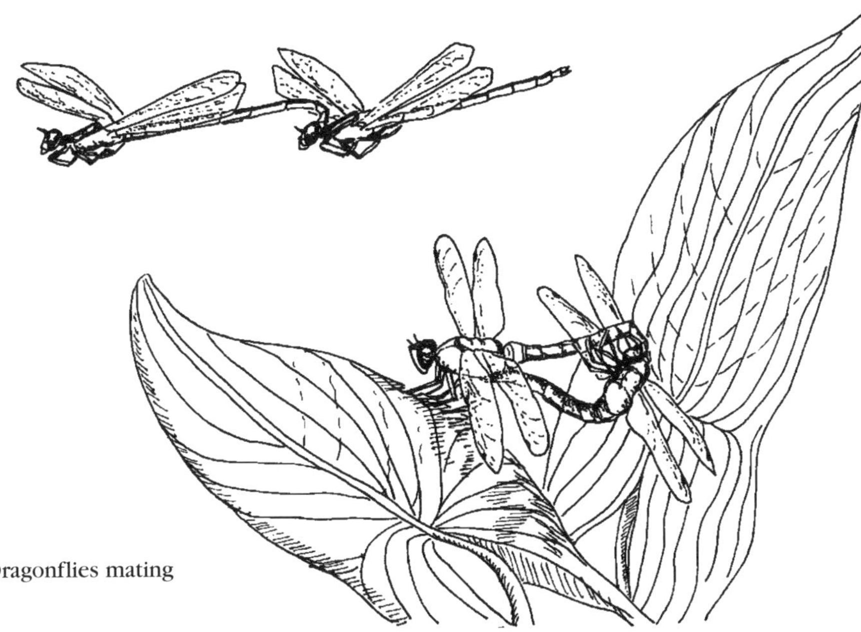

Dragonflies mating

female. The male organ is equipped with an array of attachments evolved to accomplish this task. In some species the male's penis is blunt and is used to squash down any sperm already inside the female, thus assuring that his sperm has top priority in fertilizing the female's eggs. Males of other species have a spiked, bristly penis that is used to scrape out the competing sperm before replacing it with their own.

This process doesn't guarantee that some other male won't sneak in after the current male's handiwork. So some male dragonflies grasp the female's head after copulation, and then guide her to an egg-laying spot, holding her until she completes her task. Males of other species simply hover near her, driving away competitors. Their problem is that many other males, seeing the female about to oviposit, will zoom toward her in an effort to be the last male inseminator. This leads to a melee in which males are buzzing all about the place in a frenzy. If a new male is successful in fertilizing the female, he instantly changes roles. He becomes the guarding male trying to drive the others away, just moments after being one of the rogue males.

Whew!!

Fortunately, dragonflies and damselflies can't argue paternity in courts of law.

Anyway, the likely reason for most people to care about Odonates, besides their beauty, rests in their pseudonym—"mosquito hawk." All Odonates are pure predators, whether in larval form capturing other larval insects, or as adult flying insects capturing other flying insects, like our friendly mosquitoes, deerflies, and blackflies. A dragonfly can nab upwards of 300 mosquitoes a day, so a dragonfly resting on the brim of your hat brings a personal measure of insect protection to your life on the river.

Read more about dragonflies and damselflies at the end of the chapter.

≈ ≈ ≈ ≈ ≈

Let's change the aesthetic pace from the sublime (the beauty of dragon and damselflies) to the bottom slime, and talk about leeches. I may be a naturalist, but I sure don't like these critters. Yet, they're as fascinating as any other animal you may see on this paddle.

I clearly remember the July afternoon when our oldest daughter Eowyn, then 13 years old, took a little trip on inflated rafts with two friends down the Manitowish. We dropped them off, went home, and an hour later realized we'd forgotten to pick up some groceries in Manitowish Waters. We were traveling along Highway 51 when we saw three people in the distance running out of the woods. We slowed down to see what this story might be, and quickly identified our daughter and her two companions. Recognizing us, they sprinted for the car. Their first breathless, and hysterical, word upon reaching us was "LEECHES!!" This mantra was repeated exclusively for a while, and was interjected with wide-eyed recall for many hours, regardless of the topic at hand.

It seems they were jumping on and off their rafts, swimming and splashing and having a good old time, when at one point they jumped into a swatch of aquatic plants, then back up on their rafts. At that moment, they noticed their bodies were covered with hundreds of tiny leeches. Worse yet, they were skinny-dipping.

Not good.

One girl, the one, of course, most inclined to hysteria about bugs and such, had mostly little ones on her, but leeches nonetheless. She likely has nightmares to this day about this event.

Why did this happen to them? As best as I can figure, they jumped off into a recent hatch of leeches. The young of some leech species attach to the parent and are brooded by the parent, so the young may have reattached to our daughter and her friends who managed to step among them. Or maybe the girls stumbled into an important town meeting of leeches. It's hard to say. To the hysterical girls, my natural history cogitations were rather immaterial.

At first glance—and at second glance for that matter—it's a bit difficult to say much about the body structure of leeches. They're segmented animals, narrowing and depressed toward the ends, without a distinct head to use as a ready identifier of a front and back end. An anterior (front-end) sucker feeds, while the posterior (back-end) sucker attaches onto rocks, vegetation, and people.

Many species have toothed jaws, the teeth radiating in three directions. Once the leech has attached itself to its prey, the teeth, cutting like an electric jigsaw, pierce the skin so the leech can inject saliva into the wound. The saliva contains hirudin, an anticoagulant that allows the blood to flow readily. This is why a wound keeps bleeding after a leech has been removed.

A leech can store three times its own weight in blood, but since digestion proceeds very slowly, the leech may not desire another meal for nine months. That's good news for victims.

Some other good news is that not

all leeches are bloodsuckers. Among the 63 species of freshwater leeches inhabiting North America, including about 30 species that are found in the Upper Midwest, most feed on worms and insect larvae. Others scavenge carrion, performing an important recycling function. Only a few species suck mammalian blood. Three species are specific to waterfowl. Once attached to a duck, these species move toward the head and reattach themselves either to the conjunctivas in the eyes or enter the nasal passages.

Biologists identify leeches by examining their eyes under a microscope. The number of eyes varies from zero to five pairs, and they are distinctly arranged.

Leeches have both male and female organs of reproduction, possessing testes and ovaries. As hermaphrodites, they are both potential mothers and fathers. But the sperm of one leech must still be transferred to another leech in order to fertilize eggs.

Most leeches inseminate one another though "hypodermic impregnation," an incredible method that goes beyond bizarre. Sperm-containing capsules are deposited on the partner's body. The capsules then secrete an enzyme containing a substance that eats away the leech's flesh, enabling the sperm to enter the body cavity. Some spermatophores are killed by defensive cells, but others become suspended in the blood and are pumped to the ovaries, where they pass through the walls and fertilize the eggs inside. Three days typically pass before the holes bored into the leech's body heal.

I'm not sure how the hermaphroditic leeches decide which gets to be the male and who gets to be the female, but if choice exists rather than random occurrence, then I think the male choice is a no-brainer. Of course, leeches may themselves be literally no-brainers.

The skin of the freshwater leech serves as its breathing organ. Like snakes, leeches must also shed their skins periodically, rubbing their old skin off on rocks and woody plants or debris.

Historically, leeches were used extensively to bleed patients. For infant teething problems, leeches were placed behind the ears. For a cold, half a dozen leeches were applied to each temple. For a bad stomachache, 20 or more were attached to the pit of the sufferer's stomach.

As a result, an enormous leech trade flourished. In 1850, the French bought about 100 million leeches. Peasants collected leeches on their bare legs in order to make a buck. Russia had to pass game laws in 1848 to protect leeches from overharvest, closing the season from May through July and decreeing a length limit of 2 5/8 inches. At the height of the leech trade, leeches became a rare species in Spain, Portugal, Italy, England, and Bohemia.

Leeches have helped influence history, too. Napoleon's army invaded Egypt and suffered greatly from drinking muddy water containing leeches. The men began coughing, bleeding, and vomiting, and had trouble breathing and swallowing. A gargled solution of vinegar and saltwater was used to dislodge the leeches.

If all invaders would suffer such pain, maybe we humans would do less invading of one another.

Today, leeches seldom make their way into prescriptions from the local pharmacy, but they are used as an adjunct to specific microsurgery; by plastic surgeons when clotting can jeopardize an operation's success; for acute pulmonary edema in which the lungs fill with fluid; and even to relieve certain fevers. Hirudin has been studied for many years as a potential clot-breaker for heart patients.

Leeches also have a practical use while you're out on a river. They make a good fish bait, as most anglers know. And if you get a headache from too much sun and paddling, or a stomachache from too much camp food, well, you know what you can try.

≈ ≈ ≈ ≈ ≈

Near the pull-out on Highway 47, a few clumps of purple loosestrife in flower may appear. The plant blooms in mid-July, even though we've been working to destroy it whenever we see it. This exotic and profoundly invasive species first appeared here on the Manitowish in 1998, and more has appeared in the ensuing two years.

Invasive is the word used to describe many non-native plants and animals that find their way into ecosystems and go hog-wild in the absence of any controlling predators. The term "invasive" leaves some wiggle-room; there are little invasions and big invasions after all. Unfortunately, purple loosestrife ranks among the truly great invaders, like Genghis Khan and Darth Vader.

Purple loosestrife attained its Mongol-horde reputation fairly. Mature plants with up to 50 shoots can produce several million seeds in a summer, of which 60 to 70 percent are likely to survive. This creates a seed bank that remains viable for many years. Purple loosestrife germinates best in open, wet, warm soils, usually along river and lake edges, but even seeds submerged in water for nearly two years remain viable.

If seed germination doesn't spread the beast sufficiently, vegetative reproduction works quite well. Root or stem fragments can take root and form new plants. A broken stem will simply produce more shoots and roots. Whether the plant is submerged in a flood, baked in a drought, or trampled with a herd of bison, it will find a way to adapt. Purple loosestrife's ability to adjust to widely ranging environmental conditions gives it the competitive *oomph* to create monotypical stands. This plant tolerates a host of environmental disturbances, but doesn't tolerate competitors.

Introduced from Europe and Asia in the 1800s as a garden perennial, it now occupies all 49 states in the continental U.S., and its propagation is outlawed in at least 24 of those. Purple loosestrife arrived in Wisconsin sometime in the early 1930s, but it remained uncommon until the 1970s, when it took off. It's now found in 70 of Wisconsin's 72 counties. It's illegal to sell, distribute, or cultivate the plants or its seeds, but some folks still blithely plant it, unaware of its quest for domination. Purple loosestrife can and has overrun wetlands covering thousands of acres, and it has nearly eliminated openwater habitat in those areas. My seemingly exaggerated descriptions are not hyperbole! Destroy this plant whenever and wherever you see it, and do so with glee.

However, all is not completely lost when purple loosestrife invades. Researchers near Saginaw Bay in Michigan surveyed bird life over a two-year period on 18 wetland plots. They tracked the utilization of various types of wetlands by singing or flying birds. They found that avian density was actually higher in loosestrife-dominated vegetation types than in other vegetation types; however, avian diversity was lower in loosestrife than in any vegetation type. Swamp sparrows apparently like loosestrife, as they accounted for 65 to 95 percent of all birds in this vegetation type. The researchers also found small numbers of mallards, blue-winged teal, Virginia rails, red-winged blackbirds, American bitterns, sedge wrens, yellow warblers, common yellowthroats, and American goldfinch.

Many studies have shown that plant form and structure, rather than species composition, play the key role in determining habitat selection by marsh-nesting birds. Purple loosestrife's stout, woody growth acts much like shrub growth, so it is attractive to some birds. Still, its dense stands often replace better seed-producing species that are more attractive to waterfowl, shorebirds, and songbirds, both during nesting season and during migration.

Other studies have shown that the highest levels of avian density, diversity, and productivity occur in marshes where emergent vegetation and open-water are equally present. Dense, monotypical stands of anything, whether loosestrife or corn, lead to a reduction in wildlife diversity. If and when purple loosestrife merely

Purple loosestrife

becomes one more member of the wetland plant community, then its presence will be far less detrimental.

One last note: The recent introduction of five insect species to act as biological control agents on purple loosestrife has met with significant success. Over 100 insect species feed on purple loosestrife in Europe and Asia, but none are native to North America. The newly imported insects were tested at length to make sure they specifically fed only on purple loosestrife, and they appear to be doing just that. Two species of weevils feed exclusively on the roots of the plant, while two leaf-eating beetles and one flower-eating beetle also attack the plant and decrease its vigor.

Purple loosestrife is here to stay, but hopefully these predators will reduce its numbers so it will become simply a member of the diverse wetland plant community, and not a hulking Darth Vader in pretty purple as it manifests itself now.

≈ ≈ ≈ ≈ ≈

Pulling-out is easiest on river-right just past the Highway 47 bridge. Alternate pull-outs are river-right before the bridge, or back in the slough on river-right also before the bridge, if the river is high enough.

Under the Highway 47 bridge are the mud nests of cliff swallows. They nested in significant numbers here for at least 15 years, until 1998 when the Highway Department worked on the bridge during their nesting season and drove them away. They didn't return in 1999, but a few nests reappeared in 2000, hopefully a precursor to full colony restoration in the near future.

I recommend a walk on the abandoned railroad track here, easily accessed via the wooden snowmobile bridge. About a half mile south, extensive northern bog appears on both sides of the track, and runs for several miles.

Very little true bog occurs right along the Manitowish, because most of the river's shorelines are flood plains and are adapted to periods of inundation and drying. Bogs typically occur where there's no inflow or outflow of water, so they're seldom seen right along a river's edge. While this book is specific to river ecology, bogs often occur just back from river shorelines where they can exist out of the flow of things. Bogs occur commonly in this particular area of the Northern Highlands, offering a fascinating ecology worthy of study.

The abandoned rail-bed reveals why the town of Manitowish exists, and also why it's so small and quiet now. The Chicago-Northwestern Railroad carried passengers on four trains a day, two going to Chicago and two coming from Chicago. One stop was at the rail station in Manitowish near the present site of Chuck's Bar. Freight trains also ran through the area, and were the last to be discontinued in the early 1970s.

Given that the trains don't stop here anymore, Manitowish is perhaps fortunate to continue to exist at all, albeit as an unincorporated town with a population of only about 29 souls. Four miles south of here, at the next former railstop down the line, a few houses mark the former status of the unincorporated town of Powell. In Wisconsin, Powell provides the definition of *real* small. It hasn't a bar to its name.

I don't mind small and quiet. In fact, I'd like it even quieter. As Edward Abbey wrote, "What our economists call a depressed area almost always turns out to be a cleaner, freer, more livable place than most."

Seasonal Economic Cycles of the Ojibwa

The Ojibwa acted as fur brokers, the middlemen in the fur trade. Their permanent arrival in this area in the 1730s sparked battles with the native Sioux, Menomonee, and Fox that lasted all the way into the 1850s with the Sioux. The Sioux were driven out of the area by 1745 and the Fox by 1763. The fur trade allowed some Ojibwa to overwinter in the area by remaining close to the fur-trade posts at Lac du Flambeau. But many of the traditional people continued as seasonal migrants, harvesting the gifts that each season and place provided.

Not all bands used the same kind of resources. Some bands diversified within their territory by utilizing the unique subsistence opportunities offered within the territories of other bands. Ojibwa bands shared resources occasionally, particularly between the lakeshore and interior bands.

Bands followed an annual economic cycle based on the available seasonal resources. They moved fluidly and flexibly throughout the year, covering hundreds of miles in pursuit of what each season provided.

A typical year went like this:

Spring (March—mid-May):

Pulling toboggans, the Ojibwa would leave their winter camps along the Chippewa River in January, and travel back north on snowshoes, arriving at their sugar camps in March. The women would begin tapping the maple trees, while the men speared fish through holes they cut in the ice. Food cached the year before, including items such as twine bags of wild rice, cranberries, dried potatoes, and apples, were vital as a supplement to their diet.

When the creeks opened, the children would catch fish. Spawning runs started soon after ice-off, and the Ojibwa would spear at night using torches to illuminate the waters.

Hunting of migrating waterfowl would occur for a short period.

Gardens were planted in May with potatoes, beans, corn, and squash.

Summer (mid-May—late August):

Summer provided a variety of plentiful resources and was the easiest season of the year, allowing time for making a host of necessary tools and living commodities like mats, bags, and equipment for hunting and fishing.

Continued

Seasonal Economic Cycles of the Ojibwa (Continued)

Berries were gathered and dried as the summer season progressed. Hunting and fishing commenced as needed, but no trapping of hides took place until later in the fall when the furs were prime.

Fall (September—early November):

Wild rice came ripe at the beginning of September, and the Ojibwa would travel to the rice marshes to gather as much as possible. They'd return to their summer camps after several weeks of ricing and harvest their gardens, process their rice, and then cache the extra dried goods for their return later that winter.

The men would leave for fall trapping, and the woman would begin fall fishing, again taking advantage of spawning times, this time netting lake trout, whitefish, and cisco. When the men returned from fall trapping, the snows may already have begun, and the band would depart for their winter camp.

Winter (mid-November—January):

Some Ojibwa made a permanent residence in the Lac du Flambeau area, trapping and trading with the wintering French.

The Ojibwa who migrated south made late fall and winter hunts down along the lower end of the Chippewa River, south of present-day Chippewa Falls. This game-rich area was where the prairie and forest regions met, providing excellent hunting opportunities for bison, elk, and deer. The turbulent rapids and falls of the lower Chippewa kept the river open, permitting easy access to the water for fishing.

Sometime late in January, the Ojibwa would begin the difficult trek back to their traditional village sites and begin a new cycle with the gathering of maple sap.

A Sucker Is Born Every Minute, and It's a Good Thing

Suckers spawning

Four species of suckers were found in two fish-sampling studies done a decade ago on the Manitowish River. The white sucker (*Catostomus commersoni*) was the most prevalent, and the greater redhorse (*Moxostoma valenciennesi*)—a threatened species in Wisconsin—was the least common.

All four species come under the derogatory rubric of "rough fish" or "trash fish," an epithet given to non-game fish that is a bit like the huge box that many non-game mammals and birds are put into. Calling something "non-game" is like calling a white person "non-black", or a Catholic a "non-Lutheran." Perhaps we should likewise call our songbirds "rough birds" or "trash birds" because of their non-preferred status as a huntable species.

The species at the top of the food chain, whether fish or mammals or birds, receive disproportionate attention because of their beauty and size. Biologists often refer to them as the charismatic megafauna. We humans do a poor job of recognizing the value of all the species lower on the food chain. These species perform their ecological roles either in anonymity or heaped with disdain. We talk a lot about recognizing the little guy in our society. It's time we did the same for plants and animals that don't make the PBS specials.

So, here's a voice in the wilderness for the lowly sucker, unfortunately but accurately named for the shape of its mouth and its manner of

Continued

A Sucker Is Born Every Minute, and It's a Good Thing (Continued)

feeding. Suckers have Mick Jagger-like lips that vacuum up food along the bottom, sucking up an array of species including insects, mollusks, plankton, and higher plants, along with a few fish and fish eggs that get inadvertently mixed into the blend.

Suckers are bottomfeeders, another term we employ with derogatory zeal to folks we don't like. "He's a real bottomfeeder," we say, disparaging the person's moral fiber. Personally, I don't see any better morality being displayed in feeding from the surface than from the bottom.

While there's little romance associated with being a sucker, they are an exceptionally successful native family of fish. White suckers are probably the most widespread fish in Wisconsin, inhabiting most streams and large lakes throughout the state. They can tolerate a wider range of environmental conditions than any other fish species in Wisconsin.

In northern Wisconsin, white suckers typically spawn in late May and early June when water temperatures reach about 45°F. Spawning usually takes place over a gravel bottom in swift water or rapids. When a female moves into the rapids, she is often approached by many males to mate. But most frequently the spawning act occurs between a single female and two males, with one male on each side of her. After spawning, the female moves upstream where she spawns with other males again, so her eggs are ultimately scattered over a considerable area.

Suckers are an important forage species for muskie, walleye, burbot, bass, and brook trout. Bald eagles, osprey, loon, and heron prey on them as well, particularly during spawning runs when they are most concentrated and vulnerable.

The white sucker hosts the glochidial stage of the mollusks *Alasmidonta marginata* and *Anadonta implicata*, performing the important role of distributing these mussel species throughout rivers and lakes.

And, hey, they taste good, with a white, sweet meat that when smoked is a delicacy. The author of *Wisconsin Fishes and Fishery Management*, George Becker says that considering the white suckers, "wide distribution, its high numbers, and its delectability, the white sucker is potentially the most valuable food and sport fish in Wisconsin." While suckers won't win any beauty contests, their positive ecological role is clearly their most attractive trait.

Green Frogs—The Decline and Possible Fall of the Frog Empire

What has Happened to Green Frogs?

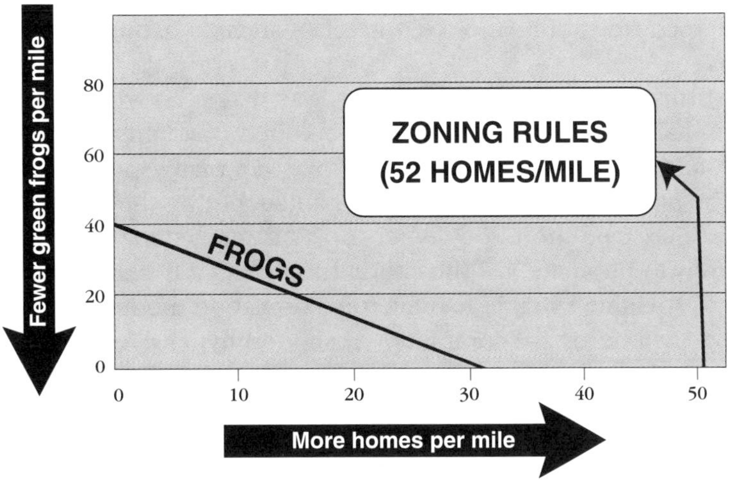

A 1997 WDNR study surveyed 28 lakes at night to map the location of calling green frogs. Researchers chose lakes with a wide range of shoreline development, selecting 12 lakes with little development and 12 with more than 50 percent developed shoreline. They also added four hyper-developed lakes, defined as lakes that were the most developed in the region.

The researchers developed cover type maps for each lake. The maps categorized each lake's shoreline habitat as excellent, adequate, or poor for breeding green frogs based on the presence of appropriate shorezone vegetation.

The researchers then looked to see if a relationship existed between the three sets of data: the proportion of shoreline classified as green frog habitat, the number of green frogs present, and the density of houses along the shorelines.

They found that 82 percent of the shoreline of undeveloped lakes was classified as suitable green frog habitat, 67 percent of the shoreline on developed lakes was suitable habitat, and only 15 percent of the shoreline on hyper-developed lakes was suitable. The average number of calling green frogs per 100 meters of suitable habitat was 2.6 for undeveloped lakes, 1.5 for developed lakes, and 0.7 for hyper-developed lakes.

Continued

Green Frogs—The Decline and Possible Fall of the Frog Empire (Continued)

They concluded that as shoreline housing density increases, the percentage of shoreline classified as suitable green frog habitat decreases, and that green frog abundance per meter of suitable habitat also decreases.

The surprise, if one could call it that, was that even where available green frog habitat occurred on hyper-developed lakes, green frogs were seldom found. This suggests that green frogs, like many species, are affected by not only a decrease in their habitat, but by the fragmenting of their habitat. One shoreline owner can't maintain suitable habitat and expect frogs to populate it if the shoreline owners adjacent to him or her aren't providing suitable habitat, too. Population models predict that green frogs will be, or are completely absent on hyper-developed lakes even though the study showed 15 percent of the shoreline to be suitable for green frogs.

Bulrushes—*Scirpus*

Bulrush belong to the sedge family, however, not all bulrushes march in step to the "sedges have edges" ditty that most of us have learned as an identifier for sedges. Instead, these tall emergents have round, smooth, dark green, spear-like stems. Their scaly brown flower clusters emerge from one side of the stem or the top of the stem. Without their flowers, some bulrushes are easily mistaken for rushes (*Juncus*). To tell them apart, bulrush flowers have scales on their spikelets. Each scale encloses a minute flower, while *Juncus* flowers have three petals and three sepals.

Bulrush is represented by 30 species in east-central North America, but the most common species along northern rivers like the Manitowish are softstem bulrush (*Scirpus validus*), hardstem bulrush (*Scirpus acutus*), three-square bulrush (*Scirpus americanus*), and wool grass (*Scirpus cyperinus*).

Softstem bulrush has bluish-green, rigid, round, spongy stems. When pressed between your fingers, the large air chambers in the stem compress easily. The spikelets of softstem appear more relaxed than the rigid posture of hardstem. Softstem grows in water up to five feet deep and seems better suited to muckier substrates and more stagnant water like that found in sloughs or sheltered coves.

Hardstem bulrush has an olive-green stem that is more solid, and is filled with smaller air chambers than softstem bulrush. The stem is firm when you press it between your fingers. Hardstem usually grows in water less than six feet deep, and seems to prefer a firmer substrate with a good circulation of water. It's usually found out on windy, wave-swept points.

The stem of three-square bulrush, or chairmaker's rush, is sharply triangular. Its bayonet-like stems have small, unstalked, tight-clustered spikelets. Muskrat, snow geese, and other waterfowl graze it extensively in shallow marshes and along river and lake edges.

Wool grass has a tall, slightly triangular stem, long leafy bracts, and most importantly for identification purposes, large, drooping, fuzzy-brown flower clusters at its top. Look for it on drier sites, in sedge meadows, around the edges of bogs, and in wet ditches.

As a group, bulrushes typically grow in dense colonial patches, have bisexual flowers, and are wind-pollinated. Their perennial rhizomes sprout new stems each spring. In fact, most of their reproduction is vegetative from rhizomes.

Continued

Bulrushes—*Scirpus* (Continued)

The bulb-like tips of the rhizome can be pulled up and eaten raw, roasted, or dried. They're rich in sugar and starch. If you're really hungry, reportedly the pollen and ground-up seeds can be used as flour.

The dried stems were woven together on frames by passing basswood twine over and under them. The bulrush mats were then tied to the outside framework of Ojibwa wigwams. The mats were also tied to the framework inside the lodge for protection against drafts. A family might use two rows of overlapping mats if they wanted greater warmth.

Tag Alder—Entanglements and Snarls

Tag alder is the octopus of shoreline plants, with branches like tentacles that either embrace you or entangle you, depending on your perspective. Individual plants may have 20 or more arching, twisted stems seemingly eager to impale you within their grip. Alder clones itself readily, forming thickets of branches and stems that literally form a living fence along many rivers. To say that a pleasant shoreline walk is not in the cards when entering an alder swamp is to engage in understatement. It can be hell.

Continued

Tag Alder—Entanglements and Snarls (Continued)

That's the bad news. The good news is that these dense snarls offer excellent cover for a host of animals. Nesting birds include yellow-bellied and alder flycatchers, swamp and white-throated sparrows, goldfinch, Wilson's warblers, yellow warblers, and common yellow-throats. Woodcock feed in the muddy bottomlands, probing the muck with their long bills in hopes of finding tasty invertebrates. Beaver eat the bark and use the stems for building their lodges and dams. Moose browse the leaves and twigs. Ruffed and sharp-tailed grouse feed on the catkins. Snowshoe hares and cottontails eat the bark. Chickadees, red-polls, pine siskins, and goldfinch forage the seeds. Deer frequently use alder for a rubbing post to remove the velvet from their antlers. While deer may browse the leaves and twigs, alder ranks quite low on their preferred food list, usually indicating tough times.

Alder contributes more than food and cover to shorelines. Its roots typically form a mesh that anchors the soil and prevents bank erosion. The leaves are rich in nitrogen, enriching the soil, while its roots have nitrogen-fixing bacteria that work to further improve the soil.

Thus, alder controls erosion, stabilizes streambanks, feeds wildlife, provides cover for wildlife, and enriches the soil—a pretty good résumé for a plant so universally reviled by humans.

To treat the causes of a difficult childbirth, the Ojibwa took scrapings from alder roots, added ground-up bumblebees, and took two table-spoons.

I suggest seeing your doctor.

If you decide to try to clear your shore of the tangle of alders, just remember that when cut, alders simply resprout in even greater numbers and density. Alder's here to stay, and we should be thankful.

Fish Sampling on the Manitowish

Two fish studies were done on the Manitowish, one in 1989 and the other in 1990. The 1989 survey sampled a section of the river nearly a half-mile long along County H just south of the County K intersection. The 1990 study looked at an equal length of stream at the state wayside on Highway 51. The researchers measured habitat parameters such as stream width, depth, velocity, substrate, and cover, all of which affect the quantity and type of fish that are found in any section of a stream.

The 1989 survey described that section of the Manitowish as 53 percent rocky substrate with very little cover. Only five percent of the streambank had woody debris, emergent and submergent vegetation, overhanging vegetation, undercut banks, and/or boulders.

The researchers electroshocked the entire section and found the predominant gamefish to be smallmouth bass, with some walleye. Yellow perch were abundant, while the predominant panfish were rock bass. Non-gamefish included brassy minnow, hornyhead chub, common shiner, white sucker, rosyface shiner, and logperch.

The overall fish assemblage was rated as fair, given the habitat problems of lack of water depth and cover. The researchers stated the area could be important spawning and nursery area for largemouth bass, smallmouth bass, and walleye.

The other fish sampling was done in July and August of 1990 at the wayside site off Highway 51. Species richness was rated as excellent, with 31 total species present, and a good diversity of trophic categories, such as omnivores, insectivores, and top carnivores. The actual counts of the adult fish and the youngof-the-year follow, as well as the ratings produced using the Index of Biotic Integrity and the Wisconsin Warm Water Physical Habitat Rating System.

Number of adult fish collected during the Fish Assemblage Survey on the Manitowish River in 1990.

Common mud minnow	21	Yellow bullhead	19
Northern pike	3	Burbot	18
Hornyhead chub	90	Rock bass	64
Golden shiner	13	Pumpkinseed	7
Pugnose shiner	2	Iowa darter	2
Common shiner	61	Fantail darter	13
Blackchin shiner	9	Yellow perch	5
Rosyface shiner	3	Log perch	44
Spotfin shiner	19	Blackside darter	6
Mimic shiner	6	Walleye	2
Bluntnose minnow	88	Mottled sculpin	2
Blacknose dace	83	Other	3
White sucker	2		
Northern hognose sucker	6	Total individuals	595
Shorthead redhorse	2	Total species	30

Number of young-of-the-year fish collected during the Fish Assemblage Survey on the Manitowish River in 1990.

Common shiner	116
Bluntnose minnow	10
Creek chub	25
White sucker	91
Greater redhorse	28
Rock bass	4
Yellow bullhead	2
Smallmouth bass	10
Yellow perch	25
Walleye	3
Other	1
Total individuals	315
Total species	11

Scores assigned to the various items in the Wisconsin Warmwater Physical Habitat Rating System (WWPHRS), 1990. The higher the score, the better the habitat. Actual values measured on the Manitowish River are in parentheses. See Appendix A for more details on the WWPHRS.

Item	Manitowish score (actual)
Bank Stability	12 (95%) = excellent
Rocky Substrate	4 (15%) = poor to fair
Available Cover for Adult Fish	25 (12%) = excellent
Average Maximum Thalweg Depth (m) (4 Deepest Depths)	24 (1.48) = excellent
Bend-to-Bend Ratio	12 (8) = excellent
Total Score (maximum score possible = 99)	77
Rating	good to excellent

From WDNR Fish Research, Paul Kanehl and John Lyons, 1990.

Calculation of the Index of Biotic Integrity (IBI), 1990. Numbers in parentheses are the score assigned to calculate the IBI: 10 = Best, 0 = Worst. The higher the score, the better the fish community (possible range: 0-100).

Species	Manitowish number	(score)
Total Number	31	(10)
No. Darter	5	(10)
No. Sucker	4	(10)
No. Sunfish	2	(10)
No. Intolerant	9	(10)
Individuals		
% Tolerant	39	(5)
% Omnivore	22	(5)
% Insectivore	54	(5)
% Top Carnivore	11	(5)
% Lithophil	51	(10)
Correction Factor	0	
Total Score		80
Rating		excellent

From WDNR Fish Research, Paul Kanehl and John Lyons, 1990.

Sense of Place—This Is A Test

In the 21st century, our lives appear to require little knowledge of the natural world. Thus, few of us have a meaningful connection to the forests and waters around us. Our physical survival, defined in the simplest terms as our ability to acquire food, shelter, and clothing, is provided for us through a worldwide trade system that allows each of us to specialize in a job and to utilize currency to provide for our physical needs. For day-to-day survival, we have no real literal need to know anything about the natural world. But while our trade has provided enormous benefits that few of us would want to give up, we have lost something essential in the bargain—our sense of place.

With that in mind, try this informal self-scoring test so that you can get a handle on your personal understanding of where you live. The questions are basic—no fancy terminology is needed. A few of the questions are specific to the Northwoods, but if you live in another geographical area, adjust these questions to fit where you live.

If you score rather low on the test, consider using your score as a springboard to learn more about the natural community in which you live.

Place in order of fruiting times the following wild species: blackberry, blueberry, strawberry, red raspberry, Juneberry.
What type of soil is in the area in which you live?
What is this soil best suited for?
What is the land-use history of where you live?
What source does your drinking water come from?
Where does your waste-water go?
How many days until the moon is full?
What planets are currently visible in the night sky?
Where does your garbage ultimately end up?
How long is the growing season where you live (the average time between frosts)?
What is the average total rainfall for your area?
When was the last major fire to burn in your area?
What species of plants and animals in your area are helped by fire?
Name five spring wildflowers that are consistently among the first to bloom in your area.
Name three nonnative plants that have become problems in your area.
From where you are now, point north.
What primary geologic event influenced the land where you live?
When did that event occur?

Continued

Sense of Place—This Is A Test (Continued)

Name two pieces of evidence demonstrating this geologic event occurred.

Of the food you ate today, what foods came from native plants and animals in your area?

Name three birds that are winter visitors to your area.

Name three birds that are winter residents in your area.

Name three birds that nest in your area and migrate south in the fall.

List three common birds in your area that are nonnative.

Name three edible wild fruits in your area.

Name an edible wild nut in your area.

Name two edible wild greens or roots/tubers in your area.

On what day of the year at noon are the shadows longest where you live?

When do deer rut in your area, and when are the fawns born?

Name three grasses and/or sedges in you area. Are any native?

Name a bird and a mammal species that have been extirpated (removed) from your area.

Name a bird and a mammal species that are more successful today in your area than they were prior to settlement.

What kind of plant community was your area originally?

Name two of the most common native trees today, and two of the most common native groundlayer species today.

What tribes of Indians lived here prior to European settlement?

How far back does archaeological evidence show people lived here?

Name an important archaeological site near where you live.

Name the river closest to where you live, and trace its connections to other rivers—from where it starts to where it enters the sea.

If you were one of the first European settlers to come to this area, where would you have picked to clear your land, and why?

Look at one of the meals you ate today—for each item you ate, where were the ingredients raised?

Does any of the clothing you wear originate from your area? If not, what geographical region does most of your clothing come from?

From what materials was your house built? Of those materials, which were from your area?

How do you heat your home? What natural source is used to create that energy, and where does it come from?

Name five local industries that are founded on utilizing area natural resources.

Continued

Sense of Place—This Is A Test (Continued)

Name three native fish, and three nonnative fish that are in the lakes and rivers of your area.
Name two bird and two mammal species that are either endangered or threatened in your area.
What is the average snowfall in your area?
In what month, on the average, does the following take place:
ice-up
ice-out
first wildflowers in bloom
swallows and shorebirds begin their migration south
eagles return
fall color reaches its peak
bears go into their dens

Scoring:

>90%	You have a true sense of place.
80-90%	You have a good understanding of where you live.
60-80%	You're paying attention now and again.
50-60%	You really need to get out more.
<50%	You hopefully just moved here.

Water Lilies—Queens of the Slough

People from a planet without flowers would think we must be mad with joy the whole time to have such things about us.
— Iris Murdock

Identifying water lilies by their leaves is quite simple. The leaf of the white (*Nymphaea odorata*) water lily comes to a right angle in its notch (remember "right-white"), while yellow water lily leaves (*Nuphar variegata*) are more oval and rounded at the notches.

The dinner-plate-sized, floating leaves of water lilies almost certainly evolved because round, smooth-edged shapes provide the least resistance to wind and wave action. The tough, waxy surface of the leaf presumably also evolved to resist water and hail penetration. Note that the stomata (breathing pores) appear on the topside of leaf rather than the underside as is common in most plants.

A strong wind will pick up the leaf and sometimes turn it over, revealing a red underside on the white water lilies. The red pigment is thought to help raise leaf temperatures slightly, thus increasing transpiration.

Flowering occurs over a short period of time. The bisexual flowers operate in a unisexual manner, the female parts opening before the male to prevent self-pollination.

Continued

Water Lilies—Queens of the Slough (Continued)

Shortly after fertilization and the release of pollen, the flower closes, its stem coils, and the flower head is pulled underwater. The seeds take about a month to mature inside a fleshy fruit. The fruit eventually breaks off, floats to the surface, and releases the seeds. The seeds sink and germinate the following spring, reaching maturity three years later.

The most remarkable historical usage of white water lilies was as a love potion. For the love potion to work, the flowers had to be picked during a full moon, though the pickers had to wear earplugs to prevent bewitchment by the water nymphs. The dried flowers were then worn as love talismans, a practice I presume to have every bit as much likelihood of success as an ad in the "Personals" section of the newspaper.

Beaver Lodges

Beavers live in a lodge or in a den in the banks of a river or lake, or sometimes both. Lodges are preferred, but sometimes they aren't feasible to construct. Large rivers that flood frequently, or areas that have few trees, present tough problems for lodge-builders. In such habitats, beaver typically live in bank dens. The bank den entrance is underwater, with the tunnel to the den dug slightly upward until it is above water level. Tunnels back to the hollowed-out chambers can be 30 or more feet long. A tangle of tree roots in the bank is sometimes utilized to provide stability for the structure.

Beavers often construct a lodge and several bank dens. A flooded-out lodge, or one sitting high and dry during a drought, obviously poses serious problems for beavers. Bank dens offer insurance against such natural disasters. Bank dens also offer lodge-dwellers optional hiding places, since lodges are very easy for predators to spot. In winter, bank dens provide a place to stop and get a breath of air while swimming under the ice.

Most beaver colonies have lodges in various degrees of repair, so beavers, like humans, may have second homes. A family or colony may build several lodges within their home range and then alternate where they stay between them. Single bachelor beavers also build lodges that are usually a bit smaller. You can't judge the number of beavers by the size or number of lodges. The largest lodges may project over 10 feet above the water, and be 40 to 50 feet in diameter.

Inside, lodges often have an initial low feeding platform where the beavers can stop and eat, groom, and dry off. Think of it like a mud-room in human homes where you take off your boots and hang your coat.

Continued

Beaver Lodges (Continued)

Above this platform, most lodges have a slightly higher level that's often carpeted with wood shavings. This room is where the beavers sleep. The inside chamber can be 5 feet across and 2 feet high, though the record beaver lodge was tall enough inside for a grown man to stand. It had beds for about eight beavers in the room, and had walls 4 feet thick. This Bayfield County, Wisconsin, lodge stood 16 feet high and 40 feet wide. The inside of an ordinary lodge usually just consists of one room, but eight or more beaver may occupy the lodge.

Most of us would be ripping at one another's throats after five wintery months inside a single dark and damp room, but a Canadian study of family dynamics within a lodge found that they got along well, even sleeping together in a friendly heap. A window was cut into the back of a lodge and the beaver were observed for two years. The study revealed virtually no familial problems that weren't resolved quickly and peacefully—a true "Leave It to Beaver" family.

Remarkably, a trapper friend of ours has seen muskrats as well as mergansers come shooting out from the exit holes of beaver lodges, indicating that some species may cohabitate with beavers within the lodges.

Velocity

To view how a river flows over time, hydrographs are used to plot a continuous record of discharge. A hydrograph depicts the passage of water from days to years. It can show the passage of a flood event, and illustrate the pathways and rapidity with which rainfall reaches a river.

Baseflow represents the groundwater input into normal river flow, while stormflow is graphed as an increase above baseflow. The stormflow has a rising limb that shows how long it takes precipitation to enter a stream, and how much water ultimately enters. The rising limb has a peak, and then a recession limb, which shows how long it takes the river to return to its baseflow.

A hydrograph can visually represent a watershed's ability to handle a storm. If a substantial overland flow of water causes a rapid rising limb, it can result in significant erosion of the land surface. Such streams are typically called "flashy," and quickly develop floods that result in dramatic damage to shorelines and associated flood plains.

In contrast, a slow-rising limb is typical of forested streams. Here, soil permeability permits the slow infiltration of rainwater. This rate is rarely exceeded because of the presence of soil litter, plant root structure, and the interception of rainfall by leaves.

Alteration of a watershed by urbanization, such as the use of street gutters, storm sewers, and acres upon acres of asphalt, shortens the lag time between when a storm occurs and when its effects are seen on a river. The peak discharge rate increases, intensifying the effects of flooding.

In a study near Berkeley, California, storms occurring on local urbanized landscapes showed an increased peak flow of as much as eight times higher than what should have occurred under natural conditions.

Stream velocity increases in natural systems due to physical characteristics other than land-use alterations.

Velocity increases:

- Proportional to the gradient of the channel.
- When a stream passes through a chute between rocks.
- With volume. The greater the volume the faster the river flows.
- When water temperature increases. Water becomes more viscous the colder it gets. Between 39°F and 68°F, water increases its velocity a half percent for every one degree rise in temperature.
- With fewer obstacles in a channel. Velocity is slowed when encountering obstacles.
- With less viscosity in general. Glaciers are really slow-moving rivers.

Stream flow conditions can be seen on the U.S. Geological Survey's Web site at <www.usgs.gov>.

Dragonflies and Damselflies—Order Odonata—The "Toothed" Ones

Insects are so structurally different from humans that they seem like aliens. There is no point of empathy. — May Berenbaum, entomologist

The "devil's darning needles," "eye stickers," "mule killers," "snake doctors"—dragonflies have been nicknamed all these things. The names arose out of various superstitious fears of this insect that seems all eyes and all speed. Dragonflies stimulate visions of a giant mosquito whose massive bite could seemingly fill a syringe with blood.

Fortunately, none of these fears have an ounce of truth behind them. In fact, if seen with the proper perspective, a case can be made that dragonflies should rank as one of the most remarkable, and beneficial insects on the planet.

We sing the praises of butterfly metamorphosis, encouraging children in elementary school classrooms to watch the development and emergence of monarch butterflies. But the metamorphosis of dragonflies? Few even know that the larvae emerge in an equally remarkable fashion from rivers and lakes every spring and summer to transform into adults.

Dragonflies spend over 90 percent of their lives as larvae, or nymphs, submerged in water. About a week before a mature nymph is to emerge as an adult, it stops eating, its internal mouthparts dissolve, its eyes turn translucent, and its wing pads swell. Adult colors become apparent through the skin, and some species move to the surface and begin breathing air.

The nymph soon crawls out of the water and anchors itself onto a rock, a plant stem or a pier post. The exoskeleton then splits down the back, and the new adult dragonfly unfolds, soft and glossy from its larval shell. It swallows air to pump up its body to triple the size of the larval skin from which it emerged. Weight lifters talk about getting "pumped up," but they can't hold a candle to dragonflies. Then, the two cellophane-like wings unfold, flatten, and are held out in the sun to dry and harden.

Emergence is a dangerous time for dragonflies. Several hours are needed for the skin to harden sufficiently before the dragonfly can weakly fly away. One motorboat wave can wash them away, and a host of predators, from birds to spiders, find them an easy and defenseless meal.

Continued

Dragonflies and Damselflies—Order Odonata—The "Toothed" Ones (Continued)

The dry, tan husk of the larval skin remains attached to the emergence site, with a hole in its back to show where the adult exited. These crunchy husks are easily found and collected.

Depending on the species, adult dragonflies may live four to 10 weeks.

A dragonfly in flight resembles the most daring of experimental pilots, performing aeronautical maneuvers not even conceived of by most flying animals. Only the hummingbird may be seen as a worthy competitor in accomplishing such aeronautical marvels. Dragonflies can hit a maximum flight speed of 60 miles an hour, can lift more than double their own weight, take off backward, come instantly to a dead stop, hover, even somersault. Their wings may look like brittle cellophane, but they are intricately cross-braced, so they can be bent to produce whatever aerodynamic effect they want.

Dragonflies are considered masters of unstable aerodynamics. They can maximize turbulence and use it to their benefit, while human aircraft typically break up in such conditions.

In the larval state, dragonflies still emulate aircraft. They come equipped with jet propulsion, using the rectum to breathe water in and out across the gills in the abdomen, and to shoot the water out as a means of propulsion. The weaker damselfly larvae have a three-finned tail assembly that fails to offer the zippiness that dragonflies enjoy.

Males typically defend a territory, fighting off intruders by utilizing their superior vision. While dragonflies are considered to have better vision than any other insect, no one is still sure what that really means. Their compound eyes are made up of 30,000 individual facets that are capable of giving them many separate images. But what exactly does the insect brain do with all these images? If we experience double vision, we're incapacitated. What would 30,000 images look like? Whatever their perception, it's clear that their eyes are excellent for seeing movement. They also can see the ultraviolet end of the light spectrum.

Dragonflies clean their eyes with a special brush built into their front legs. Their wraparound eyes bulge out so far that they have virtually 360-degree vision. Over 80 percent of their brains are given over to analyzing what they see, say research entomologists.

To court a female while on territory, the male faces away from her on his perch, drops his back wings down like a cape, lifts his hind end, and displays his colors.

Continued

Dragonflies and Damselflies—Order Odonata—The "Toothed" Ones (Continued)

If their extraordinary eyes and flight don't impress, consider their mouth. Dragonfly nymphs have an extensible lower lip that unhinges like a flat shovel. Called the labium, it stays folded underneath the head, even though it's one-third the length of their body. A nymph can shoot its labium out in a hundredth of a second, seizing a prey with the two grasping hooks at the tip of the labium. Mosquito larvae are a favorite. The labium has formidable hooks and knives that turn the dragonfly's mouthparts into what one entomologist calls "a combination of hands, carving tools, and serving table."

In Wisconsin, researchers list 152 Odonates, of which 108 are dragonflies and 44 damselflies.

Otherness is what I have always liked about bugs. — Sue Hubbell, entomologist

Chapter 6
Our Course: Highway 47 to Murray's Landing

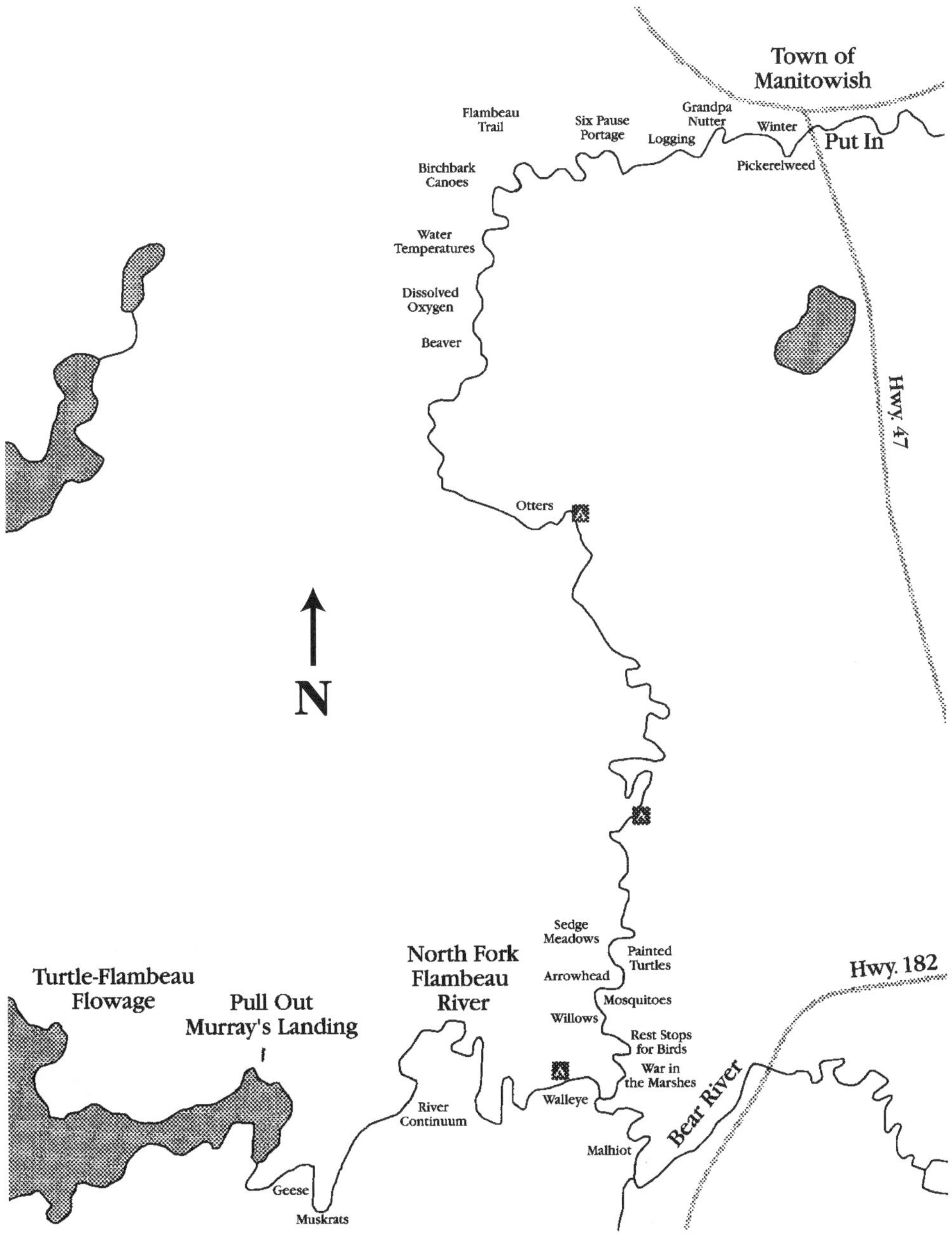

River, take me along in your sunshine
Sing me your song.
Ever moving winding and free,
You rolling old river,
You changing old river,
Let's you and me river,
Run down to the sea.

From "*River*" by Bill Staines

Chapter 6
Of Winter, Portages, Oxygen, Otters, and Mosquitoes

Flow gently, sweet Afton, among thy green braes,
Flow gently, I'll sing thee a song in thy praise;
My Mary's asleep by the murmuring stream,
Flow gently, sweet Afton, disturb not her dream.
—Robert Burns, "Afton Water"

My family and I live in the town of Manitowish in the home that Mary's grandparents, John and Ann Nutter, lived in from 1922 to 1967. Mary's mother Alyce was raised here, and Mary spent many summers as a girl visiting her grandparents. The house, built in 1907, stood empty for 15 years until we moved here in 1984. The memories that were resurrected by bringing the house back to life are priceless.

We've watched the river for many years from here and kept records of natural events, like when the water lilies first bloom, when the first geese return, when the first red-winged blackbird shows up, and so forth. The study of the timing and progression of natural events, known as phenology, keeps us attuned to the world around us, and lets us anticipate the blooming of an arrowhead or the return of hooded mergansers like visits from old friends. If we could invite them in for a cup of tea and some conversation, we would. We'd like to know how the winter went for them, and what their plans are for this year.

One event we've kept our eye on is when the river ices up in the fall and when the ice goes out in the spring. Seeing the water flow after months of winter's solid lid is like watching the genie escape from the bottle again. The magic carpet ride is back in operation.

On a July paddle down the Manitowish, the thought of winter seldom intrudes upon the average soul whose major meteorological concern is which number of sunscreen to use. But winter deserves the headlines in any book about the Northwoods, because it is the limiting factor determining who and what will live here.

How does ice form on a river whose current would seem sufficient to prevent such a phenomenon? As a rule, the faster the current, the more delayed the freeze-up. For water to freeze despite the efforts of the current, flowing water must first become supercooled, meaning the water temperature falls below freezing without solidifying. But supercooling has its limits like everything else, so scattered crystals of ice, called "frazil" ice,

Winter on the Manitowish

eventually form. Frazil ice tends to stick to whatever it touches. Some crystals stick to each other, forming floating slush, which then grows into pans of ice. These pans pinball their way down the river, banging into one another and smoothing their edges into rounded discs. Some pans freeze to the river's banks, while some freeze directly to the riverbed and become anchor ice, covering portions of the bottom with a slippery coating.

Meanwhile, ice forming along each bank grows toward the river's center. The width of the river shrinks as ice pans floating past get hung up on the projecting ice, sticking to each other and the bank ice, until the river finally freezes over.

The surface ice then thickens much like lake-ice. Frazil ice adds itself to the underside of the surface ice, and snow settles on top of the ice, some of which compacts into ice.

We stay off river-ice whenever possible. Sometimes the weight of the snow cracks the ice, or pushes the ice down until water flows onto the surface and partially freezes into slush. We've had the experience all too many times of snowshoeing across the river and getting into three inches of slush that immediately freezes onto our shoes, causing them to weigh, oh, approximately 200 pounds each.

We've also seen many open holes that come and go on what seems like a daily basis. At some point every winter, we see snowmobile tracks going down the Manitowish, a practice that well exceeds the definition of foolish. A word to the wise, or sober: If you don't know the river, stay off its ice.

If a river is very shallow, it can freeze right to the bottom, damming the river completely, and causing it to flow out onto its flood plain. Overflows may happen several times in the same area, causing sheets of ice to build up in layers on the flood plain.

Some stretches of river freeze up only during the most extensive below zero (F) periods, and then open up again when the weather warms to above zero. The Manitowish stays open nearly all winter where Highway 51 crosses it just north of Manitowish Waters.

I'm always impressed by the number of small creeks that stay open most winters in the Northwoods. Typically, a series of springs keeps their water temperatures moderate enough to prevent ice-up. Groundwater emerges as a seep or spring wherever the water table reaches the surface. Unless a spring comes from a shallow aquifer less than 10 feet down, the temperature of the water will remain constant all through the year. Even in cold climates like the Northwoods, a spring will flow all winter, though lakes nearby may be frozen tight.

Springs

Springs occur where aquifers intersect the soil surface. Small springs are little more than seepages or trickles, though many small seepages together can form the headwaters of a river. Most springs are unseen, emanating from below lakes and rivers. When we speak of a spring-fed lake, we mean the lake receives its water directly from the groundwater. Rivers too can be perched on top of springs and receive a constant supply of groundwater.

The word "spring" derives from the Old High German *springan* meaning "to jump." Spring is the word given to the place where groundwater appears on the surface, to the season where life is reborn, to events that leap forth suddenly, to the elastic coil of wire that constantly recovers its shape, and to the first stage and freshest period of something—the spring of life.

Water nymphs lived in the spring called Cassotis on Greece's Mount Parnassus. Lonely souls, they would drag young men into the water and drown them, so they would have eternal companions.

Not all nymphs worked such negative magic. The naiads of Greek mythology were mortal water nymphs who had the power to heal. They supposedly inhabited every sizable spring. Sick people would visit springs to drink, or bathe in the water for healing. That was the good news. The bad news was that any person who gazed at a naiad would be driven insane. And woe to anyone who offended a naiad—this indignity always resulted in some misfortune.

We live today with a transformed Greek mythology in our springs and streams. We see naiads and nymphs in nearly every stretch of water, though not as beautiful maidens, but in reference to the immature stages of many aquatic insects.

Life under the ice slows down for most organisms. Metabolisms wind down to the point where many fish move at a snail's pace.

If clean, powder snow accumulates on top of the river, it reflects nearly all the sunlight back into the sky in a blinding glare called the albedo effect. Ninety-nine per cent of the solar radiation may be reflected back, necessitating the use of sunglasses to prevent snowblindness. However, all snow and ice are not created equal. Transmission of light through older, crystalline snow is increased sixfold compared to through new powder snow, while clear ice actually transmits light better than the water beneath it.

Light, of course, is the engine for photosynthesis. Without available sunlight, most photosynthesis, and thus plant growth, stops. Without photosynthesis, oxygen production also stops. Over the course of a long winter, a series of snowstorms can so darken the water of a shallow lake that oxygen depletion may become drastic and result in a winterkill of organisms. The winters of 1995-1996 and 1996-1997 dumped so much snow in northern Wisconsin that numerous shallow lakes experienced die-offs of fish.

Rivers tend to withstand oxygen depletion much better than lakes because their flowing waters often remain open, permitting oxygen exchange with the atmosphere. The icy, canning-jar lid that covers lakes seals them tight, but the lid on rivers comes off enough times during the winter that rivers get a chance to literally breathe, while shallow lakes gasp.

One last note on ice-up, ice-out, and oxygen depletion. In deeper lakes (but very seldom in rivers), these processes are usually preceded in fall or followed in spring by an event called turnover. Lake turnover occurs when a complete mixing of the water takes place. In the summer, deeper lake water seldom mixes because water temperatures typically stratify the water into layers. Colder, heavier water sinks to the bottom, while warmer, lighter water remains closer to the surface. Not only is water at the bottom of a lake colder, but it typically holds far less oxygen. Less photosynthesis takes place in deep water because little or no sunlight reaches the bottom.

As surface waters cool in the fall, the differences in temperature between the surface and deep waters declines, until the temperatures even out. At that point, no density difference exists between water layers, so nothing prevents the free movement of the water. What follows is lake turnover—or overturn, if you wish. Currents, set in motion by wind, stir and mix the water. The whole body of the lake becomes thoroughly blended. Dissolved oxygen, dissolved nutrients, and microorganisms, which all had been stratified in different layers, mix together. This allows the whole lake to "take a breath and have a bite to eat."

The process is a bit like shaking one of those snow-globes. Snow starts falling throughout the scene. Likewise, fall turnover briefly distributes oxygen and nutrients throughout the lake until colder temperatures cause stratification to occur, and the scene becomes unshakable, at least until spring turnover occurs.

Not all lakes stratify. Lakes in southern regions with mild winters may not stratify, but simply remain

The When and the Why

Ten inches fall today
The roads are empty
Silent.
The town closes down
The world retreats.

We snowshoe into the big pines leading
to the camp by the river.
Heavy pine arms sag
like strings on a violin
loosened and thrumming low, low.

Whiteness, dizzyingly bright,
contours the ground in gentle swells
gracefully lapping onto downed trunks,
softly clothing sudden steep hills
of tipped up roots.

We look for tracks,
for browse, for signs
of life in February,
and try to guess
the why and the when.

Tracks cross, veer, amble,
some straight and purposeful,
others a curious anarchy.

Deer have browsed every young maple,
every cedar and hemlock,
each stem gnarled
from years of losing
spring buds to the hungry.

The trails take on the look
of starvation,
of desperation.
The deer have circled
deeper into the forest and out to the river's edge,
extending their shoreline,
each wave lapping further into the treeline.

Coyote prints overlap deer.
The distance between each track leaps,
and we try to guess
the when and the why.

mixed all year-round. Spring turnover is not an event for them.

On the other hand, small and very deep lakes in the Northwoods may remain stratified for years on end. Called meromictic lakes, these lakes have a top layer that is very warm and a bottom layer that's very cold. The bottom layer (the hypolimnion) in these lakes is stagnant and lifeless, because dissolved oxygen is so limited near the bottom. Winds can't get enough fetch to stir these little lakes up. The lake water may partially circulate at times, but it does so incompletely, never fully experiencing fall turnover. Mary Lake near Winchester in Vilas County, Wisconsin, is one such lake. With an area of only 3 surface acres but a maximum depth of 67 feet, it's a good example of a meromictic lake.

Read more about winter at the end of the chapter.

≈ ≈ ≈ ≈ ≈

Immediately downriver from the town of Manitowish are large beds of pickerelweed that typically come into bloom in mid-July. Mary and I own a few acres of land on either side of the river here, and one summer we discovered that a fellow had taken thousands of the pickerelweed from the sloughs where they grow in such great numbers. We reported the theft to the warden thinking that it was a state forest issue, but found out that we were the owners of the plants! By dint of owning the land on either side of the river, we were by state law also the owners of the riverbed. By extrapolation, we also owned whatever grew out of it. So now it became a theft of "our" pickerelweed.

Pickerelweed

The long and short of it is that the market for aquatic plants has grown, and pickerelweed may sell for $4 a plant. The problem is the fine for stealing such plants from state or private land is minimal, so the profits far outweigh the penalty, and stealing plants takes a lot less work, space, and time than growing plants.

Back in the lumbering era, stealing trees and/or logs was relatively common. A few thieving lumberjacks lost their lives pulling a log from the river, cutting off the end with the owner's mark on it, and pounding in their own mark. A century later, Northwoods theft has evolved into stealing aquatic plants. There's a market for selling aquatic plants to all the people who build ponds on their property. There's also a market for selling plants to the companies that are required to build wetlands in mitigation exchanges for destroying wetlands.

Pickerelweed (*Pontederia cordata*)

In late July, pickerelweed distinguishes itself from its slough-mates by its colonies of long, violet-blue flower spikes rising from glossy, heart-shaped leaves. Usually only a few pickerelweed flowers bloom at a time on the spike, coming into blossom initially at the bottom of the spike and progressing upward as the season advances. Insects pollinate the flowers. Each nectar-rich flower only lasts a day or so. After pollination, the upper petals close, and a single seed develops. When all the flowers are finished, the spike bends and releases its seeds into the water. Pickerelweed seeds require about two months of cold water in order to germinate.

Pickerelweed emerges in monotypical beds, spreading from their rhizomes, which are found several feet deep in the sediments of marshes and sloughs. Most reproduction occurs vegetatively by means of these creeping rhizomes. Pickerelweed can sprout into colonies so quickly that dense beds are often perceived as a nuisance. Pickerelweed thrives in water less than 3 feet deep and is often found on the inside curve of a river where sediments are deposited.

The leaf stalks have aerated chambers that function as interior flotation devices, holding the leaf upright in the water. In drought years, the leaves of pickerelweed lay stranded on the river's mudflats, left prostrate without the support of the water.

Pickerelweed stems provide sites for the emergence of dragonflies and damselflies into adults. The dried-out exoskeletons of the nymphs can often be found still attached to the leaf stem. Some dragonfly and damselfly adults, like the green darner dragonfly (*Anax junius*) and the black-winged damselfly (*Calopteryx maculata*), return to the plants to deposit their eggs in the stem tissues just below the water surface.

Beds of pickerelweed stabilize shorelines and buffer waves, just like their wetland associates arrowhead and bulrush.

Pickerelweed seeds are eaten, but not necessarily relished, by surface-feeding waterfowl like gadwalls, black ducks, mallards, woodies, green-winged teals and pintails. Muskrats eat the leaves and seeds, as do deer. The stems are used as lodge material by muskrats. But overall, pickerelweed has low wildlife value when compared to many other aquatic plants. It does provide shade and shelter to aquatic organisms. The jury appears split on whether pickerelweed provides spawning habitat for its namesake, the northern pike. Some biologists say it's important spawning grounds, but others say that there's no special association between pickerelweed and northern pike, or any other fish.

Mary and I laugh about becoming pickerelweed farmers. It sounds like there's money to be made, but we like our wetlands just as they are.

Around a few bends on river-left, a DNR campsite appears under some large pines. This is the site of Mary's grandparents' former summer home. They moved here in the summer in order to raise an acre or two of potatoes. They built a small barn and raised a few cattle, too, all in what is now part of the 5,400-acre Manitowish River State Wilderness Area.

Grandpa Nutter even built a bridge across the river here, though no hint of it remains. One night in 1934, Grandpa sat with his gun at the base of the bridge waiting for Al Capone or one of his gang to try to cross. The famous shootout between Capone and the FBI at Little Bohemia in Manitowish Waters had occurred that night, and the FBI called for all the locals to be on the lookout for Capone.

We're rather glad Grandpa never saw a soul that night. If he had, chances are pretty fair that none of us would be here today, though John was a marksman and no one to trifle with. Neither though was Al Capone.

John was a logger, and many of his crews stayed in the house we now live in. We still have the blankets he gave the crew, many of the tools he used, some of the cookware they used in the camps, and a few other odds and ends.

Historically, logging outfits that cut along river systems showed little respect for the environmental costs of their operations. Clearcutting along the shorelines caused direct erosion into the rivers, loss of habitat for a host of species, siltation of spawning beds, and loss of the forest's value as a sponge and storage cellar for rainwater and snowmelt.

Today, we have best management practices (BMP) to prevent such disregard. No commercial logging occurs directly along the river's edge. Nearly 85 percent of the Manitowish is in public ownership, and while state law does not require best management practices, state foresters generally implement BMP as a rule along waterways. Water quality issues must be addressed in the narrative descriptions that are part of the write-ups accompanying all timber sales.

Many studies have looked at the effects along shorelines of timber harvests that do not meet best management practices. Increased streamflow and sedimentation are among the most serious consequences. By removing standing vegetation along shorelines, the important function of leaves as interceptors of raindrops is lost. In a hard rain, where would you rather stand—under a forest canopy or out in a field? Rain hitting the ground head-on moves soil far more easily than rain hitting the ground as scattered smaller droplets. After a rainstorm, more water carrying more soil gets into a river from a logged shoreline than from an unlogged one.

Transpiration is also reduced by timber harvest along a stream. This simply means that much of the water that was once taken up by plant roots and transpired through the leaves back to the atmosphere now runs directly into a stream or lake.

When an experimental forest in New Hampshire was clearcut and subsequent regrowth suppressed with herbicides, runoff into the stream increased on an annual basis by 40 percent. This additional runoff represents water that in an intact forest would primarily have been returned to the atmosphere via transpiration.

Runoff not only includes water, but also the soil it may carry with it. Major sources of sediments include landslides on deforested slopes, surface runoff from logging roads, and greater erosion of the streambanks and streambed. As discussed in an earlier chapter, sedimentation fills in the spaces in spawning gravels, smothers organisms and other substrate materials, and reduces the delivery and removal of various gases and nutrients.

The leaf canopies of trees also shade the forest floor and extend the period of snowmelt in the spring. Removing the forest canopy typically increases the streamflow at some times of the year (the winter and spring), and decreases it at others (the summer and fall). During a major storm, increased runoff heightens the speed and depth of a stream surge, increasing the likelihood of flooding.

There are limits to the rate at which soil can absorb rainfall. If the maximum infiltration rate is exceeded, water will begin to move across the soil surface. Machine logging tends to compact soils, decreasing the infiltration rate and increasing overland flow.

The logging of shorelines also exposes the streambed to more sunshine, warming the water and increasing plant growth. A study of 38 trout streams in southern Ontario found that trout numbers were greatly reduced by water temperature increases that followed the removal of the riparian canopy. Streams that maintained adequate trout populations were characterized by summer water temperatures less than 71°F. The amount of forested streambank upstream of a site can be effectively used as a predictor of downstream water temperatures. Thus the quality of upstream shorelines can be used as a partial predictor of where the best trout populations might be found downstream.

The good news about poor logging practices along waterways is that the biological effects are short-term due to the relatively quick regrowth of vegetation. Recovery of shading typically occurs in 10 to 30 years in temperate climates like ours. However, one study compared amphibian abundance in headwater streams. Some of the streams were located in uncut forests; others were located in forests that had been logged 14 to 40 years earlier. The researchers found species richness was significantly lower in the cutover sections of stream. So, timber harvests along streams can also have long-term consequences.

It's very important to note that the magnitude of all of these effects depends on a host of variables, including the amount of annual rainfall, the steepness of the slopes leading into the waterway, the type of soil and vegetation, and the specific timber management practices. As is typical in the natural world, generalizations seldom accurately describe the reality found at one site at one particular time.

Read more about logging practices along rivers at the end of the chapter.

≈ ≈ ≈ ≈ ≈

A quarter mile or so downstream from the DNR campsite, the original portage trail from Mercer Lake entered the Manitowish. We have a general sense of where this 4-mile-long trail likely would have met the river, but we believe the actual trail has long since grown over. In the 1700s and 1800s, the voyageurs and American Indians entered the last leg of their journey from Lake Superior to the trading posts and village at Lac du Flambeau.

In 1847, J.G. Norwood paddled this route and describe it as follows:

"The outlet from Little Turtle Lake [Grand Portage Lake in Mercer] is through a very narrow channel connecting it with another lake [Mercer Lake], which we crossed, and came to the beginning of what is known as 'Six Pause portage.' As the voyageurs had to make a double portage, we took our packs and walked on to its termination, at the east branch of the Chippewa River, or as it is commonly called the Manidowish, where we arrived at noon. The trail runs over a sand barren, with the exception of the last half mile, which runs through one of the worst tamerack swamps I have ever seen. A few stunted pines, with the occasional patches of course grass, is the only vegetation supported on the high grounds.

"The Manidowish River at this point comes from the northeast, is deep and clear, about thirty feet wide, and winds through the centre of a broad wet meadow, with grass from two to five feet high. After the portage was made, we descended the river four miles, though probably not more than one mile in a direct line from the portage to a favorable place for a camping ground."

(See appendix B for Norwood's complete description of his 1847 journey from Long Lake to the Wisconsin River.)

While the Six Pause Portage sounds a bit rough, it was duck-soup compared to the first part of the famed Flambeau Trail. At 45 miles in length, the trail had to be portaged from Lake Superior to Long Lake, north of Mercer.

Maps may be called the light or eye of history.
— Hulsius, 16th century geographer

The Flambeau Trail was not for the faint of heart or frail of body. Francois Victor Malhiot, a 28-year-old Frenchman appointed to take charge of the North West Fur Company's trading post in Lac du Flambeau, crossed the Flambeau Trail in July of 1804. He's most often quoted for this observation of the trail:

"Of all the spots and places I have been in my thirteen years' of travels, this is the most horrid and most sterile. The Portage road is truly that to heaven because it is narrow, full of overturned trees, obstacles, thorns, and muskegs. Men who go over it loaded and who are obliged to carry baggage over it, certainly deserve to be called 'men.'

"This vile portage is inhabited soley by owls, because no other animal could find a living there, and the hoots of those solitary birds are enough to frighten an angel or intimidate a Caesar."

Malhiot first arrived at the Flambeau Trail on July 25, 1804. Three Frenchmen, who had overwintered in the area, arrived on July 27 to help him portage, but they were "thin and emaciated like real skeletons," so he gave them each "a drink of shrub [species unknown], two double handfuls of flour, and two pounds of pork and they began to eat with such avidity that I was twice obliged

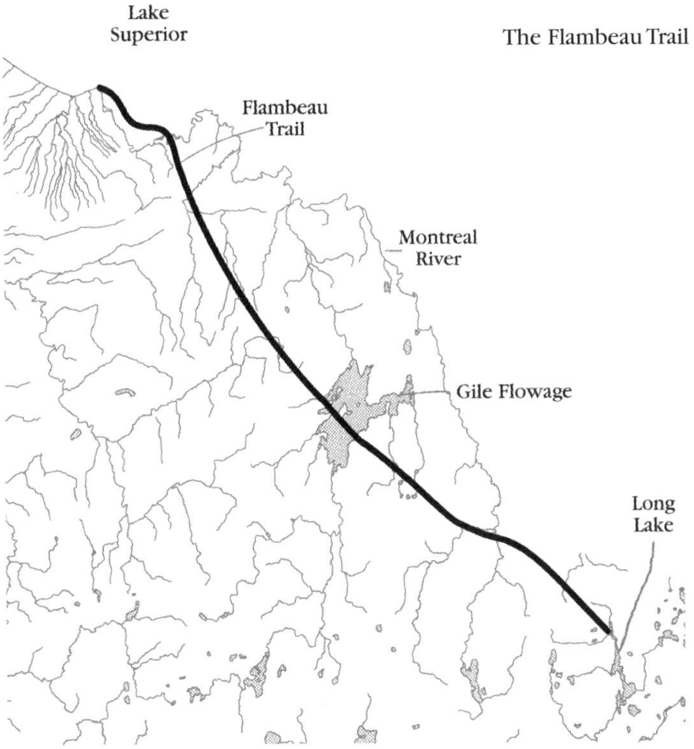

to take the dish away from them . . ."

He began the portage on July 28. "I ordered the men to get ready to enter the portage tomorrow. I gave each one a double handful of flour, a pound of pork and a drink of rum as a treat." Here he also gave an Indian he called 'le Genou' (the Knee) "16 plus credit ... I gave nearly as much to his brother 'La Pourceline.'" (Plus was a monetary unit roughly equal to one good beaver pelt.)

"I took with me a bale of merchandise, a roll of tobacco, 20 pounds of shot, 20 pounds of bullets, three quarters of a sack of corn, a barrel of rum double-strength, and all my baggage. Today we did forty pauses."

Distances were measured during the fur trade in pauses, or the times that the voyageurs paused to rest. A pause could be anywhere from 600 yards to half a mile depending on the terrain. Bogs and hills were a good cause to pause. The Flambeau Trail was considered a 120-pause portage. Each voyageur customarily carried two packs, each pack weighing 80 to 90 pounds, for a total of 160 to 180 pounds.

One of the voyageurs, Durocher, had to stay behind because of a bad case of poison ivy. His duty was to take care of the remainder of Malhiot's goods.

On July 29, they did only 20 pauses because Malhiot was suffering from a terrible toothache. He had to get his "head sweated this morning which soothed the pain a little." Malhiot's toothache was so bad that two weeks earlier on his paddle across Lake Superior south of Grand Marais, "after trying every imaginable remedy," he took

50 drops of opium without any effect, and then decided to take some rum, swallowing in one gulp half a pint of the raw spirit.

No record exists for July 30, but on July 31, Malhiot and his men began at 7 a.m. and arrived at Long Lake at 1 p.m. One man still had severe pains in his legs. The men didn't get any rest, because Malhiot immediately sent them to get the canoes that were cached. He told the men to gum them, and then make paddles so they could begin the trip to Lac du Flambeau in the morning.

They began the paddle on August 1 at 4 a.m., arriving the next day at Fort du Flambeau at 3 p.m. the next day. The fort stood on the north side of the lake, probably near the original Indian village, but the actual site is still unknown.

Portaging then was little different than today. The usual array of tormentors—bugs, rain, heat, high water, storms, injuries—plagued the voyageurs, just as they plague us today.

Malhiot retraced the Flambeau Trail in May 1805, and experienced an additional array of problems:

May 26. — Yesterday we crossed the Portage des Six Poses [Manitowish River to Mercer Lake] and that of the village of la Tortue [on Grand Portage Lake] and, at one o'clock this afternoon we reached the Grand Portage of the Montreal River where my canoe was broken, and we are obliged to camp in order to allow four packs to dry that got wet. The two portages we crossed are exceedingly bad and the Savages tell me this one is still worse.

May 27. — It rained all last night . . . having come to camp at the Petite Riviere, this side of the Riviere des Pins [probably what was then known as the Pine River flowing from Pine Lake and meeting Layman's Creek]. The portage was never so bad and the flies are eating us up.

May 29. — My people did sixteen pauses today although the water was frequently up to their knees, and they complained a good deal. We are camping at the Riviere des Sapins [possibly Layman's Creek].

May 30. — The road is so bad and there are so many overturned trees that I was lost for an hour, and should still be so, had I not had a gun.

May 31. — Today we are obliged to make a small raft to cross the Riviere du Milieu [undoubtedly the west branch of the Montreal River], and we camped there. I have not seen the water so high for a long time.

June 2. — The rain prevented us from carrying. We have done only seven pauses since Friday [May 31]. There are billions of flies! We are weak owing to bad food, and we shall have none at all unless the weather changes.

June 3. — [Yesterday] we did ten pauses with one half the baggage. Today, it is raining hard and we are completely weather-bound.

June 4. — The only remaining food is ten quarts of corn not treated with lye. [To soften the corn, French-Canadians put a small quantity of lye in the water while boiling the grain. The corn was usually eaten with pea soup or with milk and sugar].

June 6. — We have at last finished the portage at a quarter past twelve, all very tired. . . I was lucky enough to get four sturgeon from the Savages today, which will, I hope, last me to la Pointe, where I left a sack of corn in a cache last autumn."

They had begun on May 27, 1805, and reached Lake Superior 11 days later.

When I was a younger man of stronger back and weaker mind, I often said that if I had a time machine, I would use it to join up with the voyageurs. I'm older now, and better read. I now only wish to hover over them in my time machine, watching, appreciating, and heading for home in one piece. The leading cause of death for the voyageurs was stangulated hernias. That tells me all I need to know about whether or not I should disembark from my time machine.

Read more about the Flambeau Trail at the end of the chapter.

The movement of trade goods by canoe over 2,000 miles from Montreal to Lac du Flambeau staggers the imagination. The 35-foot "Montreal" canoes were the freighters of the time, hauling up to 4 tons of goods, provisions, and voyageurs from Montreal across Lake Superior to the Flambeau Trail. On inland waterways like the Manitowish River, the 25-foot "North" canoe was utilized as a compromise vessel that could turn and maneuver, carry a large load, and be light enough to slip through rocky shallows. Still smaller birchbark and dugout canoes were utilized by individual trappers and traders.

While on the 1820 Cass Expedition, the famous 19th-century Indian agent, explorer, naturalist, and writer Henry Schoolcraft described the canoes in which they traveled throughout the wilderness of what is now Michigan, Wisconsin, and Minnesota:

"The northwest canoe is . . . constructed wholly of bark, cedar splints, the roots of the spruce, and the pitch of the yellow pine [red pine] . . . and these articles are fabricated in a manner uniting such an astonishing degree of lightness, strength, and elegance, and with such a perfect adaptation to the country and the difficulties of the northern voyages, as to create a sentiment of mixed surprise and admiration. Those of the largest size, such as are commonly employed in the fur trade of the north, are thirty-five feet in length, and six feet in width at the widest part . . . They are constructed of the bark of the white birch tree (*Betula papyracea*) which is peeled from the tree in large sheets, and bent over a slender frame of cedar ribs, confined by the gunwales, which are kept apart by slender bars of the same wood. Around these the bark is sewed by the slender and flexible roots of the young spruce tree, called wattap, and also where the pieces of bark join, so that the gunwales resemble the rim of an Indian basket. The joinings are afterward luted, and rendered water tight, by a coat of pine pitch, which, after it has been thickened by boiling, is used under the name of gum. In the third cross bar from the bow, an aperture is cut for a mast, so that a sail can be employed when the wind proves favorable. Seats for those who paddle are made by suspending a strip of board, with cords, from the gunwales in such a manner that they not press against the sides of the canoe.

"A canoe of this size, when employed in the fur trade, is calculated to carry sixty packages of skins, weighing ninety pounds each, and provisions to the amount of one thousand pounds. This is exclusive of the weight of eight men, each of whom are allowed to put on board a bag or knapsack of the weight of forty pounds. In addition to this, every canoe has a quantity of bark, wattap, gum, a pan for heating the gum, an axe, and some smaller articles necessary for repairs. The aggregate weight of all this may be estimated at about four tons. Such a canoe, thus loaded, is paddled by eight men, at the rate of four miles per hour in a perfect calm—is carried across portages by four men—is easily repaired at any time and at any place, and is altogether one of the most eligible modes of conveyance that can be employed upon the lakes, while in the interior of the northwest—for river navigation, where there are many rapids and portages, nothing that has been contrived to float upon water offers an adequate substitute. Every night the canoe is unloaded, and with the baggage, carried ashore . . .

"The canoes, when laden, are hauled out in deep water; the men then catch up the sitters on their backs, and deposit them in their respective seats. When this was done, they struck up one of their animated songs, and we glided over the smooth surface of the lake with rapidity . . ."

Today, such canoes are difficult to visualize, much less have a feel for how they would paddle. There are a few examples in the area, however. Camp Manitowish in Boulder Junction has on display in one of its dining halls a smaller birchbark canoe that was found above Fishtrap Dam. The George W. Brown Cultural Museum in Lac du Flambeau has a dugout canoe on display that was found under water near Strawberry Island in Flambeau Lake. The North Lakeland Discovery Center in Manitowish Waters has a 22-foot authentic birchbark canoe they use for a few trips every year. The canoe was built in 1998 by master birchbark builder Ferde Goode and his students.

Hopefully, these descriptions of the extraordinary hardships of the voyageurs and the Ojibwa inspire you to forget your aches and pains as you make your way downriver. Misery loves company, but your misery is likely nothing compared to that of the voyageurs.

≈ ≈ ≈ ≈ ≈

In passing through this section of river, the voyageurs and Ojibwa would certainly have swum in the river a few times, though water-based recreation was hardly a recognized concept in their time. The river water is warm, only slightly stained in color, the bottom sandy more often than not. The inside banks on curves frequently provide a fine, albeit small, sand beach.

The Manitowish is classified as a warmwater river, and is certainly not to be confused with a coldwater trout stream. That warmth has far greater importance to the biota of the river than it does to us as a good place to swim.

Water temperatures play a major role in the life of a river. Most organisms have evolved a range of temperatures that are optimal for their growth, as well as a range of temperature extremes that are barely tolerable. Temperature influences many biological processes, but most important may be its influence on the amount of oxygen available in the water and its rate of use.

Temperature affects the amount of oxygen that can be dissolved in water. Cold water can hold much more oxygen than warm water. For example, water near freezing can hold over 14 milligrams per liter (mg/l) of oxygen. Water at 55°F can hold 10.5 mg/l, while water at 75°F can only hold 8.4

The Relationship of Temperature to Dissolved Oxygen

°C	mg/liter	°C	mg/liter
0	14.62	21	8.91
1	14.22	22	8.74
2	13.83	23	8.58
3	13.46	24	8.42
4	13.11	25	8.26
5	12.77	26	8.11
6	12.45	27	7.97
7	12.14	28	7.83
8	11.84	29	7.69
9	11.56	30	7.56
10	11.29	31	7.43
11	11.03	32	7.30
12	10.78	33	7.18
13	10.54	34	7.06
14	10.31	35	6.95
15	10.08	36	6.84
16	9.87	37	6.73
17	9.66	38	6.62
18	9.47	39	6.51
19	9.28	40	6.41
20	9.09		

mg/l. For a brook trout, temperatures over 78°F usually are fatal, while optimal growth occurs between 55°F and 65°F.

Trout typically need a minimum of 6 mg/l of oxygen to survive. Given that 78°F water can hold up to 8.1 mg/l of oxygen, obviously the amount of oxygen isn't the only variable in the survival equation. Another factor is the metabolic rate of aquatic organisms. Metabolic rate increases with rising water temperatures, which results in greater oxygen demand. A trout uses five times more oxygen while resting at 80°F than at 40°F.

A trout in warm water must feel like someone with severe emphysema. No matter how deeply and quickly the trout breathes, it can't quite get enough air. Activity exacerbates the need for air, so the only recourse is inactivity—hardly an ideal solution for a fish that needs to actively pursue prey. The combination of not enough oxygen in the water and a much higher need for oxygen is fatal to trout.

Rising temperatures also affect oxygen in the water by increasing the rate of plant photosynthesis. On the one hand this sounds good, because photosynthesis produces oxygen. On the other hand, more plants growing means more plants eventually dying, and decomposition of plants consumes oxygen. When photosynthesis

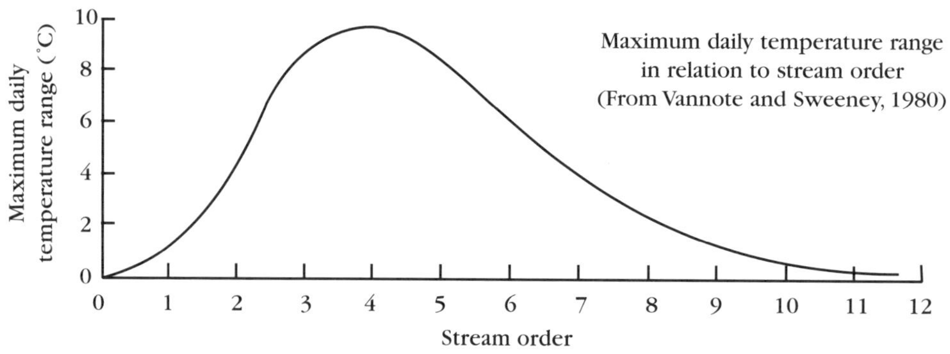

goes up, the need for oxygen (the biological oxygen demand, or BOD) typically goes up also. Too much of a good thing can be a bad thing, even where photosynthesis is concerned.

One last major effect of warmer temperatures on an organism is the stress that is caused by lower oxygen and higher metabolic rates. Like humans, aquatic organisms exhibit increased sensitivity to toxins, parasites, and disease when they are stressed.

Water temperatures obviously fluctuate seasonally. They also typically rise and fall in response to daily air temperatures, though not dramatically. Water is much better than land at holding onto heat or cold. Thus, rivers (and lakes) are natural storage sinks for warm or cold water, and act as temperature moderators on the surrounding land. Planting gardens on the uplands next to a large river or lake makes as much sense today as it did to American Indians in the past. The native people understood that early autumn frosts are less likely to occur on uplands next to water because of the moderating atmospheric effects of the warm water.

Daily temperature fluctuations do occur in rivers , but most commonly in small, unshaded streams. Because of their much greater volume of water, larger rivers tend to have fairly constant water temperatures. Spring-fed brooks also remain relatively constant in temperature, because of groundwater influence and streamside shading. The temperature of groundwater, which feeds the springs, is usually within one or two degrees of the mean annual air temperature and stays constant throughout the year.

When all is said and done, daily temperature variation is usually greatest in natural streams of intermediate size, like the Manitowish. Intermediate streams are usually much less shaded than headwater streams, but have far less water volume than large rivers.

In lakes, water temperatures often vary significantly from the surface to the bottom. Organisms often adapt their life cycles, partly in an effort to find optimal temperatures. River-water temperatures, however, seldom stratify. Due to their relatively shallow water and moving currents, rivers are usually well-mixed, the blending of waters keeping temperatures nearly constant from the surface to the bottom.

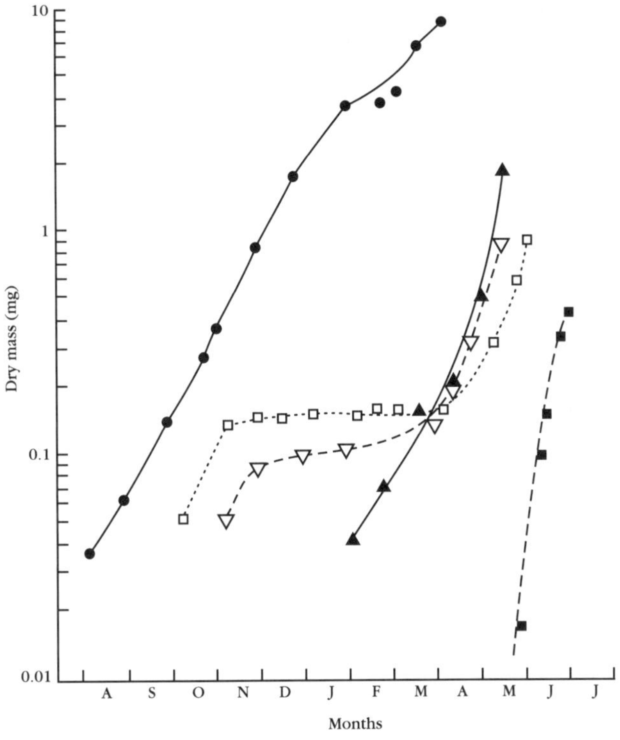

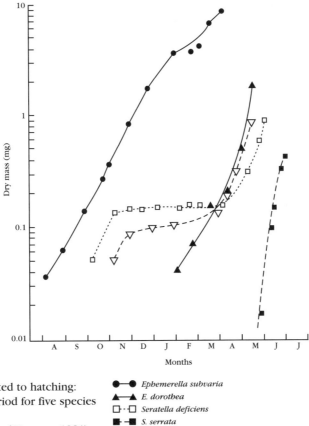

Temperature related to hatching: Larval growth period for five species of mayflies (From Sweeney and Vannote, 1981)

As discussed above, temperature unquestionably sets limits on where a species can live, but it's not just cold-water fish like trout that are affected. The timing of an aquatic insect's life cycle is often cued and regulated by water temperature. Temperature determines when insects hatch, when they most actively grow, and when they feed. Closely related species often hatch, grow, and emerge as adults in a staggered sequence over many months. This life cycle separation, a form of resource partitioning, appears to reduce competition between some species.

Microorganisms like bacteria are also affected by temperature. The decomposition of leaf litter on land proceeds at a rate 20 percent faster at 50°F than at 40°F. The processing of leaves in water also occurs more rapidly at warmer temperatures.

It's important to note that even though all plants and animals have upper and lower tolerance limits to temperature, most can gradually acclimate to seasonal temperature changes. A wintering trout would die if suddenly exposed to the same warm water it tolerates in summer. Rapid swings of temperature that don't provide acclimation time are a source of great stress for most aquatic organisms.

It's equally important to remember that warmer temperatures aren't always bad. Shortened seasons due to cold reduce the number of insect generations that may occur in a year. (I personally like this idea for mosquitoes and their ilk). A given species of insect may only produce one generation per year in colder latitudes, while producing multiple generations per year in warmer latitudes. The caddisfly *Rhyacophila evoluta* requires two to three years to complete its life cycle in cold waters, but only one year in warmer waters.

Frog life cycles provide a good example of how cold water slows metabolism and curtails growth. Bullfrog tadpoles in the frigid waters along the Apostle Islands in Lake Superior often take three years to mature into adults, while their brethren in the warmer inland waters of northern Wisconsin typically only require two years.

Still, the extent to which cold-adapted aquatic insects can remain active and grow at near-freezing temperatures is surprising. Stoneflies are noteworthy for their ability to grow in very cold water. *Taeniopteryx nivalis*, a species of stonefly, grows significantly between November and March in temperatures only a few degrees above freezing. Sometimes the adults hatch when the ice is just going off a river or lake. We've seen adult stoneflies appearing to hatch right out of the ice, an occurrence that we always find remarkable and slightly insane.

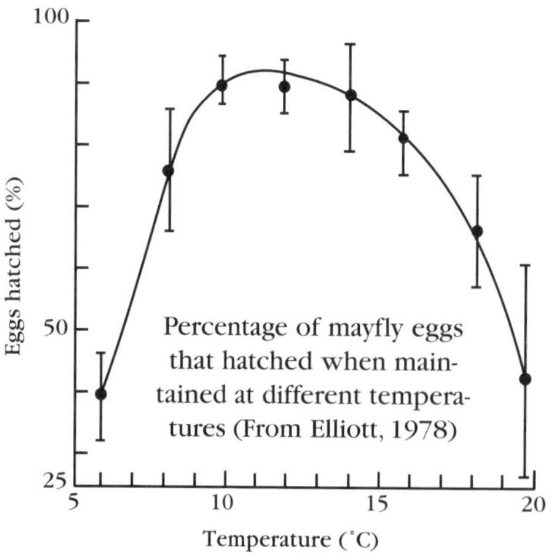

Percentage of mayfly eggs that hatched when maintained at different temperatures (From Elliott, 1978)

Insect eggs are also adapted to hatching at certain temperatures. Sometimes cold temperatures are necessary; much like some garden seeds, the eggs of a few mayfly species must undergo low temperatures before they can develop and hatch, a process known as obligate diapause. When a dam was built on the Saskatchewan River, winter water temperatures were maintained at 39°F, preventing the insect fauna from experiencing a prolonged period of near-freezing followed by a rapid temperature rise in spring. The egg cycle of the mayfly *Ephoron album* was interrupted, and it then disappeared because the temperature regime was modified. Virtually all other aquatic insects disappeared as well. Twelve orders, 30 families, and 75 species were reduced to only the midge family *Chironomidae*.

≈ ≈ ≈ ≈ ≈

The Manitowish meanders for the next 4 miles as the crow flies, but probably 8 miles or more as the river turns (a good title for a Northwoods soap opera). The second DNR campsite is on river-left about one and a half hours down from Highway 47, depending on one's paddling haste or lack thereof. Beaver lodges are common in this stretch (read more about beaver physiology at the end of the chapter).

Dissolved Oxygen (DO)

Measuring DO is probably the most significant water quality test one can do to determine the suitability of a stream for aquatic life. However, a DO test is only a snapshot taken at that moment and place on the river, because DO can vary throughout the day and the year, and from site to site.

DO is removed from water by:

- Respiration of fish, bacteria, fungi, and protozoa.
- Decay processes of organic matter.
- Chemical reactions.

Aquatic plants produce oxygen through photosynthesis during the day, but the plants also respire and use oxygen at night. Respiration by plants and bacteria removes DO. A particularly heavy level of plant growth can lead to a depleted DO content. When insufficient oxygen is available for decay processes, anaerobic decay results, producing a rotten egg smell from the production of hydrogen sulfide gas.

DO levels change over the course of a 24-hour day. DO is at its highest in late afternoon after many hours of active photosynthesis, especially from algae. DO is at its lowest at sunrise, because respiration continues at night when no photosynthesis can take place.

Waters with high levels of organic matter that slowly decay have more respiration taking place than waters with less organic matter.

There's also a positive relationship between water velocity and DO. The faster the water, and the more it churns, the higher the DO. Rapids are like a mechanical blender, mixing oxygen into the water.

Wind also churns water, mixing in oxygen. But wind varies from lake to lake. Its velocity is affected by the shape of the lake and the surrounding topography. Large lakes situated in flat topography obviously experience stronger winds than small lakes tucked in between hills.

Water depth reduces mixing of oxygen, so waters near the surface typically have more DO than waters near the bottom.

Seasonal changes affect DO level, too. Late winter DO levels are typically very low, because of the effects of five months of ice cover. However, the colder the water temperature, the more oxygen the water can hold.

DO levels mean life or death to some aquatic animals. Trout do best in at least 8 parts per million (ppm), while bass and bluegills do well at 5 ppm. Carp and bloodworms can survive in water with less than 1 ppm of dissolved oxygen.

Otter family

Many years ago, Mary and I drifted around the last bend before the second campsite, and came upon an otter family in the river. The otters initially dove and surfaced repeatedly, snorting at us whenever they popped up. Eventually, they swam back into some willow shrub, and all was quiet for a minute. Then we heard birdsong coming from the willow. We wondered why birds would be singing from within a small shrub complex when six otters were hiding there. The birdsong wasn't familiar to either of us, and we couldn't identify the species. With our binoculars, we panned the shrubs for any sign of songbirds, but saw nothing. Several minutes passed. The songs continued, and we decided to move on, concerned that we might be stressing the otters, even though we were on the other side of the river.

The next day we looked up otters in Hartley Jackson's *Mammals of Wisconsin*, and found to our amazement that otters emit "an oft-repeated birdlike musical chirp, evidently a call note." We'd never known that was possible, but his description fit what we heard. We had yet another remarkable experience to add to our appreciation of otters.

Otters leave clear sign of their presence if you don't actually get a chance to see them. They deposit an oily black scat that usually contains orange crayfish scales. Look on the inside curves of the river where sand deposits accumulate. These sites offer warm, dry tables for otters to eat their meals in peace.

The largest member of the weasel family at 3 to 4 feet long (including the tail) and 15 to 25 pounds, the otter's long, muscular body is a swimming machine. Otters can outswim all other land mammals and most fish, utilizing their webbed hind feet, a slight sculling motion from the tail, and a back-to-belly flexing undulation for propulsion. They are able to stay under water for 6 minutes or more, dive to 40 feet, and outmaneuver much smaller fish. Otters can capture trout if they wish, though trout are a minor component of their diet. Their straight-away speed under water is estimated at up to 7 miles per hour.

Like most animals, otters feed opportunistically on whatever's available at the wetland buffet, which ranges from crayfish to frogs to muskrats to fish, and even a few birds. However, research studies in Wisconsin, Michigan, and Minnesota show that otters feed primarily on slower, rough fish like suckers. Presumably, the otters prefer not to exhaust themselves by chasing faster fish like trout.

Otters breed in March or April, and utilize delayed implantation to give birth a year later. A typical litter consists of two to four pups, fully furred but with eyes closed for the first month. The pups are weaned at two months, but stay with Mom until the following spring.

Otter dens are found in an array of places, including confiscated animal burrows, hollow logs, brush piles, beneath tip-up mounds, and abandoned beaver lodges. Otters utilize beaver lodges extensively in winter as access points to water and fish. Otters often evict beavers from their lodges, given that the lodges are located on prime shoreline real estate. Typically about 3 square miles of lakes and streams lie within the territory of one otter pair.

But otters are best known for their beautifully developed sense of play. They appear to play familiar games

Ask Me

Some time when the river is ice ask me
mistakes I have made. Ask me whether
what I have done is my life. Others
have come in their slow way into
my thought, and some have tried to help
or to hurt; ask me what difference
their strongest love or hate has made.

I will listen to what you say.
You and I can turn and look
at the silent river and wait. We know
the current is there, hidden; and there
are comings and goings from miles away
that hold the stillness exactly before us.
What the river says, that is what I say.

like Tag and Follow The Leader, engage in races, perform wild versions of tag-team Wrestlemania, go up and down slides, and in general cavort around as if life was meant for joy and not the hardship of survival. The purpose of their play remains controversial. Animal behaviorists see it as a means of sharpening reflexes and hunting skills, while others are hard-put to assign survival training to hours on end of sliding down mud banks.

In winter, their slides can make a river edge look like an Olympic toboggan venue. Otters usually bound for a few steps and then belly slide up to 20 feet, moving along at speeds near 15 miles per hour, and covering miles of ground. We've seen their slides miles inland from water, and often wonder where they're going and how they find their way.

In the Northwoods, otter populations are doing well, but because they are at the top of the food chain, they will always remain relatively uncommon. Given their uncommon approach to life, that's only fitting.

≈ ≈ ≈ ≈ ≈

Sitting in the canoe along the shoreline often brings you eyeball to eyeball with the extensive sedge communities that array the river's flood plains. Sedges provide identification difficulties that usually overcome the most ardent efforts of non-botanists. In fact, the sedges offer a host of challenges even for professional botanists, so don't feel bad if you wouldn't know a sedge from a grass.

More important than knowing the specific identification of sedges is the knowledge of their importance in the aquatic community, particularly as a site for birds. Sedge meadows support a significant number of bird species and individuals. Their lack of vertical

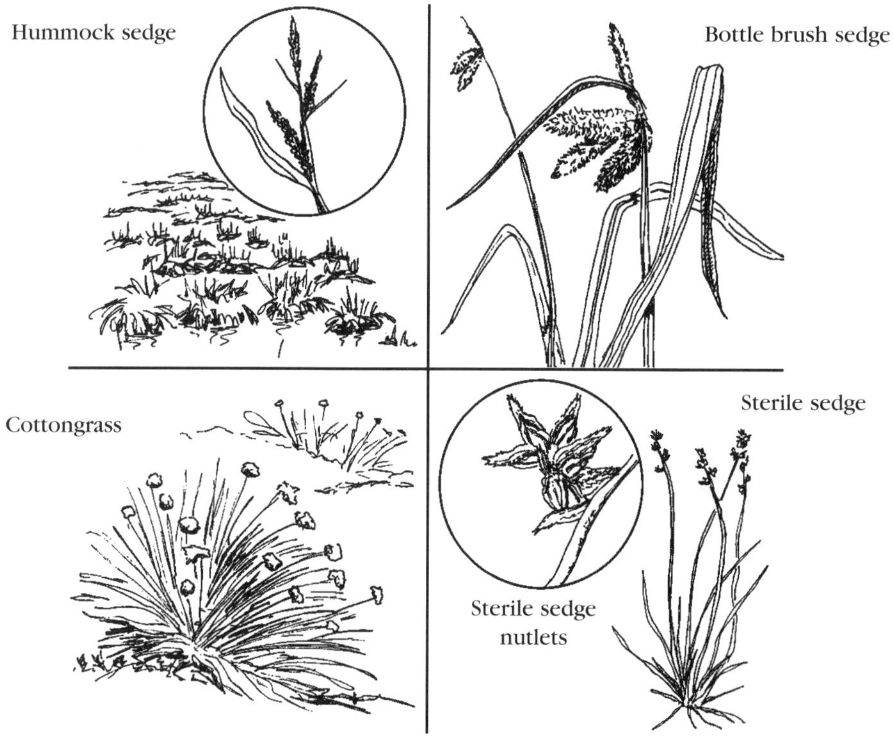

structure—a meadow only grows a few feet high after all—limits nesting opportunities to species that nest low to the ground. Still, sedge meadows are a rare plant community these days in the Upper Midwest, and because of their rarity, offer a rich birding experience.

What should you look for? In 1997, researchers studied the relative abundance of nesting birds in the sedge meadows of Necedah National Wildlife Refuge, Wisconsin. The species found most commonly included pied-billed grebe, Canada goose, Virginia rail, sora rail, common snipe, sedge wren, yellow warbler, common yellowthroat, red-winged blackbird, swamp sparrow, and song sparrow. Less common were sandhill crane, least bittern, great blue heron, northern harrier, blue-winged teal, American bittern, gray catbird, bobolink, warbling vireo, and rufous-sided towhee.

In her book *Birds and Forests*, Janet Green lists the most common birds of sedge fen as sedge wren, bobolink, savannah sparrow, and Le Conte's sparrow.

If I had to pick the bird I hear most commonly in local sedge meadows, it would be the diminutive sedge wren. This little 4 1/2-inch dynamo sings throughout most of the day from the top of a waving sedge. Its song is usually described as beginning with a few single notes, followed by a staccato trill. It sounds a bit like a weak machine gun to me, starting slowly before giving you a full burst.

Because sedges are infrequently identified and poorly understood by the general public, their conservation dangles at the low end of species checklists. That's unfortunate. A sedge meadow has a prairie grass aspect to it, a willingness to bend in concert with the wind that reminds me of ocean waves. Few people have written poetically about sedge meadows—prairies get all the press—but the following quote by Willa Cather could just be applied to a large sedge meadow rather than a field of big bluestem: "More than anything else I felt motion in the landscape; in the fresh, easy-blowing morning wind, and in the earth itself, as if the shaggy grass were a sort of loose hide."

When I look at sedge meadows, I see and feel beauty, a sensation that I experience often in this wilderness section of river where the natural community is preeminent.

≈ ≈ ≈ ≈ ≈

Pick a river, any river. If you sit beside it long enough you will hear many things, and most of them are worth waiting for. After a while you will hear a voice out on the waters whispering, saying, in a tone as soft and low as a mother singing to her restless child, that each of us is flowing, flowing like a river, and that one day we will all, from gossamers to galaxies, flow out of time into something even greater than a river. — John A. Murray

Rivers bring out people's stories. A white water lily slough, tranquil, sweet-smelling, and warm, helps to draw the words out. Sloughs are places that encourage drifting in mind and spirit. On a typical river trip, we spend as much time talking about our lives as we do about the life of the river. Quiet rivers evoke an intimacy, a meditative quiet that allows you to reflect on your life. Often in the course of river conversation, we talk of where we're from and how we came to this river at this time.

I'm often asked how I became a naturalist. Was I born in the Northwoods? Did my folks teach me about the woods and rivers? I tend to cringe a bit when asked, because I was raised a city kid, and somehow I feel my naturalist badge gets tainted when I reveal my roots. Not only was I raised in a city, but I was born in Gary, Indiana, a city renowned for its level of pollution. My family moved when I was 1 year old to Pittsburgh, Pennsylvania. As you might guess, my father worked for U.S. Steel.

On the surface, it's an anomaly. I was raised with more concrete under my feet than soil, more cars than wild animals. Yet, here I am a naturalist. I'm sure my former city friends wonder at this transformation, as have my parents.

But I remember natural influences in my life. Three rivers come together in Pittsburgh: the Allegheny, the Monongahela, and the Ohio. Three rivers that, as a boy, I recall were only slightly less chocolate than the black kids I saw swimming in them. I envied those black kids in the summer, their freedom, their fun, when I was so hot. But I also wondered how they could possibly swim in that stuff. The rivers were little better than gutters then, carrying debris of every shape and kind, a garbage truck in perpetual motion.

One spring, the rivers flooded over the wharfs and parking lots, coming up close to the freeway. I was amazed at the power of the water, and frightened of what might be in that water. God only knew what combination of human pollution and natural river life (snakes and the like) might have been swirling by.

I never saw the rivers as desirable, except from a distance. They were beautiful as landscapes, but not as living entities.

"The Point" is a famous landmark in Pittsburgh, the site where the three rivers come together. It was a romantic place for me in my adolescence, with lit fountains and bridges, beautiful at night while surrounded by the lights of the downtown.

"The bridge to nowhere" was there, too, an uncompleted new bridge spanning most of one of the rivers, offering a footbridge from which to dream at night, where the lights reflected off the water and the frenzy of road traffic couldn't reach. My girlfriend and I would walk there and gaze down at the river. We'd hold hands and talking would come easy, the sound and feel of the water beneath us eroding our harder edges.

Sometimes we'd take the trolley ride up the side of the steep hillsides that overlook the rivers to the "Top of the Hill," as it was called. Here the city unfurled far beneath us. Pittsburgh gleamed, but in my mind not as brightly as the rivers.

We took an elementary school trip once on the Good Ship Lollipop, a brightly colored tugboat of sorts that for a price ran people up and down the river. All I recall now were the stark, industrial shorelines occasionally enlivened by a copse of trees or a forested island.

Other rivers came into my consciousness. In high school, some friends asked if I wanted to go rafting with them. "Sure, why not?" I responded. The reason "why not" should have been profoundly obvious—I simply hadn't seen the Youghiogheny River before. Later, when I took off my glasses in order not to lose them in the wild whitewater, I experienced a chaos of blurred plunging through enormous rock gardens, holding on for dear life, white-knuckled beyond anything I had ever experienced. The bravado, fear, thrill, and camaraderie were indelible.

The water slowed at times, the glasses went back on, and we floated by Appalachian homes along the shoreline, revealing rural poverty beyond anything I had ever seen. I wondered what they thought of us, passing by their lives on a rubber raft, city kids without a clue. I felt like an invader.

We returned the next spring and found the river in flood, and still in our inestimable ignorance, nearly put-in. The brown, boiling water would have swept us to our deaths in short order. The power of that water, its terrible strength . . . only one other time have I seen water like that. We were traveling somewhere in the Midwest when I was quite young, and we came to a washed-out bridge over a large river. The water surged by carrying trees and parts of houses. Then an entire house came rushing by, with two people on its roof. I couldn't believe my eyes. We were helpless to help them. I was a small boy, and I clung to my mother. I can't remember anything else about the incident.

The most anticipated fun of every year was our vacation on Lake Wawasee back in Indiana. We rented a cottage for a week or two, and swam and swam and swam. As much as anything, I loved to dive down in the water and just go limp, hanging there in the water until my lungs would just about burst. I savored the sense of weightlessness, of being in another world, of losing my body, so different—so pleasant.

We often borrowed a neighbor's little sailboat, a wooden SailFish that was little more than a flat board with a rudder and a sail. That craft was perfect. If it went over, there was no consequence. We just stood on the centerboard, flipped it back up, and went on our way again. We would lean that boat out to its absolute edge, hanging in the balance, until we'd go over, yelling and laughing, and ready to do it again.

We learned to read the wind a bit, be attuned to it so the boat would use it just right, to its very maximum. We'd just fly, or so it seemed, skimming waves, our muscles taut from holding the ropes and the tiller.

Whew! I'm ready to go again right now.

I was always a marginal swimmer, and a terrible diver, while my brother Dave was like a fish—smooth, inexhaustible. He could literally swim for miles. I envied him that, and I always wondered why we were so different.

Still, we played all day in the water. We'd stop only to eat. There's nothing like the hunger one gets from being in and out of the water all day.

My mom liked best to just float on her back, or to sit in one of those Styrofoam lounge-chairs with a drink. She would just be. I think she liked the coolness, the undulation of the waves, the quiet when it happened, the sounds of children when they happened.

My grandmother Mimi would

come and join us from her Indiana home and show us how to catch fish. One day I fished with her for about an hour, and she caught 18 fish, while I got one bite. She was uncanny, and always so sweet about out-fishing the rest of us, though we could tell she was proud of it. She'd drawl in her Hoosier accent, "Well, Johnny, the fish seem to just like me better than you."

We'd sometimes fish in a canal that connected some lakes, catching catfish that were mud-brown and beyond ugly. I didn't want anything to do with them, but everyone else seemed thrilled by them.

It was always hot in Indiana in the summer. We'd often watch the heat lightning at night over the lake, marveling at the light show, while hoping for a cool breeze to come across the lake, through the screens, to take away the sweat. To my mind, the aluminum screen ranks as one of the greatest inventions of humankind.

My brother Dave, the fish, was an All-American high-school swimmer. We spent a great deal of time at indoor swim meets. I have strong memories of the smell of chlorine, the heat, the bang of the starting gun, the screaming for every race reverberating and pounding in your ears until suddenly the race was over, and it was quiet.

The outdoor pool was different, with the hum of people around the pool, the shouts of children, the whistle of the guards, the smell of suntan lotion and warm skin, and the ever-present chlorine.

While holding our breath underwater at the pool, we used to sing songs to one another and try to guess what was being sung, popping up out of breath and laughing and laughing at how silly the other person sounded.

I dove a few times off the high dive, never quite hitting the water at the proper angle, and always on the way down in total fear for the hour or so it took until I finally reached the water. The low board held its own fear, because you had to hit the spring of the board just right, and then calculate the exact proportion of up and out. Miscalculation meant you could come right back down on the board and break your neck. I saw a guy do that once, and I can still see him being carted away.

I liked to dive off the side into the deep part of the pool, swim down to the 9 foot bottom marker and hang out there as long as I could. I still like going down as deep as I can in a lake, swimming through the stratification layers and hitting the cold water, then just staying there and watching.

My father bought a folding double kayak one year, for no reason that any of us could fathom, since none of us canoed or kayaked. It was a Klepper, a German boat that folded down into four bags. We still have a scratchy 8mm home movie of unpacking and putting together the whole thing in 18 minutes, our world record.

We had a main sail and a jib for that kayak, but since you had to sit down inside a cockpit, you couldn't hang off the side to counterbalance the wind. Over we'd go. The kayak would fill with water, and we'd have to swim it back to shore, which was like pushing a whale because of the weight of the water it held.

So we mostly paddled, trying to stay in rhythm, because if you didn't, you would clang paddles. As might be expected, my brother and I failed the

Aging

Through the sweep of wind and rain
a young headwater stream,
born from topographical chance,
drains away the forest duff.

I too have traveled in this way,
as part of the sweep,
and then been jettisoned on sand or rock
to fend for a while
until forces arose again.

And as that wild stream grows older,
so must I.
Aging, my shores widen,
ferment slows.
I am less prone to emotional flooding
taking in more,
accepting more,
carrying more,
learning expediency.

I eventually delta my life,
as all rivers must evolve,
flowing in rich veins through old wildness,
home to wood ducks and egrets,
deepening color and softer light.

I spread to form vast webbings and channels.
I cry crane callings
from instinctual pulses,
lift wings
to believe air or ground
of equal limitation and expanse.

The sea awaits
salty and vast
embracing the finality of land and its souls.
I am relentless in the pursuit,
powerless in this flight.

rhythm test repeatedly, as siblings are wont to do in virtually all early-life circumstances. It was always his fault as I still correctly recall.

But we never paddled on a river, and we never saw quiet, clean rivers like the Manitowish. It was either the extreme of pollution of the Allegheny or the extreme of power of the Youghiogheny.

Now, I can always come home to the Manitowish. The river runs right through "our land," though I have no sense of ownership of the land or river. They belong to the snipe and bitterns and yellowthroats and song sparrows, to the alder and the hummocks.

I'll never get over the fact that this water flows all the way to the Gulf of Mexico. While it's a downhill journey, we're only at 1,600 feet, and the river has to go 1,000 miles or more farther. There's barely a slope there, but the river finds one and travels down it with power and volume. Each day, every minute, every second some drop of water is undertaking the journey of a lifetime. I envy the river, because it's always on an adventure.

Whenever I first get on a river, I can feel that adventure in the possibilities of where I could go, the beauty of what I may see, the loss of the grip on life's reins, the magic.

When all is said and done, and we've measured every last thing we can measure about the river, when we've identified every plant and bird and insect and thought about every hydrological concept, the most important thing about rivers will still remain: the magic carpet ride, and its occupants, that puts every other carnival ride and mechanical ride to shame. Every one of those other rides will stop when the gas or the electricity gives out or the parts wear out or the ride's over. But the ride's never over on the river. We just flow into bigger waters, never stopping unless catastrophic drought intrudes. The river has no mechanical parts, no gears, no external energy source that will run out. Our ingenuity in tool and machine making will never match the genius of the river.

≈ ≈ ≈ ≈ ≈

Whenever I'm paddling the river, I see painted turtles sunning themselves all along the shorelines. They appear to lead a pretty good life, basking in the sun, wearing armor wherever they go so attackers can't get to them, and sleeping away the winter at the bottom of lakes. But it's obviously not that easy, or more of the 11 indigenous Wisconsin species would live up here, too. Being both a cold-blooded reptile and an air breather is a problem. Five months of winter push ectotherms like turtles to remarkable adaptive lengths in order to survive.

Painteds comprise the most common turtle species on the river. Snappers live here in good numbers, too, and with exceptional luck, you might see a wood turtle or an Eastern spiny softshell turtle, both of which are present but rarely seen in the river.

The painted turtle occurs ubiquitously throughout Wisconsin, and ranges from British Columbia to Newfoundland, and south to Florida. They're successful because they can adapt to most any kind of water, including slow-moving streams, ponds, lakes, spring-fed fast streams, marshes, and even polluted waters. An omnivorous diet consisting of about half animal life (crayfish, snails, insects, small fish, tadpoles) and half plant life (algae, cattail, duckweed) provides diverse feeding opportunities. They are pretty much assured of something to eat, at least from April to late October, after which they hibernate.

In June, the females dig a flask-shaped nest in loose soil, lay 4 to 20 eggs, cover the eggs up with soil, and then head back to the water. Hatching may occur in September, or the hatchlings may overwinter in the nest until the next spring.

If you see a painted turtle digging a nest, or laying eggs, appreciate how few of those eggs are likely to become adult turtles. Raccoons, skunks, foxes, squirrels, gulls, and a parade of others either dig up the eggs or eat the hatchlings as they scuttle toward water. One study of nests along Lake Mendota in Madison found that 70 percent (74 of 107) of the nests were dug up and the eggs devoured by thirteen-lined ground squirrels. Other studies have found nest mortalities at or nearly 100 percent. Easy-to-find eggs in easy-to-dig soft soil represent a great night-on-the-town for predators.

However, hatchlings that make it past their first year, stand an excellent chance of survival, and may live 40 years or more.

≈ ≈ ≈ ≈ ≈

Turtles often sun themselves on fallen logs along a shoreline. Several species of arrowhead typically bloom along such undisturbed shorelines. The pale-white, three-petaled blossoms modestly appoint shallow waters in late July all along the Manitowish. Like most emergent species, the arrowheads utilize clonal growth, resulting in large colonies that anchor and protect shorelines.

The male flower displays are arranged in whorls of three, placed high up on a separate stem from the female flowers. The female flowers also bloom in whorls of three, but lower on the stem and clustered inconspicuously in a head.

The leaves offer the quickest key to arrowhead identification, though their extreme variability can make differentiating between arrowhead species difficult. Submersed plants typically develop narrow, ribbon-like leaves that look like a long, flat grass, while emergent plants display broad leaves. Arrowhead can grow completely submersed in several feet of water, or up on muddy banks. The leaves appear to adapt themselves to the water levels, the narrow, ribbony leaves offering the least resistance to a water current, while the broad leaves offer the most photosynthetic surface area. Sagittarius is, of course, the archer, and the image of his sharp arrows gave rise to the genus name *Sagittaria*.

Small 1- to 2-inch potato-like tubers form at the ends of long subterranean runners that originate at the base of the plant. These can be harvested by using a hoe or rake to free them from the mud. Historically, they were gathered by the Ojibwa in the fall and hung to dry in the wigwam. The tubers were later boiled like potatoes. People were eating arrowhead tubers 3,000 years ago according to studies in the western U.S. In the Far East, arrowhead is cultivated as a crop along the margins of rice paddies.

The European settlers gave arrowheads the names "swamp potato" or

Arrowheads

"duck potato," while the Ojibwa called it *wapato*. Duck potato refers to arrowhead's delectability as a waterfowl food. Tundra and trumpeter swans dig out the tubers, and also eat the small, flat seeds, as do black ducks, gadwalls, mallards, pintails, wigeons, wood ducks, canvasbacks, and an array of other waterfowl. However, the tubers are often buried too deep for most ducks to root out. Muskrats and beavers also relish the tubers, and even porcupines venture into the water to eat the leaves and stems. Beds of arrowhead offer shelter and shade for fish.

Arrowhead leaves lift and transpire water very efficiently, and are often blamed for drawing down water levels in seasonal reservoirs and pools. However, if you're trying to restore a shoreline or aquatic plantbed in front of your property, arrowhead comes highly recommended. It has a wide tolerance of pH, grows on a variety of sediments and in a variety of water levels, and removes phosphorous from the water and sediments. Planting the tubers usually leads to rapid clonal growth. All in all, arrowhead offers a combination of beauty, wildlife value, and ecological utility.

≈ ≈ ≈ ≈ ≈

I've done a great job so far of avoiding the topic of mosquitoes. Sometimes the less said, the better. They are attracted to carbon dioxide after all. But no discussion of water in the Northwoods can forever evade the little buggers, any more than we paddlers can. Usually the mosquitoes leave you alone paddling downriver, at least during the day. I've paddled on still nights, however, and the story changes. But it's most often when you come ashore that the tiny savages suddenly discover you, swarming from the bushes in hordes.

How is it that mosquitoes can find us so quickly? Are they waiting everywhere, or just here? How long can they wait before their miserable little bodies dry up and blow away? These may be the most troubling natural history questions to have endured over the centuries.

Almost two centuries ago, on July 25, 1820, those questions were certainly on Henry Schoolcraft's mind. While camped some 28 miles south of Sandy Lake, Minnesota, on the Mississippi River, he wrote:

"It commenced raining during the night, and as we had neglected to have our tents pitched, we were first awoke by the falling rain, and during the intervals of the showers, the musquitoes assailed us in such numbers as to forbid the hope of rest. In this situation we passed the remainder of the night around our fires,

Mosquitoes—Guerilla Warfare at its Best

Mosquito

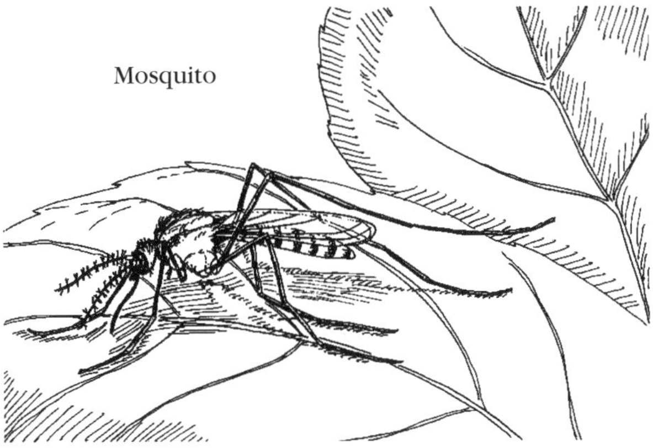

What insect repellents work on mosquitoes? Check out the June 2000 issue of *Consumer Reports* for a full-length review of their testing. But the long-and-short of it is that DEET (N,N-diethyl-meta-toluamide) remains the one proven deterrent. I usually lightly apply 100 percent DEET to the brim of my hat, and then rub some on my wrists, ears, and neck. But CR's research shows a concentration of 20 to 30 percent DEET is as effective as 100 percent—the lesser concentration simply doesn't last as long.

Given that DEET can damage synthetic fabrics like spandex, rayon, and acetate, as well as plastics like your eyeglass frames and lenses, paint, dashboards, and other materials, less can be more. I frequently watch folks tumble out of their cars, line up the kids next to the car, and start spraying them from head to foot with repellant. I wonder what their car finish looks like when they're done, and what harm they've done to their clothes, not to mention their lungs.

As a solvent, DEET takes tree pitch off your hands, so solvents have their advantages when used in the right circumstances.

A 1998 review in the *Annals of Internal Medicine* found that DEET has "a remarkable safety profile," so it's considered generally quite safe. But, as with any chemical, a small number of severe reactions can occur. Use common sense and apply it at the lowest dosage you can get by with, and then wash it off when you're back indoors.

Citronella showed virtually no repellent effectiveness, and the ultrasound machines advertised to repel mosquitoes appear equally useless. The only "natural" repellent that worked in the *Consumer Reports* testing was one that utilized a 2 percent soybean oil, but its effectiveness was very short-lived. The CR research did not examine all repellents on the market, so there may be a magic bullet out there. I sure don't know about one—I'd use it if I did.

endeavoring to divert our reflections by the interchange of anecdote, and absolutely prevented from falling asleep by the labour of brushing away the voracious hordes of musquitoes, which unceasingly beset us with their stings and poured forth their hateful and incessant buzzing upon our ears. It certainly requires a different species of philosophy to withstand, undisturbed, the attacks of this ravenous insect, from that which we are called upon to exercise upon the sudden occurrence of any of the great calamities and misfortunes of life. He who is afflicted, without complaining, by an unexpected change of fortune or the death of a friend may be thrown into a fit of restless impatience by the stings of the musquitoe; and the traveller who is prepared to withstand the savage scalping knife and the enraged bear has nothing to oppose to the attacks of the enemy, which is too minute to be dreaded and too numerous to be destroyed."

I've always wondered how the Ojibwa and voyageurs naturally protected themselves from mosquitoes without the benefit of DEET. Schoolcraft revealed the prescription (though I don't think you want to try it), writing on June 28,1820, while camped at the mouth of the Ontonagon River:

"The Indians hold this animal [the black bear] in the highest estimation, not only on account of their great fondness for its flesh, but because there is no part of it which is useless . . . The oil is, however, considered the most valuable part, whether kept for use or for the purpose of selling to the traders. They rub their bodies with it to protect themselves from the bite of the musquitoe. It has the singular property of destroying lice in the hair, and if occasionally used, of preventing their appearance altogether. They also rub their joints with it, believing with the Romans that it renders them supple."

In the absence of a good supply of bear oil, what to do? "Know thy enemy" seems to be the best strategy.

Female mosquitoes need animal blood protein in order to develop their eggs. Since females need to locate a source of protein amidst all the activity in their environment, they've developed sensing gauges that respond to increased heat, carbon dioxide, lactic acid (a byproduct of exercise), volatile fatty acids, and visual cues like dark clothing, color contrasts, and movement. Mosquitoes have receptors on little feelers that can detect chemical emanations from as far away as 115 feet. Basically, the vapor trail we exude gives us away every time, and the more abundant our body and breath vapors, the easier the trail is to follow. If you could hold your breath, and prevent all sweat and scents from escaping your skin pores, you would probably remain unbothered by mosquitoes. In all likelihood, you'd also be dead.

Staying cool, calm, and collected emotionally as well as physically definitely helps deter mosquitoes. Some people are mosquito magnets, while others seem to walk serenely among them. A person's skin chemistry does affect mosquitoes, and diet appears to be one of the determining factors. But since we exude literally hundreds of chemicals, the optimal skin chemistry is unknown. Washing with unscented, antibacterial soap seems to help deflower our personal scents, but the jury is still out on a host of folk remedies, including B vitamins, garlic, and others.

By all means, find the wind whenever possible because it disperses your chemical plume, making your track less easy to read.

A female mosquito can suck two to three times her own weight in blood. She then lubbers off, looking for a place to rest and let her eggs mature. If she has mated, she's stored the sperm, and is already fertilized. In fact, she can use the sperm for many future reproductive cycles.

Depending on the physiological state of the mosquito and the environmental conditions, she may wait two to six days to lay her eggs before resuming another search for blood protein to support a new batch of eggs. A female may live a month or more, so 15 reproductive cycles aren't out of the question. Prime mosquito real estate may contain a million or more mosquitoes per acre.

The good news is that mosquitoes become lethargic at 60°F, and can't function below 50°F. In the Northwoods, there is only a two-month window in June and July during which nighttime temperatures regularly stay above 60°.

The bottom line is that you have to learn to live with them, because you aren't going to live without them.

≈ ≈ ≈ ≈ ≈

The river meanders in more and tighter loops as it heads south to its eventual junction with the Bear River. Here, it will turn east, becoming the North Fork of the Flambeau River, and pursuing a path for the Turtle-Flambeau Flowage. The meanders bring paddlers nearly back to previous locations again and again, while uplands lay distant from the river banks.

Willow shrub may be the most dominant plant in these extensive flood plains. Some 30 shrub willow species grow in east-central North America, most growing in or along side wetlands. Willows hybridize readily and frequently show variation on individual plants, so identification can be a matter of great frustration. Unless you have a voracious curiosity or an all-consuming scientific bent, knowing a willow's genus (*Salix*) is plenty good enough.

Willows typically have long, narrow, toothed, alternate leaves. Their buds are often distinctive, pressed flat against the stem from which they emerge (a characteristic called "appressed buds"). Willow flowers usually emerge very early in the spring in long catkins, the male and female flowers occurring on separate plants. The familiar pussy willow often flowers well before the snows are gone, clearly gaining a competitive advantage by opening under such cold conditions.

Shrub willows clothe streamsides, cloning rapidly to secure a shoreline in their rooted grasp. In places where clones may not reach, the silky-haired seeds float on the softest of winds, dispersing in great numbers like their cousins cottonwood and aspen will do later in the spring. Willow seeds must germinate quickly or lose their viability—an evolution that makes sense in light of their ability to rapidly germinate on open sandbars and shorelines. The seeds have the best chance of success if they spread early in the spring as floodwaters recede and leave behind scoured shorelines.

Willows typically occur in "shrub-carr" habitats, plant communities of

tall, deciduous shrubs that grow on seasonally flooded soils. These communities are usually dominated by willows, red-osier and silky dogwood, with a groundlayer of hummock sedge, lake sedge, and Canada bluejoint grass. Mixed in are usually a variety of other herbaceous species like meadowsweet, Joe-Pye weed, marsh milkweed, and blue vervain. The Manitowish supports extensive shrub-carr habitat.

Willows do the tough work of stabilizing riverbanks, withstanding flooding and silting with ease. They grow quick, up to 7 feet a year, but don't tolerate shade, dying when other plants shade them out.

Willows often have large numbers of galls on them, produced in response to larval insects that feed on their plant tissues. Galls contain tasty insects, a fact not lost on many birds.

Dense willow thickets create excellent cover for nesting and loafing. Yellow warblers, common yellowthroats, willow and alder flycatchers, gray catbirds, song sparrows, northern waterthrushes, and American goldfinches commonly nest in these shrubs. An array of shorebirds, marshbirds, and waterfowl use willow thickets for cover.

White-tailed deer, elk, and moose browse the twigs. Moose are particularly fond of willow as a winter browse, and the enormous stands of willow shrub along the Manitowish make it optimal moose habitat. One day moose will return to northern Wisconsin and paddling will be more eventful than ever with bull moose wading in the shallows around river meanders.

Beaver, snowshoe hares, cottontail rabbits, and porcupines also gnaw on willow bark or nip off the twigs. Ruffed and sharp-tailed grouse eat the buds in winter, as do squirrels.

While little wildlife may appear at midday in this section of the river, rest assured these shrubby wetlands provide excellent bird habitat. A recent study tried to determine what Northwoods habitats were utilized by migrating birds as stopover sites. The number one most attractive site was, as you might guess, wetland shrubs along rivers. The least attractive site was pine plantation (read more about rest stops for birds at the end of the chapter).

These shrubby flood plains provide important ecological values besides bird habitat. A healthy river will absorb and digest a flood. The riparian edge holds the earth in place with its network of tree and shrub roots. A flood may scrub some areas clean of vegetation, but the vegetation will come back quickly.

Natural logjams redirect surging current, causing it to dig out pools, flow into side channels and sloughs, and slow down. Thus, flooding can dig new pools, and its load of debris brings new wood and rock for river structure, increasing river complexity. Flood events that spread out over their natural flood plain are actually key to shaping and maintaining high-quality habitat for fish and other organisms.

A look around you at this point yields a vision of water, water, everywhere. Thank goodness you're paddling, because these wetlands would be hell to try to walk through. Taking to the uplands wherever they can be found would offer some respite, but the uplands here are often islands among wetland.

That's the glacial legacy in northern Wisconsin, a legacy that was used not only for travel and trade but also for battle strategy. Fur trader Nicolas Perrot describes one such strategic battle between the Dakota Sioux and the Huron in his journal from 1656 to 1662. Where exactly the battle took place is unclear, but it was in a country that was:

"Nothing but lakes and marshes, full of wild oats [rice]; these are separated from one another by narrow tongues of land, which extend from one lake to another not more than thirty or forty paces at most, and sometimes five or six, or a little more. These lakes and marshes form a tract more than fifty leagues square . . . Consequently, the Sioux are inaccessible in so swampy a country, and cannot be destroyed by enemies who have not canoes, as they have, with which to pursue them . . . If any one of these little villages be attacked, the enemy can inflict very little damage upon it, for all its neighbors immediately assemble, and give prompt aid wherever it is needed. Their method of navigation in lakes of this kind is to push through the wild oats with their canoe, and, carrying these from lake to lake, compel the fleeing enemy to turn about and thus bewilder him."

Perrot describes some 3,000 Sioux driving 100 Huron into a swamp, where the Huron hid in the mud and water under the tall rice stalks.

"The loud noise, the clamor, and the yells with which the air resounded showed them plainly that they were surrounded on all sides; and that their only resource was to make head against the Sioux (who were eagerly striving to discover their location), unless they could find some place by which they could retreat. In this straitened condition, they concluded that they could not do better than to hide among the wild oats, where the water and mud reached their chins. The Sioux, who were sharply searching for them, and only longed to meet them in battle, found very few of them, and were persuaded that they themselves were entirely hidden by the wild oats; but they were greatly astonished at seeing only the trail made in leaving the lake, and no trace of the Huron's entrance. They bethought them of this device: they stretched across the narrow strips of land between the lakes the nets used in capturing beavers; and to these attached small bells . . . They divided their forces into numerous detachments, in order to guard all the passages, and watched by day and night, supposing that the Hurons would take the first opportunity to escape from the danger which threatened them. This scheme indeed succeeded; for the Hurons slipped out under cover of the darkness, creeping on all fours, not suspecting this sort of ambuscade; they struck their heads against the nets, which they could not escape, and thus set the bells to ringing. The Sioux, lying in ambush, made prisoners of them as soon as they stepped on the land. Thus from all that band but one man escaped; he was called in his own language Le Froid [he who is cold].

Battle often resulted from conflicts over the fur trade. The Sioux and Ojibwa battled for this region for over a century, their last battle occurring in the 1850s.

≈ ≈ ≈ ≈ ≈

The Bear River modestly joins the Manitowish on river-left, a confluence that can easily be missed. A DNR campsite on river-right just 30 yards past the junction marks this historic intersection.

The historic trade canoes would now have to push upstream for nearly 20 or more miles (as the river flows) to reach the village and trading post at Lac du Flambeau. However, the task was not as difficult as most upstream pulls, given that the current on the Bear offers only moderate resistance.

The Bear River looks very much like it always has. Only one short stretch of river near Highway 182 reveals a few homes and cottages. Otherwise, the river flows uninterrupted and undeveloped through public or tribal lands for its entire length. Woodlands and wetlands comprise nearly the entire watershed. Less than 1 percent of the watershed has been cleared or is in agriculture, leaving a remarkably untouched river system.

The Bear flows even more slowly than the Manitowish. The stream gradient drops less than 1 foot per mile, a barely perceptible elevational change. By contrast, the Manitowish drops 2 feet per mile. A mid-April flow measurement taken near Highway 182 during a bank-full stage recorded 291 cfs, a very moderate flow for a high-water event. The lightly stained brownish water matches the Manitowish in color, alkalinity, chemistry, and temperature.

Northern pike, some smallmouth and largemouth bass, and panfish inhabit the Bear's warm waters, but the river's wildlife reputation rests on its waterfowl migrations. Dabblers like Canada geese, mallards, black ducks, and wood ducks find excellent foraging on the river, particularly in the fall after the ripening of the extensive wild rice beds.

In low-water years, save yourself some serious walking and leave the Bear to its isolation. Five road crossings between Flambeau Lake and the Manitowish River provide public access, but a lot of river runs in between each crossing. The river bot-

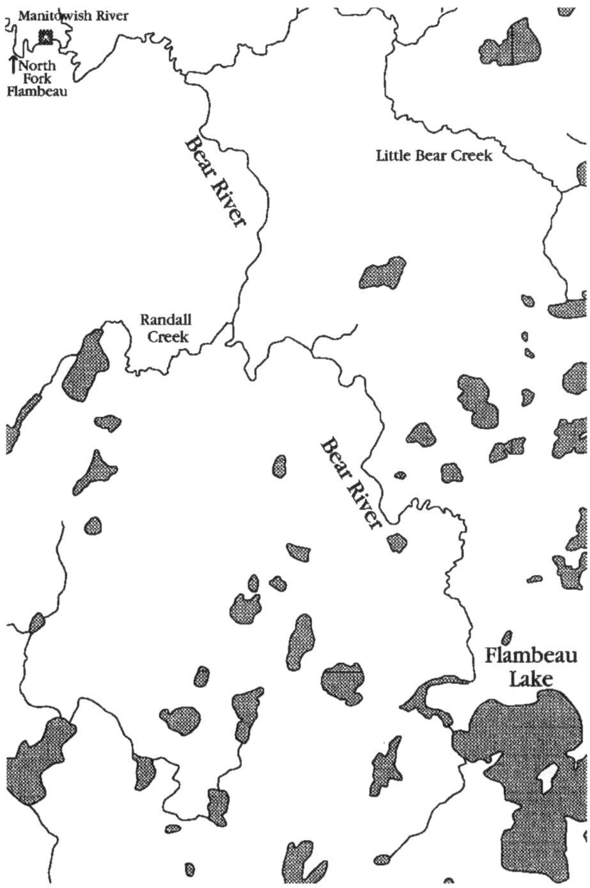

The Bear River

tom varies from sand to silt to muck, and while a walk through the silt or muck in low-water makes for good hardship storytelling later on, it's a mess in the present tense.

Geologist J.G. Norwood paddled the Bear on September 26, 1847, after camping the night before on the Manitowish. He described the terrain in his journal:

"Soon after entering Lac du Flambeau River [the Bear], which we ascended to the lake of which it is the outlet, large boulders began to show themselves, some of them of great dimensions. One which was examined, measured fifteen feet in the long diameter, twelve feet in the transverse, and stood seven feet out of the water. It was composed of mica slate, and studded with garnets of small size.

"Just before reaching a range of hills, the river runs through what was once evidently the bed of a large lake, now drained, and overgrown with aquatic grasses [wild rice today]. Through this the river flows in many channels, some of them fifty yards wide. This alternate widening and narrowing of the river occurs all the way to the lake. The trunks of hundreds of dead tameracks are standing in all the spaces between the channels, and give a peculiar air of desolation to the scene . . .

"The river is exceedingly crooked, its general course being south-southeast. We reached the Lake late in the afternoon, and crossing its northwest arm, camped near the old Trading-House of the

American Fur Company, now deserted."

Francois Victor Malhiot, the clerk in charge of the North West Fur Company's post in Lac du Flambeau in 1804, must have been having a bad day when he wrote the following about the Bear:

"As to Lac du Flambeau, it is worthier of the name of swamp than of lake and at this season [August] it would be easier to catch bullfrogs in the nets than fish. I have had the nets set three times since my arrival without catching a fish . . . With regard to the river I will never call it anything but a small stream, because in many places a mouse could cross it without wetting its belly."

It's important to note that Malhiot found his post less than idyllic, and his journal issues both-barrel complaints about his men, the Ojibwa, the food, the weather, the bugs, the fur trade, and so on. But Malhiot's description of the Bear as a small stream portrays the river fairly, though he must have seen coyote-sized mice in the 1800s.

Read more about Malhiot's journal at the end of the chapter.

A former logging dam at the outlet of Flambeau Lake today holds back a 2-foot head of water on the Flambeau Chain of Lakes, raising the chain enough to ensure better recreational access, but apparently having little effect on the river.

Today, the river's placid isolation allows dreams of Malhiot's time or the Ojibwa's era to come easily. The landscape remains the same. A trade canoe bearing Ojibwa or Frenchmen coming around a bend would seem natural and right.

≈ ≈ ≈ ≈ ≈

The campsite near the river junction of the Bear and Manitowish has very likely been a resting place for many centuries. A scan of a topo map shows virtually no other high ground along the river for several miles in any direction. It only makes sense that native people and eventually the French might stop here before proceeding farther down the Flambeau or up the Bear. While no archaeological study has taken place here, what more likely spot would there be for an ancient and long-standing campsite than the high-ground junction of two important rivers in the midst of miles of wetlands?

The riverbed turns briefly to gravel and cobble here, making a good site for spawning walleye.

If you're a non-angler like me, the life cycle of fish isn't something necessarily near and dear to your heart. The problem for non-anglers is that we can't casually encounter fish like we can a wildflower or a bird, so their lives can easily remain quite mysterious and inaccessible.

To begin to understand walleyes, we need to know about the general life cycle of a fish. Life begins for a fish when eggs are buried within the substrate of a lake or river, or broadcast over the surface of those substrates, or into the water column, or attached to plant material. The eggs incubate for a few days to a few months, mature, and hatch to produce a free-embryo phase where they rely entirely on energy from the egg yolk. Once the free-embryo phase is completed, the fish become larvae and feed externally. The larval phase concludes after the formation of the axial skeleton and the development of fully formed organs and fins. The fish are then termed juveniles. They subsequently undergo favorable and unfavorable periods of seasonal growth and migration until they reach sexual maturity and migrate to their spawning sites. There, they deposit eggs and the cycle begins anew.

That's it in a dry, scientific nutshell.

The problem with, and the beauty of, this general scheme is the tremendous diversity inherent in such a multi-step process. One of the major challenges for fisheries biologists is to first understand the life histories of each species of fish, and then establish the links between the landscape processes that may affect one or more of the stages of each life cycle. These landscape variables include things like the shape and size of the stream channel and its obstructions, the substrate materials and their size, water depth, the current velocity, riffle and pool development, meanders, slope, and other factors. These are all major influences upon the growth and survival of fish during their early life stages.

The entire terrestrial drainage basin must be taken into account in order to understand what takes place under water. This is a daunting task to say the least, and one not bargained for by the typical young fisheries student who just wanted to study what takes place in the water. The scope of such study extends well beyond this book, and my expertise. But let's take a look at what a spawning walleye might be thinking right here in front of the campsite.

The walleye's spawning period depends largely on the time of year (the photoperiod) and the water temperature. Walleyes generally spawn shortly after ice-out when water temperatures reach from 38°F to 44°F, and they're usually finished by the time temperatures rise to 50°F.

Walleyes require rather specific spawning habitat requirements. On lakes and rivers with rocky shorelines, they broadcast their eggs over the rocky, wave-washed shallows. On lakes with inlet streams, spawning takes place on the stream's gravel bottoms. And in some places, walleyes spawn on flooded wetland vegetation, demonstrating a dependence on spring flooding and protected flood plains. This is another in the long list of benefits provided by floods.

Females typically spawn out in one night, but males often spawn over a longer period. Walleyes will spawn earlier here on the river than on most lakes simply because the ice goes off the river much earlier, and the river obviously warms more rapidly without a lid of ice when exposed to the sun.

If you're an intelligent angler, you pay attention to water temperature. Spawning walleye feed only moderately, but a week of spawning uses up a lot of energy. Rising temperatures increase their general metabolism, making them more hungry. The minnow population is low at this time of year, making food hard to find. The combination of factors is a formula for optimal fishing. Walleyes typically go on a feeding binge when the water temperatures reach about 50°F. Keep a basic thermometer handy along with all your fancy lures. If the water temperature hovers around 45°F or less, or exceeds 54°F, you've missed the feeding frenzy.

Interestingly, fish biologists believe walleyes were historically confined to the larger lakes and waterways in Wisconsin. Because walleye are the most popular sport fish in Wisconsin, stocking of walleye fry and fingerlings by the DNR and sport-fishing clubs has extended their distribution into many lakes and rivers throughout Wisconsin that were not original walleye sites. This provides an ironic twist to the continued debate between white anglers and tribal governments over Ojibwa rights to spear walleye. Many northern lakes never had walleye before the 20th century.

≈ ≈ ≈ ≈ ≈

We have left the Manitowish River Watershed and have entered the Flambeau Flowage Sub-basin. These watersheds nest like Russian dolls inside the Upper Chippewa River Basin. The North Fork of the Flambeau eventually meets the South Fork to form the Flambeau, which flows into the Chippewa, which flows into the Mississippi.

To the voyageurs, the river now simply became the Chippewa, bypassing the Flambeau designation altogether. Travel clear to the Gulf of Mexico became possible, the only problem being getting back, which would require the seemingly impossible effort of paddling back upstream to where the rivers flow north again into Lake Superior.

No thank you.

The North Fork is renowned for its whitewater, but rest assured that no rapids occur until you portage around the Turtle-Flambeau Dam on the far western end of the Flowage.

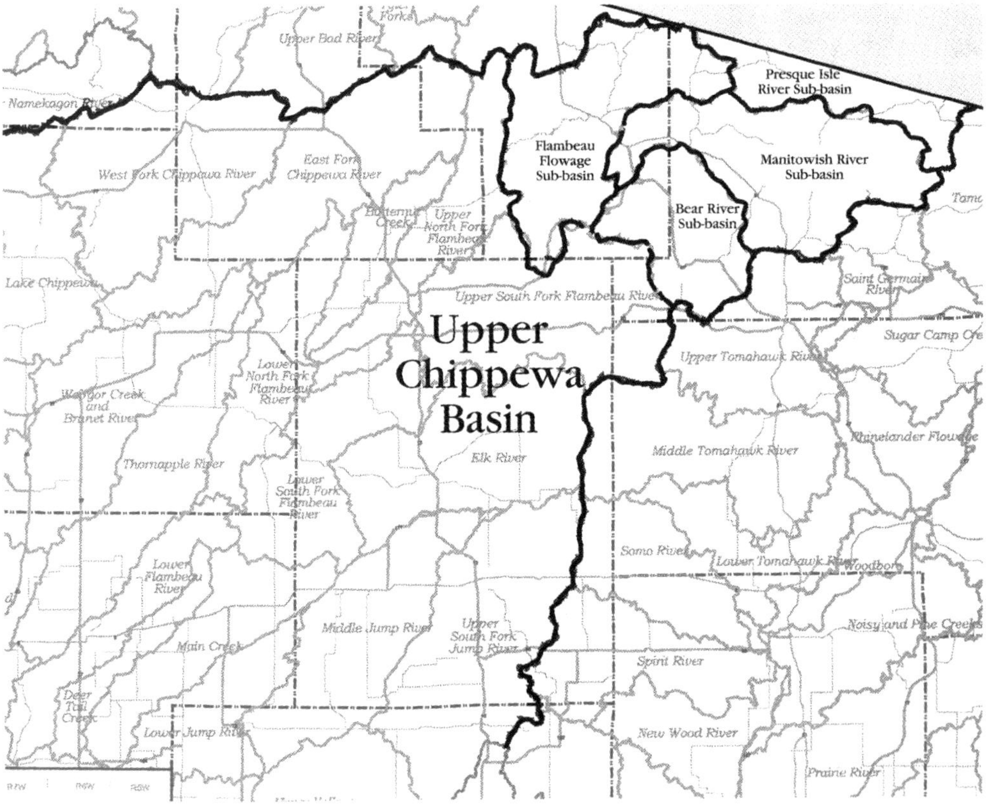

This book pulls out on the eastern edge of the Flowage at Murray's Landing.

The character of the Flambeau begins to change now from its small-stream ancestry as the Bear and the Manitowish, into a deeper, less meandering river carrying a higher volume of water at a slightly greater velocity.

As the river changes physically, it also changes biologically. This is not only true of the Flambeau, but of all rivers as they grow from headwater streams into large rivers that eventually empty into the ocean. To describe these biological changes along the entire length of a river, stream ecologists have constructed a general theory called the "river continuum concept." The theory says that a dynamic equilibrium exists over extended reaches of a river, and that biological communities become established in places that match, or fit, the dynamic physical conditions of the stream. From the headwaters of a river onward downstream, the physical variables present a continuous gradient of conditions, including such things as river width, current velocity, depth, flow volume, and temperature. Along this length, a continuous replacement of species occurs based on their functions down the continuum. However, the species community is dynamically stable at any point on the continuum.

Compare this concept to forest succession, which is the natural, gradual, and continuous replacement of one group of plant species by another at a point. A forested site changes over time, say from pioneer, sun-loving species like aspen and jack pine to climax, shade-tolerant species like hemlock and sugar maple. A stream, however, does not change over time. Instead, it maintains a dynamic equilibrium at that point on the river. So, today you might sample a specific short stretch of river for aquatic insects and find a certain variety and total number of species. This same general community of species should have been residing there 100 years ago, and should be there 100 years from now. Numbers and species will fluctuate to some extent—little is stagnant in the natural world—but the biological community in the river won't change over time like it will in a forest. The river community changes in a generally predictable manner as you go downstream, but it does not fundamentally change at any given point.

Here's how it works. In headwater streams, the forest vegetation along the stream is the main energy source, typically providing 99 percent of the energy for organisms living in and along the stream. Leaves, needles, twigs, bark, and branches drop or wash into a stream. They become the storehouse of organic material needed by stream organisms for energy. The forest along the headwater streambanks shades the river. Without sunlight, little photosynthesis can take place, so as little as 1 percent of the stream's energy may derive from photosynthesis.

This shower of organic materials from the forest provides a rich food base, but it is only useful to stream organisms that can utilize such large and coarse materials. A diverse population of insects and microbes browses and shreds the leaves, gouging tunnels into logs and branches, and rasping off algae and fungi. Nearly two-thirds of this woody debris is processed within

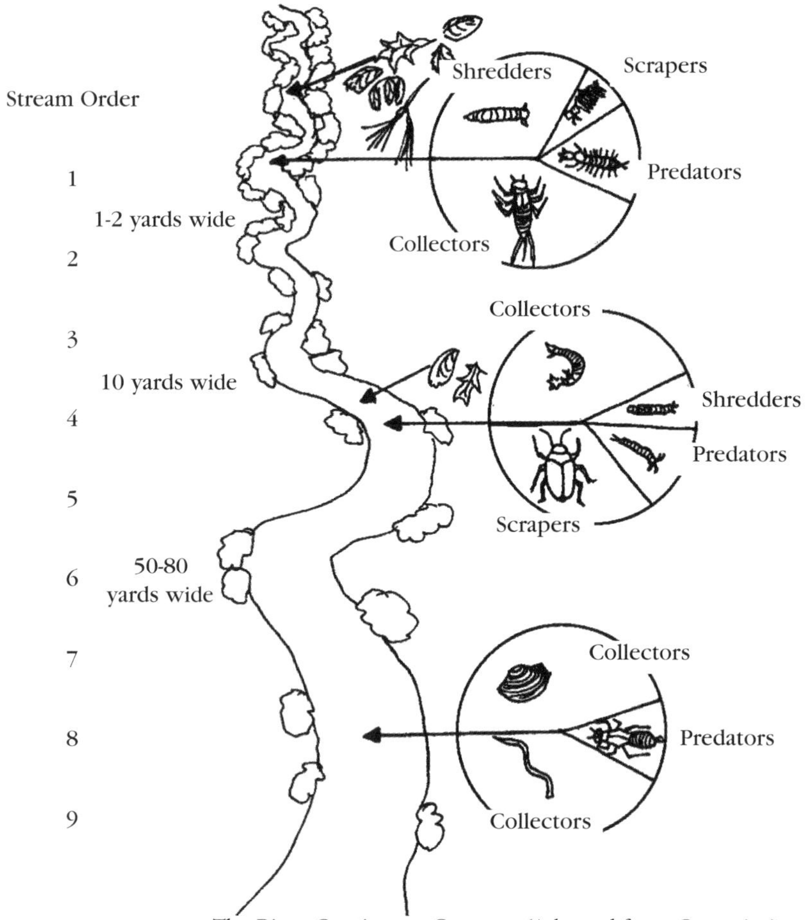

The River Continuum Concept (Adapted from Cummins)

the headwaters. Very little of it leaves the headwater system without being at least partially processed.

Aquatic organisms respond appropriately to the type and location of food. In headwaters, aquatic insects must get food from this rain of coarse particulate organic matter (CPOM), and so the stream is dominated by shredders and collectors.

Shredders use plant litter, but only do so after it has been colonized and conditioned by microbes and the soluble organics have been leached out from the leaves by the streamwater. Typically one-third or more of the dry weight of a leaf is lost through leaching in the first two days. The conditioning done by microbes like protozoans, bacteria, and aquatic fungi may take weeks or months depending on the plant species and on the stream temperature. Leaves from basswood, alder, and most herbaceous species process quickly. Maple and birch leaves process somewhat more slowly, while oaks, conifers, and most ferns break down very slowly.

If the headwaters are turbulent, the mechanical action of the water abrades the leaves and CPOM and helps tear them apart.

The shredders eventually get into the act, and skeletonize the leaves, converting about 40 percent of what they eat into their own tissue and respiration, and excreting the rest. The microbial organisms that helped condition the leaves for the shredders are also ingested, serving I suppose as the protein in the leaf sandwich.

So, shredders set the table for the collectors downstream, as well as for another group of organisms called grazers or scrapers. Collectors filter their food from the water, or gather what they need from the sediments, while scrapers shear attached algae and other materials from rock and plant surfaces in the water. The shredders turn the CPOM into FPOM (fine particulate organic matter), acting like food grinders to make the materials smaller and more digestible downstream.

In streams orders 1 to 3, like the size of the Manitowish, shredders and collectors are co-dominant. In mid-sized rivers of orders 4 to 6, collectors and scrapers are co-dominant. In big rivers of orders between 7 and 12 (the Mississippi is an order of 10), where significant photosynthesis takes place, collectors are dominant. So, as the particle size of organic matter becomes progressively smaller down the continuum, the stream community becomes progressively more efficient at processing the smaller particles that arrived at its table.

In the meantime, predatorial insects like dragonfly larvae and a host of others, feed fairly constantly throughout the stream orders. However, the predatorial species gradually change from insect-eaters to more zooplankton and fish-eaters as rivers become larger.

Mid-sized streams and rivers contain the greatest diversity of species, in large part because they offer the greatest diversity of habitats. The mosaic of habitats can include riffles, eddies, side sloughs and channels, various-sized pools, logjams, boulders, and a diverse streambed that changes from sand to gravel to bedrock. As one would expect, diverse habitats support a rich biological community.

≈ ≈ ≈ ≈ ≈

The river begins to widen as it approaches the Turtle-Flambeau Flowage, an effect of the Turtle-Flambeau Dam pushing water back upstream. Now the quality and character of the river gradually changes from that of a flowing stream to a reservoir. Eagles and ospreys become more numerous, along with Canada geese, pied-billed grebes, and other marsh birds.

Wild rice appears in section 36, as the river continues to open up into bigger water. Where the river channel narrows a bit between two points of land before opening into a huge marsh complex sits an eagle's nest, active for decades but now inactive in 2000. It can be seen on river-right, a third of the way down in a relatively smaller white pine. In years past, another eagle's nest could be seen a mile from here on the far western side of the marsh, a rather rare incidence of two eagle's nests occurring in sight of one another. The latter nest blew out in the early 1990s, and was never rebuilt. Another nest rests in a white pine on the south edge of the marsh.

This marsh houses a rich array of birds and mammals. Muskrats must find this a paradise, because in the fall of the year when they build their

Trapping

Venus glows above the landing
as we slide our canoe into the black water.
The aluminum grates on the pebble shore.
We paddle upstream through the cape
of darkness that melds river and bank.
The slough ice suddenly shatters
and the stars glimmer on the shards.

In the silence and cold of late October
life is expressed in the quiet
splashes and the widening circles
of the wakes of animals headed home.

Even our softened voices are thunderstrikes.

We pull for the muskrat houses where the traps
vise the young and the old,
who were wary of mink,
but not of metal jaws hiding
in their runways.

As the Eastern sky colors gradually in waves,
we glide to the first black hutch,
the houses eventually emerging out of the dark waters
in gray
to brown
to salmon
to the tan of cattail and bulrush.
The interior of the hummocks have been eaten
away into a room for eight
to wait out the months of ice.

We carefully count as we pull the traps
33 houses, 67 traps, 28 muskrat.
Four alive with broken legs,
each dispatched with a paddle blow
to its thin skull.
Two eaten by mink or owl,
who were surely pleased by a prey
exposed and tethered to its home.

For tomorrow's catch, we make tears
in the walls of the houses,
in the plunge-hole exits,
in the feeding platforms,
and place our traps.
Here repair, rest, and retreat are expected,
but now drowning awaits.

Three bald eagles observe from high pines.

As the morning grows, the wind follows,
sweeping the wetlands of their darkness.
The marsh awakes with a rush of air,
where sound had earlier ridden
mutely on the dark balm.

We become efficient.
Each man performs his task in context,
the script executed.

Later the trapper demonstrates to me how
to skin a muskrat in less than a minute,
the fur lifting from the skin with ease
as if it had been a coat all along.
A snip and cut from the back ankles to the anus,
and the pelt pulls over the head.
Then a snip at the front feet and jaws
and the skin peels away.
It is as easy as taking off of a sweater,
as easy as opening the housing of his outboard,
only now he exposes
the red machinery of the marsh.

Muskrat

huts, one can count well over a hundred houses. A good friend of ours has trapped "rats" here for years, and consistently takes 150 to 200 muskrat each season without any apparent impact on the population.

Muskrats eat marsh plants, and this marsh produces a prodigious banquet for them. Marshes can produce more plant matter (biomass) than a cornfield, so quantity of food is not an issue. As mentioned earlier, without muskrats, marshes can fill in and become dense monotypes of cattails, a poor habitat for most waterfowl, which prefer a mixture of open water and aquatic plants. If the lake or river you live on has too much plant life, one possible corrective measure is to import a family of muskrat and let them be your controlling agent. A family of mink is also needed to control the muskrat population.

Muskrats not only eat cattails, but also bulrush, arrowhead, sweetflag, pondweed, and various reeds. Empty mussel shells littering a shoreline are usually testament to the muskrats' omnivorous diet, which basically includes whatever the opportunity presents, from frogs to crayfish to even carrion.

Muskrats build their lodges by cutting and piling up emergent vegetation into a big heap, usually about 6 feet in diameter and 4 feet high, though only 2 feet above water. They then excavate plunge-holes and nest chambers inside the lodge, leaving walls about a foot thick. They may even dig an open porch into the outside wall to serve as a resting area.

Muskrats also build feeding platforms, feeding huts, and "push-ups." Feeding platforms are little more than piles of discarded plants that serve as a dry spot to eat above the water. Feeding huts are similar to lodges but smaller, and are constructed to serve both as secure feeding sites and as underwater air pockets. Push-ups are built during the winter on top of holes that the muskrats cut through the ice. They function much the same as the feeding huts.

Muskrats typically don't travel much in the winter, staying within a 15- to 30-foot radius of their feeding huts and push-ups. A series of huts and push-ups allows them to forage under water for submerged vegetation. It's a long winter under the ice, and because muskrats eat a pound or two of vegetation a day, they need access to a large store of food.

In a productive marsh like this one, muskrat may reach densities of 25 individuals per acre. In non-marsh habitats, muskrats rarely exceed 16 adults per mile of shoreline. More typically, their numbers are half that, simply because available aquatic plants are less abundant along a narrow riparian zone. The home range of a muskrat in a high-quality marsh may be as small as 200 feet in diameter, with most of its activities occurring within 50 feet of the lodge.

Weighing 2 to 3 pounds, and equipped with a scaly tail and tiny ears that barely protrude from their heads, muskrats look like an overgrown field mouse. But muskrats provide excellent table fare for a host of predators, including minks, otters, raccoons, bobcats, fox, and eagles. Muskrats are also prodigiously reproductive. An average female will produce two litters of five kits each per year. Trappers have learned that muskrat populations can sustain annual harvests of 75 percent of their total population. The terms "old age" and "Social Security" are probably not part of their vocabulary.

On the other hand, if their numbers explode, they can eat out a marsh, literally eating themselves out of house and home by exceeding the carrying capacity of their habitat.

Observing muskrats usually requires an evening watch, because about 80 percent of their activities take place under the cloak of darkness. But daytime sightings often occur—particularly near dawn or dusk, or on rainy days when muskrat, like most wildlife, seem to understand that humans don't like to get wet.

≈ ≈ ≈ ≈ ≈

Nesting Canada geese appear close to Murray's Landing. Paddlers from southern Wisconsin and points farther south tend to express regret when they see geese, even going so far as to say it's a shame that we have geese up here. Their experience with high numbers of geese fouling lakeshores and golf courses with their droppings has led them to see geese as a weed species. They're often astonished to learn how thrilled we northerners are to see geese, and how in fact the WDNR annually imports a few hundred from down south to release on our lakes and rivers.

The difference in our responses comes in large part from the fact that we have yet to transform the Northwoods into the Northlawns. We certainly have the capability to do so, and many owners still strive to suburbanize their shoreline properties, but overall, the grassy shorelines that geese so enjoy remain fairly uncommon in the Northwoods. And as long as we don't alter our shorelines and uplands too much further, northern geese simply won't be able to make a good enough living to come here in pestilential numbers.

Less controversial, and far more unusual, are the black terns that can often be seen in the big marsh above Murray's Landing. These birds may be present one year and inexplicably gone the next. Black terns sweep up from South America, appearing late in May and departing in mid-August, less than three months later. Their erratic, buoyant flight, dark heads and dark bodies, and tendency to hover while looking for prey make them easy to identify.

Black terns typically nest in loose colonies right on marsh vegetation, or make use of muskrat feeding platforms. In either case, they nest atop 2 to 3 feet of water and usually 50 feet or more apart. If you inadvertently near a nest, the terns will dive-bomb your head and may actually strike you. Black tern researchers often wear helmets when they're doing nest surveys.

Surveys done in the early 1980s and more recently from 1995 to 1997 show a very significant decline in numbers, so much so that black terns will likely be recommended for threatened status by the Wisconsin Bureau of Endangered Species. The reason(s) for the decline remains unclear. Young-of-the-year migrate to northern South America and spend their first year there before returning as 2-year-olds. Degradation of the wetlands in South America may be part of the problem, but the jury is out on this one.

Keep an eye out also for yellow-headed blackbirds in the center of the marsh. Their presence is even more enigmatic and mercurial than that of the black terns. While common in western Minnesota and points farther west, yellow-heads are quite uncommon for our area, and are a source of great birding pleasure. Listen for their cacophonous, creaky-screen-door song.

Murray's Landing beckons, the site of an old logging dam on the Flambeau River in the early 1900s. Murray's marks the beginning of the Turtle-Flambeau Scenic Waters Area, a mouthful, but one worth speaking. Dammed in 1926, the Turtle-Flambeau Flowage flooded 16 natural lakes, drowned the Flambeau River, and impounded approximately 14,000 acres of water, making it the fifth largest waterbody in Wisconsin.

The State of Wisconsin acquired the Flowage in 1990, and its ownership now exceeds 32,000 acres including 114 miles of mainland shoreline and 195 islands. Sixty campsites, accessible only by water, offer great camping. More importantly, the TFF has the highest density of breeding bald eagles, common loons, and ospreys in Wisconsin, though the numbers of breeding pairs fluctuates from year to year. For anglers, the TFF supports a world-class walleye fishery, and excellent muskellunge, northern pike, smallmouth and largemouth bass, black crappie, and rock bass fishing.

≈ ≈ ≈ ≈ ≈

Pull-out at Murray's Landing. Or continue on your way if the notion possesses you. New Orleans, here you come. The river goes there every day, every second. This magic-carpet ride never breaks down, seldom falters, frequently meanders, and on occasion overflows and changes its course. Rivers are the bloodstream of this earth. Keep them healthy. Read and accept their philosophy. Follow their course. And let them course through you.

You cannot step into the same river twice.

— Heraclitus

Winter Ice

In the autumn, mists often rise from the Manitowish in a perfect tunnel over the river, giving the channel an atmospheric depth and an otherworldly aspect. Paddling the river within that tunnel is like being in another universe, or another time. Voyageur or Ojibwa canoes could easily glide into view.

The mist rises when warm riverwater evaporates and condenses as it comes into contact with the colder air above. The river is cooling, heading toward the inevitable winter ice-up.

Rivers freeze later than lakes because the faster the water current, the more delayed the freeze-up. But rivers differ most dramatically from lakes in that they are typically so shallow that temperatures in riverwater remains completely mixed. The water at the bottom is usually very close in temperature to the river surface, so riverwater is at freezing or below freezing all the way to the riverbed.

The first ice to form on a river is paper-thin and clear. It's often called black ice, because it's so transparent. The ice turns more and more white as new ice is added on top of the original layer. New ice arrives as snow, or as "overflow ice" when water seeps through the cracks of the ice, floods the surface, and freezes.

After the black ice is formed, heavy snowfalls will push down on the ice and force water through cracks up onto the ice. This begins a process where the river water mixes with the layer of snow and forms a layer of slush. If it snows again, the first layer of slush may freeze and create an inferior ice under the new layer of slush. Over time, numerous alternating layers of slush and ice may form on top of the river. By the end of winter, if a lot of snow has fallen, lakes and rivers may be covered with 2 to 3 feet of this crumbly ice.

Unlike a lake whose freeze-up date is determined by local temperatures, the freeze-up date of a river depends on the history of the river. Where has the river come from, and how long has it taken to come? If the river has been cooled by cold tributaries, it freezes earlier. If it flows northward, it's slower to freeze, because it's bringing warmer water within it.

The wind often sweeps snow from the center of a river or lake and drifts it along the shorelines, so the center ice may appear black, while ice around the shores appears white.

Though the ice may appear thick on a river, don't bet your life on it. Currents may be flowing if there is warmth coming from the river bottom. Heat may be stored in the river sediments, accumulating via the Earth's internal heat or from sunlight penetrating the ice.

Continued

Winter Ice (Continued)

Ice isn't a benign substance. When it's pushed up on shore in the spring, ice can uproot trees, move rocks, and rip apart piers and even houses.

When a frozen lake makes new ice you can hear the growing pains ... The sounds vary. Sometimes they are continuous and steady, like an enormous bowling ball rolling across 20 miles of wooden flooring. Sometimes they are sudden and short, a series of abrupt electric cracks. There are random shots as piercing as artillery shells, followed by intermittent sonorous thumps, like giants stomping across the bay. Sometimes it is dull and thudding, sometimes metallic and reverberating. It can sound like distant jets or noisily burning fires.

— Jerry Dennis

Logging Practices Along Rivers

Timber cutting has significantly altered the nature of many forests, and thus has inadvertently altered the nature of many streams that bisect the forests. Studies of the effects of logging along streams typically show that total fish biomass initially increases after clearcutting. Increased growth occurs because more sunlight can reach the river, which fuels more algae growth, which fuels growth all along the food chain.

But soon the growth spurt in the river declines. The loss of streamside trees deprives streams of their most important food source—leaves, needles, and twigs. Temperatures change, because without trees to shade a stream, the water warms up, making survival difficult for coldwater fish. Streams lose much of their complex shoreline and underwater structure, because large trees are no longer available to fall into the stream. Stream sediments, like gravel and dirt, increase in the river due to erosion from the exposed soils, smothering the bottom substrates.

Streams also lose 50 to 75 percent of their pools after a clearcut. In one study, streams that had little timber harvested along their shorelines averaged more than 15 large pools for every stream mile. But after moderate logging, the same streams averaged fewer than 7 large pools per mile. The pools decreased, because the streams lost the logjams and meanders so typical of an unaltered river. Logjams and meanders are the structures that create and maintain pools.

After a clearcut, the temperature of large rivers is unlikely to be affected much by the loss of shading. But losing shade trees is highly important to small streams. A study in northern England examined the impacts of logging the shorelines along a small stream. Researchers recorded water temperatures before and after the clearcut took place, and found that the maximum summer temperature of the water increased from 59°F before cutting to 70°F after cutting.

In other studies examining small watersheds with clearcuts and extensive roads, water discharge after rainstorms increased by 50 percent during the first five years after logging. Nearly 25 years later, runoff in those same clearcut watersheds was still 25 to 40 percent higher than before the cut. It's important to note that many variables affect this rate, like slope, geology, climate, and road design, as well as the actual logging practices.

Logging streamside forests also amplified the affects of floods. In one study, watersheds with clearcuts had 14 to 60 percent more runoff per unit of area than unlogged watersheds, even when the harvest had occurred 20 to 30 years earlier.

Continued

Logging Practices Along Rivers (Continued)

Logging can also minimize the in-stream refugia, or the hiding places necessary for animals to survive a churning current during a flood. In flooding streams with lots of large logs in the water, fish survive better.

Forest roads can also be a major factor impacting stream ecology. Roads capture water draining off slopes and route it into ditches, or through culverts and down roads, channeling it into streams much faster than it would ordinarily have traveled.

The bottom line is that streamside logging, road building, and other poor forest practices often simplify rivers, reducing the number of deep pools, and the presence of coarse woody debris, riparian forests, and complex habitats. Changes then cascade through the aquatic communities, from insect larvae to larger fish, ultimately changing the life of the river community.

It's not difficult to avoid these problems. A simple logging plan should include:

• Retaining a majority of the shade trees that project over the stream in order to maintain normal water temperatures.

• Retaining a majority of the original tree canopy to provide organic material essential to the stream food chain.

• Retaining all downed timber in an aquatic and riparian area to provide shoreline and in-water structure.

• Retaining all dead, standing trees to provide habitat for insects, birds and small animals.

The Flambeau Trail

The Montreal River posed an impossible navigational barrier between Lake Superior and the Chippewa River. A series of falls roared just a short distance above the mouth of the river. Extreme rapids continued above the falls and made further canoe travel impossible. The Ojibwa called the Montreal River *Kawasiji-wangsepi* or *Kawasidjiwong*, meaning either "White Fall River," or "where there is a strong foaming current in the river."

During his expedition with Schoolcraft in 1820, James Doty wrote:

"This stream is very rapid, and at its mouth where we landed a beautiful fall is seen of about 70 ft.—the banks are 100. From the fall the banks widen, forming a fine bottom through which the river meanders 1/4 of a mile, in the middle of which the Indians have erected a weir for the purpose of taking whitefish and sturgeon . . . A little above the river on the lakeshore there are several lodges of Indians on a piece of level ground bounded on 3 sides by mountains, through which a small creek runs . . .

"The Montreal River is not navigable; but at its mouth, on the east side, a portage is made of one hundred and twenty pauses to a small lake; in which distance the Montreal River is crossed twice, the first time at eleven pauses, and the second at eight."

Schoolcraft wrote of the same journey in 1820:

"About 800 yards above its mouth it [the Montreal] has a fall of eighty or ninety feet, where the river is precipitated over a rugged barrier of vertical rocks, by several successive leaps, the last of which is about forty feet perpendicular. This view is highly picturesque."

Schoolcraft repeated a common misconception of the time, stating that the Montreal connected with the headwaters of the Chippewa and Wisconsin rivers. When Wisconsin Territory was established in 1836, the boundary established between Michigan and Wisconsin was a continuous water line from the Bay of Green Bay, up the Menomonee River, to Lac Vieux Desert, then down along the Montreal to Lake Superior. The only problem with this notion was that the Upper Peninsula of Michigan would then have been an island!

T.J. Cram's survey in 1841 redefined the boundary, in part, by correctly mapping the Montreal. But the dispute on where the real boundary belonged carried on until Michigan filed a law suit in 1923 claiming that the west branch of the Montreal was the main stream of the Montreal, and therefore was the actual starting point for the state line. The U.S. Supreme Court finally confirmed Wisconsin's claim to the disputed lands, and a good thing, because Wisconsin would have lost thousands of acres of land, as well as Hurley, though some disputed whether

Continued

The Flambeau Trail (Continued)

we gained much by retaining Hurley at the time. The *Ironwood* (Michigan) *Times*, February 1924 headlines read, "WE'RE THREATENED WITH HURLEY AGAIN. Legal Action over Boundary Progressing. Wisconsin Wants to Keep Hurley and She's Welcome to the Burg."

Ah, Hurley. "Hurley, Hayward, and Hell!" was the rallying cry of many an inebriated miner and logger. But that's another story. There was hell enough along the Flambeau Trail.

An overland trek, nearly 42 miles in length, had to be made in order to reach the inland Ojibwa nations. The portage trail commenced on the east side of the mouth of the Montreal, went 6 to 7 miles over the eastern hills, and crossed the river at a point above a falls. The path continued up the southwest bank at some distance back from the river, crossed the river again, and ended at Portage Lake (Long Lake), where canoes were kept en cache, ready to load for the inland trade.

The Flambeau tested even the endurance of the legendary French voyageurs. The trail easily earned its description as a "120-pause portage." Distance was measured during the fur trade in pauses, or the times that the voyageurs paused to rest and smoke. A pause could be anywhere from 600 yards to half a mile depending on the terrain. Bogs and hills were a good cause to pause.

In 1680, Daniel Greysolon Sieur Du Luth paddled along the southern shoreline of Lake Superior, and is reputedly the first white traveler to see the Montreal. Either he or Jesuit priests gave the Montreal its name, probably deriving it from the likeness of the bluffs at the mouth of the river to the mountain at Montreal. The earliest reference to the Montreal is on a 1688 French map of Lake Superior.

An American Fur Company trading post operated at the mouth of the Montreal in the first half of 19th century.

Logging began in the 1880s, and the perpendicular walls of Superior Falls were eventually blasted down by the lumbermen to improve the ease of their log drives.

To experience the Flambeau Trail today, join the annual Flambeau Trail Trek, an Iron County Heritage Festival Event that takes place in August. Unfortunately, the actual historic trail runs mostly through private lands today, so the trek follows only a portion of the original trail.

Beaver Physiology

Beavers are the largest rodent in North America, and actually the second largest rodent in the world, exceeded only by the South American capybara, or water hog. At 3 to 4 feet from head to tail, and averaging 40 pounds (record weight of 110 pounds!), their compact body is hardly built for speed. The hind feet are large and webbed for swimming. The front feet are small and delicate with long claws that allow them to manipulate food, carry materials, and dig. Beavers don't have an opposable thumb, but they can rotate a twig while eating it, like eating an ear of corn. Their hind legs are longer than their front legs, giving them a hunched appearance when walking on land. Two inner claws on each hind foot have split nails that are used for grooming, the nails combing the fur to keep it waterproof and streamlined.

A beaver's broad and flat scaly tail is 9 to 14 inches long, 4 to 7 inches wide, and less than an inch thick. The tail is used for slapping a warning, and as a rudder when swimming. It's not used to carry mud or for propulsion. The tail has special circulatory characteristics that dissipate excess heat during hot weather and conserve body heat during cold weather.

The beaver's small ears barely protrude from the back of its head. Its nostrils open to the side instead of to the front. Both ears and nose have valves allowing them to close while swimming. A clear nictitating membrane closes over the eyeball to protect the eye when swimming. Beavers also have flaps on the insides of their mouths, which close behind their incisors, allowing them to swim under water while carrying small sticks. They can also chew wood while under water without getting water in their mouths.

Their sight is poorly developed, and thus they rely more on hearing and smell. An unfamiliar object detected by sight is usually circled while the nose is held above the water. They can hear quite well above and below water.

Beavers can stay submerged between 3 and 10 minutes, and are reputed to swim a half mile under water. Their lungs aren't particularly large, so a beaver slows its heartbeat and constricts its veins and arteries to conserve oxygen. Their respiratory systems can handle greater amounts of carbon dioxide build-up than other mammals. When a beaver surfaces for air, it exchanges up to 70 percent of the air in its lungs, more than 2 to 3 times what a human exchanges.

Their large, orange buck incisors are very noticeable. The orange color comes from the hard enamel plate on the surface of the teeth. The

Continued

Beaver Physiology (Continued)

incisors continuously grow, and must be worn down by chewing, otherwise the teeth will grow in a curved fashion and eventually prevent eating.

A beaver's fur is composed of a dense layer of underfur and a sparse layer of long guard hairs. The guard hairs are 10 times the diameter of the underfur hair, creating a coarse look. What the underfur hairs lack in diameter they make up in density. Over 100,000 underfur hairs occur per square inch on the belly. These hairs have a wavy texture that gives the pelt a downy softness when dry.

Scent glands located near the tail contain castoreum, an oily substance that beavers use for waterproofing their fur and scent-marking their territory. The oil was used historically as a base for perfumes because it retains any fragrance combined with it and slowly releases the aroma when warmed. Castor oil was also a "wonder cure" in the Middle Ages for rheumatism, arthritis, sexual impotency, and other maladies. The beaver's Latin name *Castor* originated due to the similarity between the sexes—the male organ appears to be missing, and so it has the appearance of being castrated.

Rest Stops for Birds

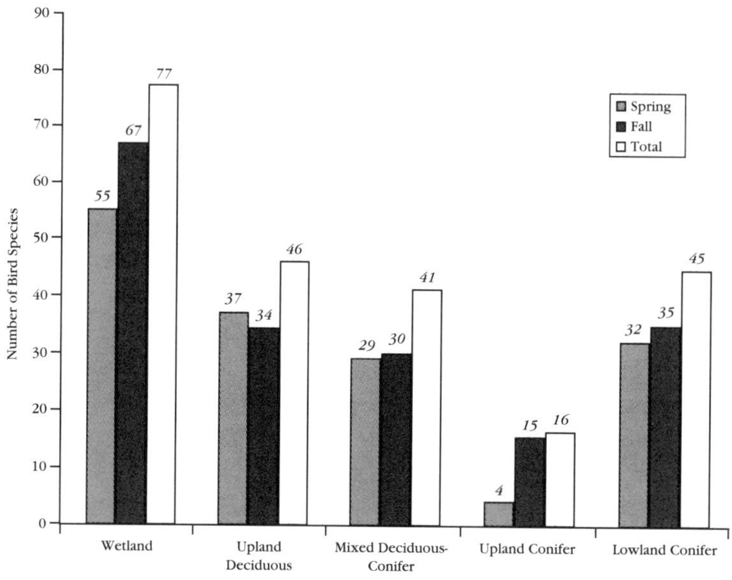

Numerous studies have documented declines in songbird populations, particularly in neotropical migrants—those species that breed in the U.S. and Canada but winter in Central and South America and the Caribbean. Almost half of the birds that breed in North America migrate to Central and South America, and nearly 100 of these species pass through or nest in the Upper Midwest. Their decline is due to issues like the destruction of their wintering grounds, fragmentation of breeding habitat in North America, loss of migration corridors, and the destruction of stopover sites used during migration. However, not enough is known about all of these concerns to definitively answer why they are in decline.

One of the problems is how little is understood about the critical habitat needs of birds during migration. With that in mind, a study was undertaken at nearby Park Falls, Wisconsin, from 1989 through the fall of 1992. Migrating birds were trapped during the spring and fall migration periods. Mist nets were set in 15 different habitats to monitor species type, their movement, their habitat use, and the condition of the birds as they moved through the forest during migration. A point count was also utilized during the spring of 1991 and 1992 in order to compare totals of birds heard singing in these habitats to those that were actually trapped.

Continued

Rest Stops for Birds (Continued)

The study found that all community types were used by a wide variety of species, but some habitats were used far more than others. The highest capture rate occurred in the wetland community type (170 birds per 1,000 net hours). Comprising only 7 percent of the study area, this community was used by migrating birds far in excess of its acreage proportion. Fifty-four percent of all birds were netted in the wetland communities. These wetlands contained dense shrubs and accessible water. In contrast, those habitats with little or no understory, such as red pine habitat or hemlock habitat, had low capture rates.

Wetland also had the highest species richness with 77 of the 89 species represented. Over half of those species are classified as long-distance neotropical migrants.

The three wetland habitats studied—alder shrub, willow shrub, and sedge meadow—all contributed to the high capture rate within the wetland community type. These three habitats each attracted more individual birds and more species than any of the other 12 habitats.

Many of the birds normally found in the upper canopy, such as red-eyed vireos and blackburnian warblers, were frequently caught in wetland, often in mixed species flocks. Foggy periods would often bring down migrating birds. Wetlands acted like a magnet in attracting birds to concentrate there. The birds probably concentrated their activities in these shrubby wetland habitats, because there was an abundant food supply of insects and berries, accessible water, and dense vegetation that protected birds from bad weather and predators. Even the agile, bird-eating sharp-shinned hawk had a difficult time penetrating the dense, shrubby wetland habitats.

A mosaic of habitats is used by migratory songbirds in a northern Wisconsin forest, but wetland habitats attract disproportionately more individual birds and species than any other habitat type studied. The Manitowish flood plains are alive with birdsong in the spring, much of which emanates from birds just passing through. The bottom line is that dense shrubbery near water is good medicine for songbirds. Luxuriant foliage provides essential protective cover and a ready food supply for the needs of migratorial birds.

In past years, wetlands have often been considered wastelands. Their value to wildlife has sometimes been underestimated by land managers. We need to learn to look at these wetlands not as wastelands, but as important songbird migration refuges.

Malhiot's Journal, 1804-05

Francois Victor Malhiot's journal offers fascinating reading, though as with all history, one must take into account the writer's bias. Indeed, Malhiot had his fair share of bias. He never seemed comfortable with living in a wilderness, or in trading with "savages," though he did marry an Indian woman in 1800. He resigned his post with the company in 1807, left his Indian wife with her own people, and took his biracial son back with him to France where he lived until his death in 1840.

Malhiot's journal may be read in its entirety on a Web site set up by teacher and historian Jim Bokern. Bokern has also reprinted the journals of Norwood (1847), Cram (1841), Doty (1832), Gray (1846), and others on this Web site—<www.marshfield.k12.wi.us/socsci/discovery/historic>.

A sampling of Malhiot's journal offers a look at some of the life's worries as seen through the eyes of a French trader in the wilderness of Lac du Flambeau:

Hardships of Fur-Trading

8/17/1804. O! wretched people of Lac du Flambeau, everything is against you! Little to eat, much work to do; sometimes ill, uncertain of obtaining returns, with reproaches to be dreaded from the Partners, anxiety about the goods out of the fort, Savages to satisfy, and adversaries to watch. What a life!! Poor Malhiot, when will you be relieved of such a heavy burden?

5/10/1805. My people have not had a day's rest since my arrival here last autumn. Of all the men who may be in the upper country I do not think there are any who have worked as hard as mine; a house twenty feet square, of logs placed one on the other made by four men; 70 cords of fire-wood chopped; pickets sawn for a fort; a bastion covered; a clearing made for sowing 8 kegs of potatoes; and all the journeys made here and there.

Rum

8/28/1804. Several of Chorette's Savages came here last night to get rum and to use violence . . . I was alone with Gauthier and they were at least 15 rascals all armed; those who had no knives or spears, had sticks or stones. Fortunately we all got off with calling one another names and threatening one another.

9/4/1804. We had quarrels all day with the Savages of Lac du Flambeau; spears, knives, hatchets, etc., all were brought into play. They made a breach in the Fort, broke one of the doors and had it not been for the aid of l'Outarde and two or three young men who were quite sober at the time, there would certainly have been bloodshed, and even somebody killed on one side or the other.

4/15/1805. "Rum flows like water on both sides. . ."

[Note that while Malhiot complains bitterly of their drunkenness and his lack of safety, he continues to trade them rum!]

Continued

Malhiot's Journal, 1804-05 (Continued)

Wild Rice

8/24/1804. We are threatened with famine because the Savages absolutely want to go on the war-path; consequently they will put the greater portion of their rice in caches, and we shall find ourselves with very little, which we shall have to purchase at its weight in gold.

9/29/1804. I also gave a laced capot to Barsaloux with a half keg of rum and a large keg to l'Outarde to be distributed in his village in exchange for rice.

Fishing

8/14/1804. We caught nothing this morning in the nets. One day of abundance and ten days of famine!

Gardening and Squirrels

8/10/1804. The squirrels are doing much damage in the corn fields; they ate 77 ears last night.

Starvation

3/1/05. I learned from two young men who have just arrived that one of the Savages I gave a coat to last autumn, starved to such an extent that he had to eat his pack, his dogs, and even his gun-cover.

Fort

4/26/1805. I thank God every day for having inspired me with the idea of making so good a fort, impregnable to bullets and to all attacks.

Furs

10/5/1804. I have just taken an inventory of the furs I have traded since my arrival here and I counted: 528 deer skins, 840 musk-rat skins, 107 lbs. beaver, 44 otter skins, 16 bear skins, 7 marten skins, 1 mink skin—the whole making probably sixteen packs.

Relationship with the Ojibwa

10/30/1804. 'L'Epaule de Canard' is . . . a brave, sober Savage, liked by the others, liking the French, capable of sacrificing himself for them; a good man for errands; he does not ask for things, is satisfied with everything that is given him and is a famous hunter . . .

There are some others whom I might include in the number of good Savages, but, as a rule, if I could put them all in a bag and know that Lucifer wanted them, I would give them all to him for a penny.

Appendix A

WISCONSIN WARMWATER STREAM PHYSICAL HABITAT RATING FORM

STREAM:______________________________ WATERBODY ID:____________________

Year: ________ Month: _______ Day: ________ Entire Stream Miles: ______________________

Evaluators: ____ Total Score: ___________________

____

CATEGORY

Rating Item	Excellent	Good	Fair	Poor	Score
Bank erosion, failure and bank protection.	No significnt bank erosion, failure. ≥ 90% of bank protected by plants or stable rock. 12	Limited amount of bank erosion, failure. 80% of bank protected by plants or stable rock. 8	Intermediate amount of bank erosion, failure. 60% of bank protected by plants or stable rock. 4	Extensive amount of bank erosion, failure. ≤ 50% of bank protected by plants or stable rock. 0	_______
Main channel rocky substrate (% of area).	≥ 65% of the bottom covered by rocky substrate (BE+BO +RC+GR). 25	45% of the bottom covered by rocky substrate. 16	25% of the bottom covered by rocky substrate. 8	≤ 5% of bottom covered by rocky substrate. 0	_______
Available cover for adult gamefish.	Extensive cover, (woody debris, rocks, or macrophyte beds) ≥ 12% of total surface area. 25	Adequate cover, 8% of the total surface area. 16	Cover limited, 4% of the total surface area. 8	Little or no cover, 0% of the total surface area. 0	_______
Average maximum Thalweg depth (4 deepest depths).	≥ 1.5 meters. 25	1.2 meters. 16	0.9 meters. 8	≤ 0.6 meters. 0	_______
BB or RR Ratio (distance between bends or riffles/avg. main channel width).	BB or RR Ratio ≤ 12 12	BB or RR Ratio = 18 8	BB or RR Ratio = 24 4	BB or RR Ratio ≥ 30 0	_______
99 = Excellent 66 = Good 33 = Fair 0 = Poor				Total Score:	_______

Appendix B

J.G. Norwood Journal, 1847—From Long Lake to the Wisconsin River via Lac du Flambeau.

September 23. At this point the long portage ended [the Flambeau Trail from Lake Superior to Long Lake], and, after discharging the extra packmen, and furnishing them with provisions, we set about preparing for our journey to Lac du Flambeau. We expected to obtain here a canoe belonging to a man living on Wisconsin river. He was at La Pointe when we left, and obligingly offered us the use of it as far as the mouth of Maple river, stating, at the same time, he would send an Indian who knew its place of concealment, and would discover it to us. Early in the afternoon the Indian arrived, with the intelligence that our obliging friend had instructed him not to show us the canoe, but to cache it in a new place, where we could not find it. We were thus deprived of the means of transportation upon which we had relief from the moment of leaving lake Superior. We had not left the alternative of making a further portage to Lac du Flambeau, over a region only traversed by the Indians in the winter, when the rivers, swamps, and lakes are frozen and passable, or of waiting until some straggling Indians should arrive with a canoe which could be purchased.

Fortunately, however, the men, in examining the lake shore, discovered a small canoe concealed among the bushes, and, under the circumstances, we determined to follow the custom of the country, in like exigencies, and appropriate it to our own use without waiting to consult the owner, who was supposed to be an Indian left sick at Madeline island. The canoe, though entirely too small for our purpose, being intended for only two persons, was perfectly new, and of excellent model, and by judicious stowage, it was supposed capable of answering our purpose until we could procure a larger one.

September 24. The early part of the morning was spent in arranging our provisions and luggage to the best advantage in the small canoe. About 8 o'clock, we left the head of the Portage lake [Long Lake], which is from two hundred to four hundred yards wide, and four miles long. We had proceeded about two miles, when we observed a canoe approaching us containing a young Indian. Contrary to the usual custom, he gave us no salutation as he approached, but, paddling swiftly along side, grasped the canoe, and claimed it as his property. After being made acquainted with the circumstances which induced us to take it, he expressed himself satisfied, and, after considerable hesitation, agreed to sell it for about three times its value, which we declined giving. Whatever may be thought of the simplicity of the Indians when bartering in the frontier villages with the whites, I have always found them not only acute dealers on their own soil, but ever ready to seize the slightest occasion for extorting money or provisions. In the meantime two other canoes came up, and we finally succeeded in purchasing an old one, a little larger than the one we were in, for the price of a new one. It took but a few minutes to exchange loads, and we were soon floating down the lake in our own vessel, secure for at least as long a time as the bark would hold together.

After leaving Portage lake, we passed a series of small lakes, connected by shallow, winding streams, with numerous granite boulders in their beds, and finally entered Big Turtle lake [Echo Lake], from the east side of which there is a portage of about six hundred yards to Little Turtle lake [Tank Lake]. At this place we camped just in time to escape the rain, which had been threatening to fall all day, and now came down in torrents.

September 25. Turtle portage is an excellent one, over the plain lying between the two Turtle lakes. At the east end of it is an Indian village, inhabited during the summer months by one of the Chippewa bands. At present it is deserted, the band having gone north to their winter hunting grounds. Potatoes and corn are raised at this village. The soil is underlaid by fine drift, with occasional large granite boulders disseminated through it. Along the shores of the lakes, sections of drift from ten to twenty-five feet in thickness are exposed . . .

September 26. The river is exceedingly crooked, and from forty to fifty feet in width from the camp to the mouth of Lac du Flambeau river, a distance of about three miles. Where the bends of the river approach the margin of the meadows, the banks are from four to six feet high, and composed entirely of yellowish coarse sand resembling very much that found on the Chippewa below the Dalles. Soon after entering Lac du Flambeau river, which we ascended to the lake of which it is the outlet, large boulders began to show themselves, some of them of great dimensions. One which was examined measured fifteen feet in the long diameter, twelve feet in the transverse, and stood seven feet out of the water. It was composed of mica slate, and studded with garnets of small size.

Just before reaching a range of hills, the river runs through what was once evidently a large lake, now elevated and overgrown with aquatic grasses. Through this the river flows in many channels, some of them fifty yards wide. This alternate widening and narrowing of the river occurs all the way to the lake. The trunks of hundreds of dead tameracks are standing in all the spaces between the channels, and give a peculiar air of desolation to the scene, only partially relieved by the evergreens on the distant highlands.

About three miles above the mouth of Lac du Flambeau river, in a direct line, we came to a range of low hills on either side of the wide meadows through which it flows, which gradually recede until they reach the height of from forty to eighty feet . . .

Shortly after passing this range the swamps again show themselves, and continue on either side of the river up to Lac du Flambeau. The river is exceedingly crooked, its general course being S.S.E. We reached the lake late in the afternoon, and, crossing its north west arm, camped near the old trading house of the American Fur Company, now deserted.

September 27. From this point we had to find our way to the head waters of the Wisconsin river, without a guide, or the slightest knowledge of the country through which we were to pass. The aid afforded by Mr. Nicollet's map not being of a reliable character for this region, I sent Baptiste to the Indian village to procure such information as would enable us to reach Vieux Desert lake, or its vicinity, with as little delay as possible. While he was gone, an Indian whom I knew very well, having met him at Madeline island in July, and afterwards at Fond du Lac, came to our camp; and from him we learned that there are three routes from this lake to Wisconsin river. One of them is by a chain of lakes south of this point, and leading into the Little Wisconsin through White Squirrel creek; another by way of Leech, Kewaykwodo and Swamp lakes; and a third through a series of lakes towards the head waters of Manidowish river, and thence, via Trout lake and a series of small lakes, to Vieux Desert lake; which last route might be changed about twenty miles southwest of Vieux Desert, so as to enable us to strike the Wisconsin ten miles south of that lake. The first route is the one usually followed by the traders in their journeys from the posts on the Wisconsin to La Pointe. As it would, however, lead us into the Wisconsin too far south to subserve the purposes of the survey, we decided upon the route by Trout lake.

Finding "Puzigwingis," the name of the Indian alluded to, intelligent, and willing to impart any information we might desire with regard to the country, in order to test his abilities I got him to draw a map of our route from Lake Superior to this place, and finding it agreed in every respect with our own observations, I determined to remain in camp to-day, for the purpose of procuring from him an outline map of the surrounding country, and particularly of the part to be traversed by us. He spent about half the day in executing our wishes, and so far as I am able to judge of the whole by the part that fell under my immediate observation, it is quite as accurate as it could possible have been made by any one having no knowledge of the principles upon which maps are constructed. The only valuable purpose which it can subserve, however, is that of a guide through a very intricate wilderness, until accurate surveys are made; but even that is a desideratum in a country where there is such a multitude of lakes and impassable swamps to impede and turn one aside from a course which might otherwise be followed from a knowledge of the general bearing of known points.

Lac du Flambeau is the largest body of water we have seen in this region. It is exceedingly irregular in its outline, resembling rather an assemblage of several small lakes, united at one point by short narrow channels. It has a number of thickly wooded islands dotting its surface. The shores recede with a gentle slope, to the height of twenty and thirty feet, and are covered at some points with bushes and grass, and by a dense forest at others. The soil, like that in the neighborhood of Turtle lake, is a light sandy loam; and, judging from its general appearance, would hardly attract the attention of a cultivator. The Indians, however, who have a village on one of its shores, raise excellent potatoes, better indeed, that are usually grown, with all the aids of cultivation, in the valley of the Ohio. The arm of the lake, near which we encamped, is called by the Indians, Pokegoma; a name, given to any lake connected with another, or with a running stream, but a short outlet.

September 28. The Pokagoma, arm of Lac du Flambeau, which we crossed this morning, is about three and a half miles long by half a mile in width. It abounds with fine fish, which the Indians take in great numbers in gill nets and with the spear. From the northeast shore of this lake a portage of half a mile, over sand hills, covered with small pines and elevated about thirty feet above the general level of the small lakes, leads to Lake Wepetangok [White Sand Lake], which we crossed in a high wind. This lake is about two miles long, and our course across it was northeast to a small channel, four feet wide and eight yards long, which led us into another small lake three-fourths of a mile long and half a mile wide [Little Sand Lake], which we crossed northeast to a portage of one mile in length, leading to Mashkegwagoma lake [Ike Walton Lake]. This portage passed over hills of the same character as those seen in the morning.

We waited sometime on the shore of this lake for the wind to subside, and at noon started across. By the time we had made two-thirds of the passage the wind increased to a perfect gale, and wave after wave, which ran almost as high as I have ever seen them in Lake Superior, broke over our canoe until it was more than half full of water and in momentary danger of sinking. By great exertions the men succeeded in reaching the borders of a small island, and we dragged the canoe into a marsh. Everything was thoroughly soaked, with the exception of my notebooks, which, very fortunately, were secured on my person. A fire was built in a spruce thicket, the highest part of the island, and we set about drying our persons, clothes, maps and instruments. As the wind continued high all the afternoon we were forced to camp on the island. The lake is about two and a half miles long and one mile and a half wide, a very small sheet of water to afford so heavy a swell. Our misfortune is to be attributed, however, more to the size of our canoe than the roughness of the lake.

September 29. Crossed to the main shore, and made a portage of a mile and a half, to the Chippewa or Manidowish river. The trail, for nearly the whole distance, leads through swamps flooded with water almost ice cold. The river at this point is about forty feet wide, winding to the northwest through marshes like the one just passed.

Had it not been desirable to visit Lac du Flambeau, we might have reached this point by ascending the river from "Six Pause portage," through "Cross" and other small lakes; and this was the route pursued by Mr. A. B. Gray and party in 1846, as I have since learned. I knew nothing of the route, however, until I reached Lac du Flambeau, when I learned it from Puzigwingis. It is the one commonly followed by the Vieux Desert and Trout Lake Indians in passing from their villages to La Pointe, and is in every respect preferable to the one pursued by us, for persons wishing to pass from the head of Wisconsin river to the neighborhood of Montreal and Bad rivers, or to any point northwest of Lac du Flambeau.

While the men were sent up the river with the canoe, Mr. Gurley and myself took the trail for Trout lake. The portage is an excellent one, about four miles and a half long, and passes for the distance over a sandy plain supporting a few scattering pines. The surface of the ground is literally covered with the wintergreen, and the general features of the landscape resemble very much those seen in the neighborhood of Lac Court Oreille. About half way the portage we ascended a hill of drift between forty and fifty feet in height, with a great number of crystalline boulders and a few large fragments of sandstone scattered over it. From the top of this hill a range of highlands were seen in our rear, distant eight or ten miles, bearing northeast and southwest. From their position and course we judged them to be a continuation of the range seen in ascending Lac du Flambeau river. The drift continues on the Trout lake.

About one mile before reaching the lake, the river becomes very shallow, and is so much obstructed by boulders as to require a portage to be made. There is an Indian village at Trout lake which is only occupied, however during the summer and fall months. They have gardens for corn and potatoes at this place, through their principal dependence for food is upon the lake, which yields them a plentiful supply of fine fish. The few Indians now here were preparing to depart for their hunting grounds. Oshtawabanis, head chief of the Wisconsin band, came to our canoe and begged some flour, in return for which he sent us a lot of very fine potatoes, a most acceptable present, as more than two-thirds of the provisions which we had brought with us from La Pointe were consumed, and we had not yet performed more than one-third of our journey.

Trout lake is seven or right miles long by about four miles wide, and contains a number of small islands. It is surrounded by drift hills, from twenty-five to forty feet high, supporting a sparse growth of small pines and birch. Our course across it was northeast, to a trail leading to lower Rock lake. We camped on the trail a short distance from the lake. At six o'clock, p.m., the thermometer stood at 31°F, and our tent and baggage, which had got wet in crossing the lake, were frozen.

September 30. Ice formed one-fourth of an inch thick last night. The portage between Trout and Lower Rock [Pallette Lake] lakes is about two miles and a quarter in length, and runs along the base of drift hills. These lakes are connected by a small stream, not navigable for canoes. The lower lake is about half a mile in diameter. A portage of three hundred yards leads to Upper Rock lake [Escanaba Lake], which is one mile in its largest diameter, and contains a number of small islands. These lakes are also connected by a small stream. They derive their name from the immense number of boulders which line these shores, and show themselves above the water in the shallow parts. The islands in the upper one are made up almost entirely of boulders, with a thin soil covering them, and supporting hornblende, and greenstone, with smaller ones of amygdaloid, were seen near the east end.

We had great difficulty in finding the portage from this lake. It begins on the northeast shore, and is about two and a half miles long. Its course is nearly due east, passing a good part of the distance along in the first two miles. They are connected by a small stream flowing into Upper Rock lake, and which is navigable for canoes up to the second pond. From this point a portage of everything has to be made to Lower White Elk lake [White Birch Lake]. The country passed over yesterday and to-day is made up of drift hills, from twenty to sixty feet high. The sand is white and coarse, while the boulders, which are disseminated through the upper part, were derived almost entirely from granitic rocks. The soil is thin, but supports a growth of small pine, birch, spruce, hemlock, fir, a few oaks, and some basswood; the swamps, as usual, being filled with tamerack, or, where that is wanting, overrun with cranberry bushes.

Lower White Elk lake, where we camped, is about three quarters of a mile long, and a quarter of a mile wide. Here we found a number of deserted wigwams and the remains of a garden. The lake affords great numbers of fish, and the quantity of their remains scattered around shows they are the principal article of food among the Indians who occasionally inhabit it.

October 1. A very heavy frost this morning; the thermometer standing at 25°F. at half past six o'clock, we crossed first White Elk lake, and, by a stream twenty feet wide and a quarter of a mile long, passed into second White Elk lake [Ballard Lake], which is about two miles long and one mile wide. From this we passed into third White Elk lake [Irving Lake], by a river ten yards wide, and three hundred yards long. This lake is nearly circular, and about one mile in diameter. It is very shallow, not having a depth of more than three feet at any point, with a mud bottom. We noticed here a phenomenon, not hitherto observed in any of the great number of small lakes we have seen in the territory. The whole surface of the lake was covered with bubbles of light carburetted hydrogen gas, which was constantly ascending from the bottom.

From this lake, a portage of a quarter of a mile brought us to the fourth White Elk lake [Lake Laura]. The portage leads due east, over drift, covered with a better soil than any met with for several days past. It supports a tolerably good growth of sugar maple, birch, oak, poplar, and a few pines. This lake is a beautiful sheet of water, about one mile long and three-fourths of a mile wide. The bottom is covered with pebbles and the shores with boulders, some of which are very large; one of them being over fifty feet in circumference. This is the source of the east or Manidowish branch of Chippewa rivers; all the lakes and streams beyond this point, which send their waters to the Mississippi, being tributaries of the Wisconsin. The hills, bounding the north and east shores, are about one hundred and fifty feet high, and are composed of white sand, and occasional boulders scattered over the surface. Almost all the boulders seen, for the last three days, were granitic, and small.

To-day, however, at the fourth Elk lake, boulders of other rocks were plenty, and, from the size of some of them, I infer that the source from which they were derived is not very distant.

The portage to the head waters of Wisconsin river starts due-east from this lake. In about half a mile the trail divides, the left hand branch leading directly to Vieux Desert lake, the other to a small lake which discharges its waters into the Wisconsin, about ten miles in a direct line south of Vieux Desert. We determined to take the shortest route, principally on account of the little provisions we had remaining, and the certainty that they would be exhausted before we could reach any point where supplies could be had.

The portage is about six-miles long, over a high, rolling pine country, which does not afford a drop of water, from the upper White Elk lake to within a quarter of a mile of the end of the portage, where a small stream, ten feet wide, from the northwest, crosses the path. I did not reach Muscle lake [Upper Buckatabon Lake] until sunset, and before I came in sight of it I heard the voyagers singing and firing guns. They were rejoicing on account of having reached a tributary of the Wisconsin, and that long portages were over for this year.

The high and broad strip of land which divides the waters of the Chippewa from those of the Wisconsin is made up of white sand, with small boulders thinly scattered over the surface. The pines with which it is covered are small, but very tall and straight, many of their trunks rising fifty or sixty feet without a branch. On some of the higher hills a great many small birch were seen; and in the vicinity of Muscle lake, the sugar maple began to appear.

October 2. The ground was whitened by a heavy frost, and the atmosphere cool and bracing. Muscle lake, upon which we began our voyage to the Mississippi, is about one mile long and rather more than half as broad. A small stream, about one hundred and fifty yards in length, led us into another lake, rather more than half a mile in diameter. It discharges its waters into the Wisconsin river, through a small creek, from one to five yards wide, running east. The creek is very shallow, very crooked, and much obstructed by drift wood, but without a rock of any description. Its whole course is through swamps, bordered by sand banks, covered with pine. The banks have quite a reddish appearance, although the sand in the bed of the river is white.

At half-past 12 o'clock we entered Wisconsin river, which is twelve yards wide at the junction and from three to four feet deep.

Selected Resources on Rivers

Web sites

American Canoe Association
www.acanet.org

American Heritage Rivers
www.epa.gov/rivers

American Rivers
www.americanrivers.org

Canoe and Kayak Magazine
www.canoekayak.com

Clean Water Action Council of Northeastern Wisconsin
www.cwac.net

Digital Time Travelers - Jim Bokern
http://marshfield.k12.wi.us/socsci/discovery

Environmental Protection Agency - Wisconsin Information
www.epa.gov/surf/stinfo/WI

Freshwater Society
www.freshwater.org

Great Lakes Declining Amphibians Working Group
www.mpm.edu/collect/vertzo/herp/Daptf/daptf.html

Great Lakes Environment
www.epa.gov/glnpo

Groundwater Foundation
www.groundwater.org

Hydrogeologist
www.thehydrogeologist.com

National Millennium Trails
www.millenniumtrails.org

National Organization for Rivers
www.nationalrivers.org

Native Trails
www.nativetrails.org

North American Lake Management Society
www.nalms.org

Project WET - Water Education for Teachers
www.ProjectWET.org

River Alliance of Wisconsin
www.wisconsinrivers.org

Rivers Council of Minnesota
www.riversmn.org

Rivers/Trails Conservation Assistance Program—National Park Service
www.ncrc.nps.gov/rtca/

Sierra Club River Touring Section (Wisconsin Chapter)
www.sierraclub.org/chapters/wi/rts

State Historical Society of Wisconsin
www.shsw.wisc.edu

United States Geological Survey (USGS)
www.usgs.gov

Water Resources of the United States (USGS)
http://water.usgs.gov/

Wisconsin District Home Page (USGS)
http://www.dwimdn.er.usgs.gov

University of Wisconsin Center for Limnology
http://limnosun.limnology.wisc.edu

University of Wisconsin Extension—Lakes
http://uwexlakes.uwsp.edu/

Wisconsin Association of Lakes
www.wisconsinlakes.org

Wisconsin Department of Natural Resources
www.dnr.state.wi.us/org/water/fhp/lakes

Wisconsin Herpetology - Milwaukee Public Museum
www.mpm.edu/collect/vertzo/herp/atlas/welcome.html

Wisconsin Lakes Partnership
www.uwsp.edu/cnr/uwexlakes

Wisconsin Land and Water Conservation Association
www.execpc.com/~wlwca/

Resources—Addresses

Adopt-A-Lake
UW Stevens Point College of Natural Resources
UW-Extension
Stevens Point, WI 54481-3897

Freshwater Society
2500 Shadywood Rd.
Excelsior, MN 55331
(612) 471-9773

Midwest Aquatic Plant Management Society
P.O. Box 100
Seymour, IN 47274

North American Lake Management Society
P.O. Box 5443
Madison, WI 53705-5443
(608) 233-2836

North American Water Trails, Inc.
56 Pease Town Rd.
Appleton, ME 04862-6455

Self-Help Lake Monitoring
WDNR WR/2
P.O. Box 7921
Madison, WI 53707
(608) 266-8117

University of Wisconsin Extension
College of Natural Resources
UW Stevens Point
Stevens Point, WI 54481-3897
Robert Korth—Outreach—Lakes
bkorth@uwsp.edu

U.S. Geological Survey Water Resources Division - Wisconsin
8505 Research Way
Middleton, WI 53562-3581
(608) 828-9901

Volunteer Monitoring Factsheet Series
Extension Publications
630 W. Mifflin St.
Madison, WI 53705
(608) 264-8948

WaterWatchers Program
Dane County Extension
Water Quality and Stream Biology
57 Fairgrounds Drive
Madison, WI 53713-1497

Wisconsin Association of Lakes
One Point Place, Suite 101
Madison, WI 53719-2809
(800) 542-5253 or (715) 346-3424

Wisconsin Department of Natural
Resources
Bureau of Fisheries and Habitat
Protection - Lakes and Wetland Section
P.O. Box 7921
Madison, WI 53707-7921

Wisconsin Wetlands Association
222 South Hamilton St., Suite 1
Madison, WI 53703
(608) 250-9971

Selected Bibliography

Albright, James. 1902. Exploration of a mound on Fox Island in Rest Lake, Vilas County. *The Wisconsin Archeologist,* 2(1):14-15.

Allen, James. 1832. *The Journal of Lieutenant James Allen, Expedition of 1832*. U. S. House Executive Documents, No. 323, 23rd Congress, 1 Session.

Allen, J.D. 1995. *Stream Ecology: Structure and Function of Running Waters*. Chapman and Hall, London.

Bates, John. 1995. *Trailside Botany*. Pfeifer Hamilton Press, Duluth, MN.

Bokern, J.K. and Stiles, C.M. 1993. Report of the 1992 shoreline survey, Manitowish Waters Chain of Lakes, Vilas County, Wisconsin. *Nicolet National Forest*, State Region 2 Archaeology Center, Rhinelander, WI.

Borman, S., Korth, R. and Temte, J. 1997. *Through the Looking Glass*. Wisconsin Lakes Partnership, Stevens Point, WI.

Brown, G.W. and Krygier, J. 1970. Effect of clearcutting on stream temperature. *Water Resources Research,* 6(4):1133-1140.

Caduto, Michael. 1990. *Pond and Brook*. University Press of New England, Hanover and London.

Chadde, Steve W. 1998. *A Great Lakes Wetland Flora*. Pocketflora Press, Calumet, MI.

Cram, Thomas Jefferson. 1840. Report of the survey of the boundary between the state of Michigan and the territory of Wisconsin. *Senate Document 151,* 2nd Session, 26th Congress, The National Archives, Washington, D.C.

Christensen, D., et al. 1996. Impacts of lakeshore residential development on coarse woody debris in north temperate lakes. *Ecological Applications*, 6(4):1143-1149.

Cummins, K. and Mayer, C. 1992. *Field Guide to Freshwater Mussels of the Midwest*. Illinois Natural History Survey, Chicago.

Cummins, Kenneth. 1974. Structure and function of stream ecosystems. *BioScience,* 24:631-641.

Curtis, John. 1959. *The Vegetation of Wisconsin*. University of Wisconsin Press, Madison, WI.

Cvancara, Alan M. 1989. *At the Water's Edge*. John Wiley and Sons, New York.

Dennis, J. 1996. *The Bird in the Waterfall*. HarperCollins, New York.

Densmore, Frances. 1928. *How Indians Use Wild Plants for Food, Medicine and Crafts*. Dover Publications, New York.

Densmore, Frances. 1929. *Chippewa Customs*. Minnesota Historical Society Press, St. Paul, MN

DiDonato, G.T. and Lodge, D.M. 1993. Species replacements among Orconectes crayfishes in northern Wisconsin lakes: the role of predation by fish. *Canadian Journal of Fisheries and Aquatic Sciences,* 50:1484-1488.

Doty, James Duane. 1820. Papers of James Duane Doty. Wisconsin Historical Collections, 13:163-219.

Eastman, John. 1995. *The Book of Swamp and Bog*. Stackpole Books, Mechanicsburg, PA.

Eastman, John. 1999. *Birds of Lake, Pond and Marsh*. Stackpole Books, Mechanicsburg, PA.

Eggers, S. and Reed, D. 1997. *Wetland Plants and Plant Communities of Minnesota and Wisconsin*. U.S. Army Corps of Engineers, St. Paul, MN.

Elias, Joan. 1997. Avian species richness and abundance levels in different habitats along the Bad River corridor, northern Wisconsin. *Passenger Pigeon,* 59:21-44.

Engel, S. and Pederson, J. 1998. *The Construction, Aesthetics, and Effects of Lakeshore Development: A Literature Review*. WDNR, Madison, WI.

Graetz, J., et al. 1997. Status anddistribution of marsh and sedge meadow birds at Horicon, Necedah, and Trempealeau National Wildlife Refuges in 1995. *Passenger Pigeon,* 59:119-130.

Green, Janet. 1995. *Birds and Forests*. Minnesota Department of Natural Resources, St. Paul, MN.

Hauer, F.R. and Lambert, G.A., 1996. *Methods in Stream Ecology,* Academic Press, San Diego.

Heat-Moon, William Least. 1999. *River Horse*. Houghton Mifflin Co., Boston.

Henderson, C., Dindorf, C. and Rozumalski, J. 1999. *Lakescaping for Wildlife and Water Quality*. Minnesota Department of Natural Resources, St. Paul, MN.

Heinselman, Miron. 1996. *The Boundary Waters Wilderness Ecosystem*. University of Minnesota Press, Minneapolis.

Hilsenhoff, William L. 1980. *Aquatic Insects of Wisconsin*. Natural History Council, U.W. Madison, Madison, WI.

Hoagman, Walter J. 1994. *Great Lakes Coastal Plants*. Michigan State University Extension, Tawas City, MI.

Hoffman, R. 1990. Birds of Wisconsin's deep marshes and shallow open-water communities. *Passenger Pigeon,* 52:259-272.

Hoffman, R. and Mossman, M. 1993. Birds of Wisconsin's northern swamps and bogs. *Passenger Pigeon,* 55:113-138.

Hoffman, R. and Sample, D. 1988. Birds of wet-mesic and wet prairies in Wisconsin. *Passenger Pigeon,* 50:143-152.

Holaday, S. 1995. *Wisconsin's Forestry Best Management Practices for Water Quality*. Bureau of Forestry, Wisconsin Department of Natural Resources, Madison, WI.

Johnson, Charles. 1985. *Bogs of the Northeast*. University Press of New England, Hanover and London.

Kalinich, J. 1991. Cranberry Growing and Wisconsin's Bird Diversity. *Passenger Pigeon,* 53:126-136.

Kappel-Smith, Diana. 1984. *Wintering*. Little, Brown and Company, Boston.

Leopold, Aldo. 1949. *A Sand County Almanac*. Oxford University Press, New York.

Leopold, L.B. 1994. *A View of the River*. Harvard University Press, Cambridge, MA.

Leopold, L.B. 1997. *Waters, Rivers, and Creeks*. University Science Books, Sausalito, CA.

Lodge, D.M., Beckel, A.L. and Magnuson, J.J. 1985. Lake-bottom tyrant. *Natural History,* 94:32-37.

Lodge, D.M., Kershner, M.W., Aloi, J.E. and Covich, A.P. 1994. Effects of an omnivorous crayfish (Orconectes rusticus) on a freshwater littoral food web. *Ecology,* 75:1265-1281.

Lyons, J. and Jordan, S. 1989. *Walking the Wetlands.* John Wiley and Sons, New York.

Malhiot, François Victor. 1910. A Wisconsin fur-trader's journal, 1804-05. *Wisconsin Historical Collections*, 19:163-233.

Martin, Lawrence. 1977. *The Physical Geography of Wisconsin*. University of Wisconsin Press, Madison, WI.

McCafferty, W. Patrick. 1998. *Aquatic Entomology*. Jones and Bartlett Publishers, Sudbury, MA.

Meeker, J., Elias, J. and Heim, J. 1993. *Plants Used by the Great Lakes Ojibwa*. Great Lakes Indian Fish and Wildlife Commission, Odanah, WI.

Merritt, R. and Cummins, K. 1996. *An Introduction to the Aquatic Insects of North America*. 3rd Ed. Kendall/Hunt Publishing Co., Dubuque, IA.

Meyer, M., Woodford, J. and Gillum, S. 1997. Shoreland zoning regulations do not adequately protect wildlife habitat in northern Wisconsin. *Bureau of Integrated Science Services*, Wisconsin Department of Natural Resources, Madison, WI.

Mitchell, M. and Stapp, W. 1997. *Field Manual for Water Quality Monitoring*. Kendall/Hunt Publishing Co., Dubuque, IA.

Moore, Kathleen Dean. 1995. *Riverwalking*. Harcourt Brace and Co., New York.

Mossman, M., Epstein, E. and Hoffman, R. 1991. Birds of Wisconsin pine and oak barrens. *Passenger Pigeon,* 53:137-163.

Mossman, M. and Sample, D. 1990. Birds of Wisconsin sedge meadows. *Passenger Pigeon,* 52:39-55.

Murray, John, ed. 1998. *The River Reader*. The Lyons Press, New York.

Naiman, R., Melillo, J. and Hobbie, J. 1986. Ecosystem Alteration of Boreal Forest Streams by Beaver (Castor Canadensis). *Ecology,* 67(5):1254-1269.

Nicholls, T., Egeland, L., Elias, J. and Robertsen, M. 2001. Songbird migration habitat relationships in a northcentral Wisconsin forest during spring and fall. In press.

Nichols, Stanley. 1999. *Distribution and Habitat Descriptions of Wisconsin Lake Plants.* Wisconsin Geological and Natural History Survey, Madison, WI.

Oliver, Mary. 1992. *New and Selected Poems.* Beacon Press, Boston.

Pearson, T., Ostergren, R. and Vale, T. 1997. The Wild Rice Harvest at Bad River. *Wisconsin Land and Life.* University of Wisconsin Press, Madison, WI.

Peterson, R.C. and Cummins, K.W. 1974. Leaf Processing in a Woodland Stream. *Freshwater Biology*, 4:343.

Pielou, E. C. 1991. *After the Ice Age.* The University of Chicago Press, Chicago.

Pielou, E. C. 1998. *Fresh Water.* The University of Chicago Press, Chicago.

Rapp, Valerie. 1997. *What the River Reveals: Understanding and Restoring Healthy Watersheds.* The Mountaineers, Seattle, WA.

Seno, William, ed. 1985. *Up Country: Voices from the Midwestern Wilderness.* Round River Publishing Co., Madison, WI.

Stiles, C., Bokern, J. and Kolb, M. 1995. Report of the 1993-94 shoreline survey, Manitowish Waters Chain of Lakes, Vilas County, WI. *Nicolet National Forest,* State Region 2 Archaeology Center, Rhinelander, WI.

Stoddard, J., et al. 1999. Regional trends in aquatic recovery from acidification in North American and Europe. *Nature,* 401:575-578.

Vannote, Robin L., et al. 1980. The River Continuum. *Canadian Journal of Fisheries and Aquatic Science,* 37:130-137.

Ward, J.V. 1991. *Aquatic Insect Ecology: Biology and Habitat.* John Wiley and Sons, New York.

Warren, William. 1885. *History of the Ojibway People.* Minnesota Historical Society, St. Paul, MN.

Winn, Vetal. 1924. The Minocqua lake region. *The Wisconsin Archeologist,* 3(2):40-58.

Index

Also by John Bates:

Trailside Botany © 1995, Pfeifer-Hamilton Publishers

Seasonal Guide to the Natural Year: Minnesota, Michigan, and Wisconsin © 1997, Fulcrum Publishing

A Northwoods Companion: Spring and Summer © 1997, Manitowish River Press

A Northwoods Companion: Fall and Winter © 1997, Manitowish River Press

Contributing author to:

Harvest Moon: A Wisconsin Outdoor Anthology © 1993, Lost River Press

A Place to Which We Belong: Wisconsin Writers on Wisconsin Landscapes © 1999, 1000 Friends of Wisconsin

Praise for *Trailside Botany:*

"I found it fascinating . . . Bates crafts his language to reflect the beauty he sees in each plant . . . Trailside Botany *is as lively and diverse as a patch of woods."* — Tom Hastings, *The Minnesota Volunteer* (Minnesota's DNR magazine)

"Trailside Botany *is a perfect addition to any hiker's backpack . . . a natural for everyone who loves the outdoors."* — The Nature Conservancy, Minnesota Chapter

Praise for *Seasonal Guide to the Natural Year:*

"There are the rare outdoor books that are so jampacked with interesting stuff that you find yourself reaching for it whenever you get a spare minute to yourself. You find yourself taking it with you camping, hiking, fishing, even just walking. And the darn thing is so well-written that you don't mind—in fact, you pick it up as much for pleasure as you do for information. Wisconsin outdoor writer John Bates has published just such a book." — Russell King, President, Council for Wisconsin Writers.

Praise for *A Northwoods Companion: Spring and Summer,* and *Fall and Winter:*

"Exquisite phenology. Bates has provided the Harpers Index of North Country phenology. This North Country naturalist reveals the depth of experience necessary in knowing one's home or sense of place." — Clayton Russell, Outdoor Education Faculty, Northland College

"The books are sprinkled with surprises and a good deal of humor that keep the pages turning and make one feel that they have gotten to know the author and the people he mentions. A delightful two-volume set that captures the mood of each season." — Lake Superior Magazine

Order Form

Telephone: Call (715) 476-2828. Have your Visa or MasterCard ready.
Fax order: (715) 476-2818
E-mail order: manitowish@centurytel.net
Postal order: Manitowish River Press, 4245 Hwy. 47, Mercer, WI 54547
Web site: www.manitowish.com

Check the following books that you wish to order. You may return any book for a full refund, no questions asked, as long as it is still in good salable condition (in other words, still like new—thank you).

Title (Books by John Bates)	Price	#Ordered	Total
Trailside Botany	$12.95	_____	_____
Seasonal Guide to the Natural Year: Minnesota, Michigan, Wisconsin	$16.95	_____	_____
A Northwoods Companion: Spring and Summer	$14.95	_____	_____
A Northwoods Companion: Fall and Winter	$14.95	_____	_____
River Life: The Natural and Cultural History of a Northern River	$24.95	_____	_____
North Country Moments by Cliff Wood and Neil Long (A Manitowish River Press book)	$14.95	_____	_____

Sales Tax: Please add 5.5% for books shipped to Wisconsin addresses _____

Shipping: Book Rate: $2.50 for the first book; $1 for each additional book
Priority Mail: $4 for the first book; $1 for each additional book _____

Total: _____

Payment:
Check ___ Credit Card: Visa ___ Mastercard ___
Card Number______________________________
Name on Card __________________________ Exp. Date _________

If you would like to receive a copy of the current schedule for Trails North, John Bates's naturalist guide service, please check here _____.

Your Name ___________________________Street/P. O. Box___________________________

City ______________________________________ State ___________ Zip _____________

Phone ______________________ Fax ____________________ E-mail ____________________